SYMPOSIUM ON THE ORION NEBULA
TO HONOR HENRY DRAPER

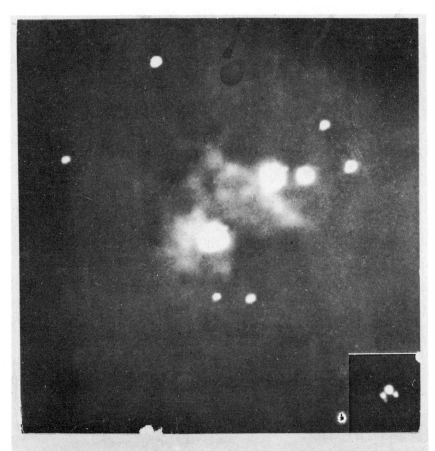

FIRST PHOTOGRAPH OF THE NEBULA IN ORION.

TAKEN BY PROFESSOR HENRY DRAPER, M.D.

September 30th, 1880, Exposure 51 minutes.

The picture is an Artotype enlargement by Harroun & Bierstadt from the original
negative. The large stars, being much brighter than the Nebula, are greatly overexposed. In the lower right hand corner is a photograph of the Trapezium alone,
with only 5 minutes exposure.

ANNALS OF THE NEW YORK ACADEMY OF SCIENCES

Volume 395

SYMPOSIUM ON THE ORION NEBULA TO HONOR HENRY DRAPER

Edited by A. E. Glassgold, P. J. Huggins, and E. L. Schucking

The New York Academy of Sciences
New York, New York
1982

Library of Congress Cataloging in Publication Data

Symposium on the Orion Nebula to Honor Henry
    Draper (1981 : New York University)
    Symposium on the Orion Nebula to Honor Henry
Draper.

    (Annals of the New York Academy of Sciences ;
v. 395)
    Primarily papers from the symposium held by
the Dept. of Physics of New York University.
    Bibliography: p.
    Includes index.
    1. Orion Nebula—Congresses.   2. Stars—
Formation—Congresses.   3. Draper, Henry, 1837–1882
—Congresses.   4. Astrophysicists—United States—
Biography—Congresses.   I. Draper, Henry, 1837–1882.
II. Glassgold, A. E.   III. Huggins, P. J.
IV. Schucking, E. L. (Engelbert L.)   V. New York
University. Dept. of Physics.   VI. Series.
Q11.N5 vol. 395 [QB855]   500s [523.1'135]   82-14549
ISBN 0-89766-180-X
ISBN 0-89766-181-8 (pbk.)

SP
Printed in the United States of America
ISBN 0-89766-180-X (Cloth)
ISBN 0-89766-181-8 (Paper)

ANNALS OF THE NEW YORK ACADEMY OF SCIENCES
VOLUME 395
October 15, 1982

SYMPOSIUM ON THE ORION NEBULA
TO HONOR HENRY DRAPER*

Editors
A. E. GLASSGOLD, P. J. HUGGINS, AND E. L. SCHUCKING

Organizing Committee
O. GINGERICH, A. E. GLASSGOLD, P. J. HUGGINS,
E. L. SCHUCKING, P. THADDEUS, R. W. WILSON, AND D. G. YORK

CONTENTS

*This volume is the result of a conference entitled Symposium on the Orion Nebula to Honor Henry Draper, held by the Department of Physics of New York University on December 4 and 5, 1981 in New York, New York.

Contributed Papers

Integration and Interpretation

Henry Draper and His Legacy

Financial assistance was received from:
- NATIONAL AERONAUTICS AND SPACE ADMINISTRATION
- NATIONAL SCIENCE FOUNDATION
- THE NEW YORK ACADEMY OF SCIENCES
- NEW YORK UNIVERSITY
- TECHNICON CORPORATION

PREFACE

A. E. Glassgold, P. J. Huggins, and E. L. Schucking

Department of Physics
New York University
New York, New York 10003

The decision to organize a symposium on the Orion Nebula in honor of Henry Draper in the year 1981 was based on several considerations: the celebration of New York University's Sesquicentennial, the one hundredth anniversary (on September 30, 1980) of the first photograph of the Orion Nebula taken by Henry Draper, the interest in the interstellar medium that has been a part of the developing astrophysics program in the Physics Department at NYU, and the association, as student and professor, of Henry Draper with this university.

Henry Draper, whose initials are familiar to astronomers through the Draper Catalog produced by the Henry Draper Memorial Project at Harvard University, was the son of another renowned scientist of our faculty, John William Draper, who took the first daguerreotype of the moon in the winter of 1839–40 and was one of the first to make a daguerreotype of the human face, that of his sister Dorothy, a fact commemorated by a modest plaque on the University's oldest building. Henry himself started his studies at New York University at the age of seventeen, had finished medical school by the age of twenty, served as a professor of physiology from 1865–82, and became dean of the medical school.

The connection between current research and Draper's early work on the Orion Nebula is illustrated by the following brief report from the April 22, 1882 issue of *Scientific American.*

> *The Spectrum of the Nebula in Orion Photographed.* During the month of March Dr. Henry Draper succeeded in photographing four times the spectrum of the nebula in Orion. The same spectrum was photographed during the same month by Dr. Huggins in England. Dr. Draper has also taken photographs of the nebula itself so as to watch for changes in it and observe whether the process of aggregation into stars can be detected. Collated with the photographs of the spectrum, they show clearly, it is said, evidences of such condensations.

Today, of course, the case can be made conclusively that the Orion Nebula is, indeed, a site of recent star formation. It serves as the prototype for the study of massive star formation.

The timing of the symposium turned out to be opportune because a large body of new observational results was just beginning to become available. A common theme in the symposium is the great usefulness of studying the birth of stars at both high-spatial and high-frequency resolution using all available wavelength bands of the electromagnetic spectrum. As the appropriate instrumentation becomes available, Orion is the first star-forming region to be examined because of its relative closeness.

In a special historical session on Henry Draper and his work, in addition to the talks by Professors Gingerich and Schucking, we are privileged to include a manuscript by Professor H. Plotkin, Henry Draper's biographer, based on a colloquium he gave before the NYU Physics Department in the spring of 1981.

An exhibit on Henry Draper was organized at the time of the symposium by Anne Kinney of the NYU astrophysics group and Ann Volpe of the Coles Science Library (where it was shown); this included material from the Draper collection in the NYU archives that had been sorted, evaluated, and identified by Professor Tibor Herczeg during his sojourn as a visiting professor in the summer sessions of 1980 and 1981. A paper by Professor Herczeg and Miss Kinney discusses these photographs and other material from the archieves. Those materials not from the University archives were made available by the John William Draper Memorial Park at Hastings-on-Hudson and the NYU Chemistry Department. We would like to thank Professor Bayrd Still, Director of the NYU Archives, Mr. Gardner Osborn, Jr. of Draper Park, and Professor E. McNelis for their kind cooperation.

The scientific success of the symposium depends, of course, on the quality of the talks presented and we would like to thank all the speakers for their contributions. We received much valued assistance from the outside members of the organizing committee: O. Gingerich, P. Thaddeus, R. W. Wilson, and D. York. In addition to their valued advice, each of them either contributed an invited talk or acted as chairman of a session. We would also like to thank NASA and NSF for travel funds for speakers and young scientists. Nancy Boggess and James P. Wright were most helpful in this regard and also made important suggestions concerning the scientific program and its organization.

Many people at New York University were helpful to us in the organization of the symposium. Two successive Deans of the Faculty of Arts and Science, Norman Cantor and Lewis Levine, provided early and continuing encouragement and support; Jim Rathlesberger helped arrange the symposium dinner and reception; Vice President for External Affairs Naomi Levine generously allocated funds for the symposium dinner and other expenses; and President John Brademas and Provost Jay Oliva welcomed the participants. We would like to thank Dr. H. Pagels, President of The New York Academy of Sciences, and Dr. M. H. Shamos of Technicon Corporation for financial support. Professor P. M. Levy, Chairman of the Physics Department, was helpful in providing various kinds of administrative services. We are particularly indebted to Valerie Kekana and Jacqueline Downing for their assistance with all aspects of the symposium. Finally, we would like to thank the editorial staff of The New York Academy of Sciences, especially Bill Boland, India Trinley, and Frederick H. Bartlett, for their work on this volume.

The following papers show how the early work of such pioneers as Henry Draper has led to the rich activity that characterizes current research on the interstellar medium.

INTRODUCTION TO THE ORION SYMPOSIUM

Lyman Spitzer, Jr.

Princeton University Observatory
Princeton, New Jersey 08544

HENRY DRAPER AND THE ORION NEBULA

It was in 1880 that Henry Draper, for many years an enthusiast of astronomical photography, took his famous picture of the Orion Nebula, apparently the first photograph of a nebula ever taken. With an 11-in. telescope, an exposure of 51 min showed only the brighter parts of the nebula. The following year, a number of photographs with longer exposures were taken; these showed stars down to about the 15th magnitude. In his description of these photographs, Draper makes the prophetic remark,[1] "In this photograph with an exposure of only 137 minutes I have depicted stars almost the minimum visible in this telescope, and it is not unreasonable to hope that by still further prolonging the exposure and by still further study of photographic processes, stars and details entirely invisible to the eye may be secured."

This photograph marked an important milestone in the history of astronomy and opened a pathway that has led to many years of successful research. Thus, it is entirely appropriate that the present symposium, which is devoted to the Orion Nebula, should be in honor of Henry Draper, whose work on spectrum analysis as well as on instrumental techniques has given him an important position in astronomy.

During the hundred years since Draper first photographed the Great Nebula in Orion, this object, also known as Messier 42 and NGC 1976, has been at the center of interstellar matter research. Since M42 is relatively near the Sun, at some 500 pc distance, its detailed structure can be studied more fully than those of most other emission nebulae; also, its high surface brightness—unusual for ionized diffuse clouds—which extends over several parsecs, has permitted studies of its spectrum at relatively high resolution. Thus, for many years, the composition, temperature, and density structure of this extended nebula were better known than those of any other such region (excluding the much smaller planetary nebulae, whose surface brightnesses are also relatively high).[2]

Even during the last ten years, the Orion Nebula has remained in the forefront of interstellar matter research. About one-fourth of the papers published in the *Astrophysical Journal* on the subject of gaseous nebulae (excluding supernova remnants and planetary nebulae) deal specifically with M42. These researches cover the entire accessible range of the electromagnetic spectrum. Measurements in visible light of the atomic emission lines and of the continuous extinction produced by interstellar grains were made extensively in previous years; more recent work in visible light has dealt with other topics, such as linear polarization. Research in ultraviolet wavelengths has been concerned not only with the structure of the nebula, as shown by ultraviolet photographs obtained from space, but also with the extinction at different wavelengths. Photons of still shorter wavelengths have also been observed; for example, x-rays were recently measured from the vicinity of the Trapezium, the four bright stars at the center of the Orion Nebula.

1

0077–8923/82/0395–0001 $1.75/1 © 1982, NYAS

Research on M42 has been particularly active at infrared wavelengths. General infrared emission has been measured from the Becklin-Neugebauer and Kleinman-Low objects, with more recent spectroscopic measurements on the emission from O I, C II, and H_2. Extensive microwave measures of molecular emission lines have also been made during the last few years. Radiation from CO molecules has been observed from the large cloud complex, or giant molecular cloud, which apparently adjoins the Orion Nebula. Emission from other molecules such as H_2CO, NH_3, SiO, and CS has been measured from denser condensations within this cloud. Radio recombination lines have also been observed from the Orion Nebula itself, and the low-frequency continuous radio radiation has been used in yet another determination of the electron temperature. Evidently, there are very few techniques of observational astronomy that have not been brought to bear on M42 and its neighborhood.

THE ORION NEBULA AND STAR FORMATION

One of the chief reasons for our interest in the Orion Nebula is the process of star formation that has recently occurred, and is, presumably, still occurring, there. As is well known, the early-type O and B stars tend to form in groups. The ages of these groups, as determined either from the maximum stellar luminosities or from the expansion velocity of stellar associations, are relatively short, only a few million years. The Orion region is one of the closest where such groups of recently formed hot stars may be studied.

Furthermore, star formation is apparently going on at the present time in the cold molecular gas adjacent to the hot ionized gas of the Orion Nebula. Some of the infrared sources observed are also sources of free-free radio emission. These sources are presumably powered by hot stars formed very recently within dense condensations. These stars have not yet dissipated the cocoons of dust and gas in which they were born. Hence, much of the energy of these newborn stars is absorbed by the surrounding dust and reradiated at infrared wavelengths. Other infrared sources, those not accompanied by free-free radio emission, may be protostars in the contraction phase, radiating gravitational energy but not yet producing ultraviolet light. Hence, there are no photons sufficiently energetic to ionize and heat the hydrogen gas in the immediate neighborhood and no measurable radio emission results from free-free transitions.

The Orion region has apparently provided one of the best examples of what is called sequential star formation. Several stellar groups of differing ages are present.[3] The oldest ones are furthest away from the molecular cloud.[4] The youngest is the Trapezium cluster at the center of M42; it is closest to the star-forming region. This observation leads one to infer that the formation of one stellar group at the edge of the extensive cloud complex triggers the formation of another group deeper in the cloud complex. Thus, the dark matter is gradually converted into a succession of star groups as the star-forming process eats its way into the complex. A detailed theory has been developed by Elmegreen and Lada, presenting one scenario by which this process might occur.[5,6] It is probably in the Orion region that some of the fascinating problems associated with star formation can best be studied. The challenge of understanding how star formation proceeds makes this region especially interesting, and star formation will, of course, be one of the central themes of the present symposium.

PHYSICAL PROCESSES IN STAR FORMATION

As background for some of the discussions in this symposium, it may be helpful to review briefly some of the physical processes that are likely to be involved in the star-formation process. These processes may be grouped under two main headings, local equilibrium and dynamics. In the discussion of dynamical principles, I place special emphasis on effects associated with the propagation of shock waves through a cloudy medium, since this important subject has not been discussed in the astronomical literature.

Local Equilibrium

This topic includes most interstellar processes that are not associated with gas motions. Ignoring the processes that accelerate and decelerate gases produces a very great simplification and yet provides, in many cases, an adequate first approximation. Such problems as the thermal and chemical equilibrium of the gas can usually be discussed without considering the time derivatives. Thus, the processes that determine the kinetic temperature of the gas or the internal temperature of dust grains can be analyzed under the assumption of a steady state. On this basis, the gas kinetic temperature is determined by the condition that the random kinetic energy gained per unit time by the atoms in a unit volume of gas just equals the corresponding losses. In a region of ionized hydrogen (H II region), these processes are well understood. The gains of kinetic energy in H II regions are mostly produced by radiative ionization of neutral hydrogen atoms, with the ejected photoelectrons carrying away kinetic energy. The losses result, in part, from the kinetic energy given up by these electrons when they recombine after they have been thermalized by collisions. Usually, the dominant loss of kinetic energy results from electron excitation of low-lying energy levels in singly and doubly ionized atoms of C, N, and O. In both cases, the kinetic energy lost is converted into visible or infrared photons. The temperature computed for this equilibrium is in good agreement with that determined for H II regions in general and for the Orion Nebula in particular. The temperatures obtained from such varied measures as the intensity ratios of visual emission lines, the intensities of radio recombination lines, the spectrum slope of the continuous radiation shortwards of the Balmer limit, and the spectrum of long-wavelength radio emission are in relatively close agreement with each other and with theoretical predictions.[7]

Our understanding of the dense neutral gas in molecular clouds is less complete, though the general physical principles seem clear enough. Since ultraviolet radiation is heavily absorbed by dust, the gas in such a region is probably heated to a large extent by infrared radiation, and exchange of energy between the dust grains and the gas is believed to play an important role. It is possible that cosmic rays penetrating the cloud may contribute a significant heat source. The cooling processes are better understood than those responsible for the heating, since the energy radiated by a molecular gas can, in principle, be computed in a straightforward manner. One gap in our knowledge concerns the collisional cross-sections involved in the excitation of molecular levels; they are less well known than those for electron excitation of atomic levels.

One complication in the analysis of molecular microwave emission lines is associated with the radiative transfer equation. Those who have worked on ultraviolet absorption lines produced by interstellar atoms have been fortunate; the equation of radiative transfer that they must solve is essentially trivial—the intensity in an absorption line decreases exponentially with increasing column density of absorbing atoms. The situation is entirely different for microwave emission lines. The dominant stimulated emission produces a much more complicated physical situation; solving the equation of radiative transfer requires a detailed knowledge of the relative populations of the molecular energy levels involved. These populations depend, in turn, on the intensity of radiation in the molecular lines. Thus, solving for molecular excitation and microwave intensity together requires the solution of complex integrodifferential equations. It is customary to solve the problem by brutalizing it with various assumptions, such as local thermodynamic equilibrium. Such assumptions have the attractive advantage that they readily provide theoretical results, but they do not always conform very closely to reality. Fortunately, modern computers make it possible to deal with the full problem, though only with considerable effort.

Dynamical Processes

Analyzing dynamical developments in the interstellar medium is generally more difficult than solving the corresponding problems in the atmosphere of the Earth. Such problems usually involve the solution of complex equations involving three spatial dimensions, as well as the temporal dimension, and computers are not yet fully equal to this task. However, it is helpful to discuss the different types of dynamical processes occurring in different highly idealized situations in order to provide some understanding of the actual complicated problems.

The first and perhaps the simplest of these processes is that of instability. This phenomenon takes place in an idealized model in which a gas is at rest or has some steady motions; in this situation, one can compute the rate of growth of infinitesimal disturbances. The simplest of these analyses is the familiar analysis of gravitational instability in a uniform medium. The analysis of this situation by Jeans is not actually self-consistent,[7] but again offers the advantage of straightforward and definite results. Self-consistent results in those cases where they can be obtained, such as the self-gravitating infinite sheet, for example, seem to be in rough agreement with the Jeans results, and there is no question but that gravitational instability is an important physical process. On the other hand, in the actual interstellar conditions, where the medium is spatially irregular and temporally fluctuating, it is not always clear how relevant the results of such idealized instability analyses are.

Another important instability is that analyzed by Parker, who showed that the lines of magnetic force in the galactic plane tend to be unstable against bending into a sinusoidal pattern; the gas flows along the lines into the troughs closer to the galactic midplane.[8] Mouschovias has demonstrated that this instability does not grow indefinitely but leads to a new equilibrium.[9] However, recent studies by Asseo et al. indicate that this equilibrium is, in turn, unstable to yet other types of perturbations. In any case, all these analyses have been developed for an initially homogeneous gas, and it is not clear how applicable they may be to a gas composed of diffuse clouds separated by a much less dense intercloud medium. This uncertainty is increased by the fact that the kinetic energy of the gas is mainly due to random systematic motions of the clouds instead of random atomic motions within the gas.

If we leave aside this discussion of instabilities and consider what happens if and when they produce a collapsing cloud, a number of processes are relevant. One of the main theoretical questions is the extent to which fragmentation occurs in such a collapsing cloud. Analysis by Hoyle has indicated that, in an isothermal cloud, the Jeans mass, which is just unstable, becomes smaller and smaller as the cloud contracts and becomes denser and denser.[11] This result suggests that an unstable cloud may fragment into successively less and less massive condensations, ultimately producing a group of many stars.

While this theoretical suggestion has many appealing features, its relevance to star formation is still obscure. Is there much tendency for a cloud to condense into different fragments and, if so, will these fragments remain separate or will they coalesce again? There is still no agreement on the simplest question—will a rotating cloud fragment into two objects that revolve around each other and remain separate, leading, perhaps, to the formation of binary stars?[12,13] Whether such fragmentation can lead to the formation of an entire cluster from a condensing cloud is even more conjectural.

Among the many processes that will occur during the contraction of a cloud, two are mentioned here. These relate to the problems raised by the magnetic flux and the angular momentum, which tend to hinder the formation of individual stars from a large cloud. It appears reasonably definite that, under some conditions, the magnetic flux can be diminished by a mechanism known as ambipolar diffusion.[7] In this process, the plasma, which is composed of electrons and positive ions, together with the magnetic lines of force around which these charged particles are gyrating, slips through the neutral atoms constituting the bulk of the gas. While the magnetic flux through the plasma remains unchanged by this "plasma slip" process, the flux through the neutral gas can be much decreased by relative motions of the plasma and the neutral gas; the gas could contract gravitationally, leaving the plasma and its magnetic field behind, or the magnetized plasma could be pulled outwards, away from the bulk of the gas, by magnetic forces. In some circumstances, the magnetic flux can also be decreased by reconnections at singular lines, in the manner recently reviewed by Syrovatskii.[14]

There are also several ways in which a gas can go on condensing, even despite a large intrinsic angular momentum. As pointed out above, fragmentation into separate objects revolving about each other may concentrate the angular momentum in orbital form, leaving the objects freer to contract gravitationally. A number of analyses by Mestel, Mouschovias, and their collaborators have shown that magnetic braking of rotation can also serve to reduce the angular momentum. For this process to be effective, it is necessary that the magnetic forces slow down the rotation before plasma slip has taken the magnetized plasma away from the cloud. Under some conditions this may well be possible, providing at least one example of having one's cake and eating it too!

Of course, the overall theory of star formation must include the effects of many other processes, as well as those discussed here. The reviews by Mestel[15] and Larson[16] discuss possible scenarios for the formation of protostars and their collapse towards the stellar stage. Other effects will doubtless be discussed in this symposium.

Irregularities and Shock Waves

Shock waves play a crucial role in the dynamics of the interstellar gas and are generally present wherever appreciable velocities are observed. Theories of interstellar shocks, such as those around supernovae and newborn stars, have generally assumed a uniform medium in which the shock propagates.

It is well known that the interstellar gas is clumpy. In the theoretical model advanced by McKee and Ostriker, a line of sight in the galactic disc intersects a warm cloud about every ten parsecs.[17] The filling factor of about 0.2 that they take for this gas, whose temperature may be nearly 10 000 K, is a reasonably direct consequence of observations. If this warm gas forms envelopes around the diffuse clouds, whose number density is reasonably well known,[7] the short distance between these warm clouds along a line of sight is a direct consequence.

A shock propagating through the overall medium will tend to be strongly influenced by the clouds in its path. The intercloud medium is at a temperature of nearly 10^6 K and the density in a warm cloud will exceed the intercloud density by nearly two orders of magnitude. Hence, to a first approximation, a shock wave in such a medium interacts with the clouds as though they were incompressible. With a few simplifying assumptions, aerodynamic theory can be used to describe this interaction. The resulting phenomena will be briefly summarized here; details will be published elsewhere.

It turns out that the energy in the shock is scattered very effectively by the clouds and is converted into pressure fluctuations that propagate in various directions through the postshock gas. The effect is similar to the reverbation of thunder produced when clouds in the atmosphere scatter the pressure pulse produced by a lightning bolt. If the shock is a strong one, with M, the ratio of shock speed to the sound velocity ahead of the shock, much greater than 1, the scattered energy is carried away in a stationary bow shock ahead of the wave. If M is less than 2.76, the flow behind the incident shock is subsonic with respect to the cloud and no bow shock will arise. Instead, the transient increase of pressure produced when the shock first strikes the cloud generates a pressure pulse that travels away from the cloud in all directions. In both cases, an appreciable fraction of the kinetic energy in the postshock gas, in a cylinder with the same cross-sectional area as the cloud and extending through the outwards-moving gas, will be radiated away as acoustic energy; for strong shocks ($M > 2.76$), the radiated fraction is about 0.5, for weak shocks ($M < 2.76$), roughly 0.1.

Most of the modes in which this radiated energy appears are damped by the hot intercloud medium within a very few parsecs. The dissipated energy reappears as heat and helps drive the shock forward. Thus, the energy tends to be recycled and has several opportunities to appear in an undamped mode. Some modes, including pure Alfvén waves and acoustic waves of relatively low frequency, are much less strongly damped and can travel many parsecs. Hence, the energy density in those modes may increase and reach a level where important effects may be produced.

In particular, this energy may be absorbed by warm clouds throughout the galactic disc, providing, perhaps, the heat source needed to maintain these clouds at their observed temperature. Only a few percent of the total energy from supernovae would be required. Model calculations suggest that the conversion efficiency may be sufficiently great to provide the necessary heat source. In any case, it appears that this physical process, which produces acoustic disturbances propagating throughout the postshock gas, must be taken into account in any detailed theory of interstellar shock propagation. While the analysis of this process has been restricted so far to supernova shocks in the general interstellar medium, similar effects may well appear as a result of shocks passing through a molecular cloud complex, with scattering of shock energy by the individual condensations within a giant molecular cloud.

REFERENCES

1. DRAPER, H. 1882. Mon. Not. R. Astron. Soc. **42**: 367.
2. JOHNSON, H. M. 1968. *In* Nebulae and Interstellar Matter. B. M. Middlehurst and L. P. Aller, Eds.: 65. University of Chicago Press. Chicago.
3. BLAAUW, A. 1964. Annu. Rev. Astron. Astrophys. **2**: 213.
4. KUTNER, M. L., K. D. TUCKER, G. CHIN & P. THADDEUS. 1977. Astrophys. J. **215**: 521.
5. ELMEGREEN, B. G. & C. J. LADA. 1977. Astrophys. J. **214**: 725.
6. LADA, C. J. 1980. *In* Giant Molecular Clouds in the Galaxy. P. M. Solomon and M. G. Edmunds, Eds.: 239. Pergamon Press. New York. See also other papers in this volume.
7. SPITZER, L. 1978. Physical Processes in the Interstellar Medium. Wiley-Interscience. New York.
8. PARKER, E. N. 1966. Astrophys. J. **145**: 811.
9. MOUSCHOVIAS, T. CH. 1974. Astrophys. J. **192**: 37.
10. ASSEO, E., C. J. CESARSKY, M. LACHIEZE-REY & R. PELLAT. 1980. Astrophys. J. **237**: 752.
11. HOYLE, F. 1953. Astrophys. J. **118**: 513.
12. GINGOLD, R. A. & J. J. MONAGHAN. 1981. Mon. Not. R. Astron. Soc. **197**: 461.
13. BODENHEIMER, P. & A. P. BOSS. 1981. Mon. Not. R. Astron. Soc. **197**: 477.
14. SYROVATSKII, S. I. 1981. Annu. Rev. Astron. Astrophys. **19**: 163.
15. MESTEL, L. 1977. *In* Star Formation. T. deJong and A. Maeder, Eds.: 213. D. Reidel. Dordrecht.
16. LARSON, R. B. 1977. *In* Star Formation. T. deJong and A. Maeder, Eds.: 249. D. Reidel. Dordrecht.
17. McKEE, C. F. & J. P. OSTRIKER. 1977. Astrophys. J. **218**: 148.

DISCUSSION OF THE PAPER

A. PENZIAS (*Bell Telephone Laboratories, Holmdel, N.J.*): Can you say a word about the dimensions for the phenomena you discussed?

SPITZER: The applications I have made of this are entirely to supernovae in the interstellar medium. Similar applications can be made for shocks in molecular clouds. The dimensions in that case would be quite different. In the case of a supernova, the typical dimension for the cloud is about 5 pc and that for the distance of the cloud from the supernova is about 50 pc.

PENZIAS: There seems to be a strong resemblance between your scheme and the number of small circular features one often sees within the envelope of supernova shocks.

SPITZER: Yes. Two analogies are useful in this connection. One is the reflected shocks seen from obstacles in wind tunnel experiments. The other is the reverberation of thunder following a lightning bolt.

QUESTION: How good is the approximation of an incompressible cloud?

SPITZER: For a reflection of a single shock from a single cloud, as I discussed it, is a reasonable approximation. For phenomena occurring over a longer timescale, a more complete treatment is needed, *e.g.,* along the lines of the theory of Cowie and McKee.

MOLECULAR CLOUDS IN ORION AND MONOCEROS

P. Thaddeus

Goddard Space Flight Center
Institute for Space Studies
New York, New York 10025

INTRODUCTION

It is evident, even from a casual inspection of the Barnard and Palomar photographic surveys and the 21-cm surveys of Orion and Monoceros,[1-4] that there is a great deal of neutral interstellar matter in the vicinity of the Orion Nebula, and it is easy to calculate that this material must be much more massive than the young stars and ionized gas for which the region is noted—probably by a factor of at least 10. Until the surveys of CO and other molecules were made, however, our knowledge of the true amount of neutral interstellar matter in the Orion region was seriously incomplete and the relation of the dense gas to the young stars was almost entirely unknown. Orion is too far away and too far below the galactic plane for dark nebulae to stand out clearly against background stars, and the 21-cm line, owing to the wholesale conversion of H to H_2, tells little about the distribution of the high-density gas from which stars are actually made.

The first interstellar CO observations, made about a decade ago, were mainly toward familiar H II regions; it was only by a process of trial and error, with little more than the Palomar survey prints as a guide, that radio astronomers gradually discovered the true size and complexity of the larger molecular clouds. K. D. Tucker, M. L. Kutner, and I began the first large-scale CO survey of the Orion region with the 5-m telescope of the University of Texas at McDonald Observatory in January 1973, shortly after this precisely figured millimeter wave antenna had been equipped with a receiver and spectrometer for spectral line observations. Initially, our goal was to study the molecular gas in the neighborhood of the H II region NGC 2024 (Ori B), but, on discovering that intense CO emission extended south from that source in an apparently continuous sheet for more than 1° to the Horsehead Nebula and the reflection nebula NGC 2023, we decided that finding the size and structure of the cloud as a whole was a more interesting objective. We soon succeeded in showing that, although there is a fairly sharp edge south of NGC 2023, the cloud stretches more than 4° to the northeast, beyond the reflection nebulae NGC 2068 and NGC 2071, and that, in all directions, it is apparently as large as or larger than the diffuse dark nebula L1630 that can be vaguely discerned on the Palomar print of this region. From the CO intensity, the visual extinction, and the virial theorem applied to the CO linewidth, we estimated the cloud mass to lie in the range 2.5×10^4–$1.0 \times 10^5 \, M_\odot$.[5]

In 1973, there were only two telescopes equipped to observe the fundamental CO rotational transition at 115 GHz: the 5-m at McDonald and the NRAO 36-ft at Kitt Peak. At that frequency, both had beamwidths so small that many years of continuous observation would have been required to fully sample a large object of the kind we had found in Orion—and these were general purpose, heavily subscribed facilities only a fraction of whose time could be devoted to a given project. There was clearly a need for

9

0077–8923/82/0395–0009 $1.75/1 © 1982, NYAS

a smaller telescope dedicated to surveying molecular clouds in CO and other simple molecules and, on returning from our first Texas observing run, I set out to construct such an instrument with three graduate students from the Columbia University Physics Department: Gordon Chin, Richard Cohen, and Hong Ih Cong.

An aperture of four feet, for a beamwidth of about 8 arcmin at 115 GHz, was chosen as a reasonable compromise between survey speed and angular resolution. To keep the telescope accessible to graduate students, and easy to construct and maintain, we installed it on the roof of the Pupin Physics Laboratory in the middle of New York City; at 115 GHz, atmospheric conditions there are about as good as at any sea level, mid-latitude site and, during the best 50–60% of the time, the sky opacity (as we had hoped) has turned out to be about as good as the average opacity on Kitt Peak—except possibly in summer, when we normally shut down. There is no detectable radio frequency interference at 115 GHz on the Pupin roof and normal shielding precautions have prevented interference by LaGuardia airport radar and other local transmitters at the intermediate frequencies of our superheterodyne receiver.

Our first observations were made less than two years after starting the project, when Gordon Chin began a large-scale survey of the molecular clouds in the Orion region as his dissertation. He ultimately obtained well-sampled maps in both CO and ^{13}CO of the Ori B molecular cloud and of an even larger cloud associated with the Orion Nebula that Kutner and Tucker had been studying in the mean time with the Texas telescope. With this extensive data, fairly good mass estimates were now possible on the local thermodynamic equilibrium (LTE) assumptions (1) that ^{13}CO is optically thin, (2) that CO is optically thick and provides the rotational temperature and hence partition functions of both isotopic species, and (3) that the H_2 to ^{13}CO column density ratio (as determined from star counts[6]) is 5×10^5. Using this method, we determined that the mass of the Ori B cloud is $6 \times 10^4 \ M_\odot$, in good agreement with our original estimate, and that the mass of the Ori A cloud is $1 \times 10^5 \ M_\odot$.[7]

Although Chin had succeeded in finding fairly definite boundaries to the Orion molecular clouds at a level of 1–2 K in peak CO intensity, I suspected that, with greater sensitivity, these objects would be discovered to be more extensive; there were also other promising objects in the Orion neighborhood, such as the cluster of reflection nebulae Mon R2, that we had hardly observed. Therefore, with a new receiver and spectrometer on the 4-ft telescope, Mark Morris and I began a larger-scale survey of the Orion region in the fall of 1977, beginning to the north and east of NGC 2071 and near Mon R2. Subsequently, J. Moskowitz resurveyed the two main Orion clouds; we now possess a fairly homogeneous and nearly complete inventory of the molecular clouds in Orion and Monoceros to a peak CO intensity of 0.5 K and a velocity resolution of 0.6 km s^{-1}. A summary of the survey, a contour map of velocity-integrated CO intensity prepared by R. Maddalena, is shown in FIGURE 1. FIGURE 2 is a schematic key to this map and FIGURE 3 shows the locations of the 6300 positions actually observed.

Several aspects of the survey deserve comment.

STAR FORMATION

FIGURE 1 is probably unrivaled as a general exhibit of the superiority of CO as a tracer of star formation. As comparison with FIGURE 2 helps show, essentially all the

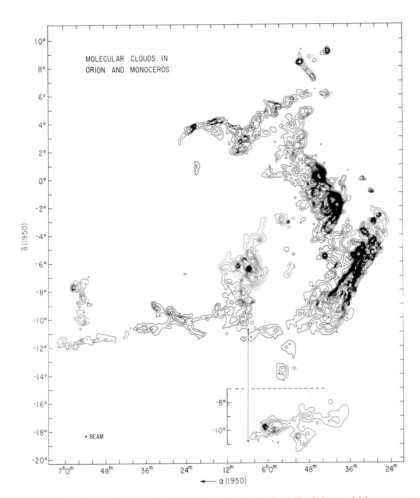

FIGURE 1. The Goddard-Columbia survey of molecular clouds in Orion and Monoceros—a map of the integrated intensity of CO line emission by the $J = 1 \rightarrow 0$ rotational transition at 115 GHz. The interval between contours is 2.6 K km s^{-1}, but the lowest contour is at 1.3 K km s^{-1} to better show the extent of weak emission. FIGURE 2 provides a schematic key to the survey. Over most of the region there is only one CO velocity component and no ambiguity in plotting integrated intensity, but there are two components in the bridge connecting the Mon R2 and Ori A clouds, one at an LSR (local standard of rest) radial velocity of about 3 km s^{-1} (shown in place) and a second at about 10 km s^{-1} (shown in the insert at the bottom of the figure). The overall velocity structure (in km s^{-1} in the local standard of rest) can be briefly summarized as follows: southern filament, left to right—15 to 11; Mon R2 cloud, bottom to top—9 to13; Ori A cloud, bottom to top—5 to 11; Ori B cloud—a fairly constant 7 to 9, but with a second velocity component at 3 near NGC 2024; and northern filament—9 ± 2.

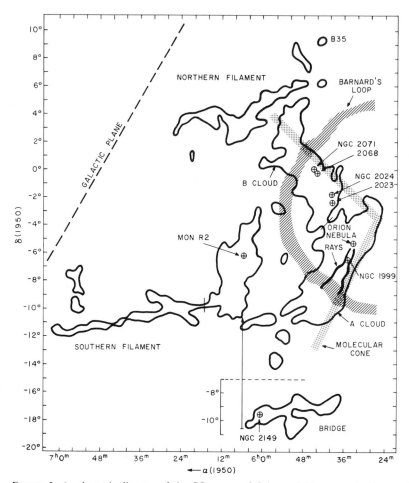

FIGURE 2. A schematic diagram of the CO survey of Orion and Monoceros in FIGURE 1, showing the location of H II regions, some of the reflection nebulae in the region, and Barnard's Loop.

well-known H II regions, reflection nebulae, infrared sources, Herbig-Haro objects, etc. that have been shown to be "signposts" of star formation in Orion and Monoceros (including, of course, the Orion Nebula) stand out clearly as regions of enhanced CO emission, which, from an observational standpoint, can be regarded as the common denominator of these disparate objects. These CO hot spots are, in turn, the result of enhanced temperature or density or both, and therefore are probably also continuum infrared sources, but continuum ir observations fail to provide the invaluable velocity information provided by CO. Since none of the molecular clouds in FIGURE 1 can be distinguished on the 21-cm surveys, the superiority of CO to H I in studying star formation is obvious.

Filaments

Carbon monoxide emission also delineates large quiescent molecular regions in which there is no evidence for star formation, and which are difficult to observe in any other way. Among the examples of such CO-specific regions in Figure 1 are the two long filaments that stretch from south of Mon R2 and north of NGC 2071 nearly to the galactic plane. The mass of these filaments may be substantial—Morris and Maddalena have observed ^{13}CO at several positions in the northern filament—but other simple molecules, (*e.g.,* CS, OH, HCO$^+$, and H$_2$CO) have not been detected and there is no

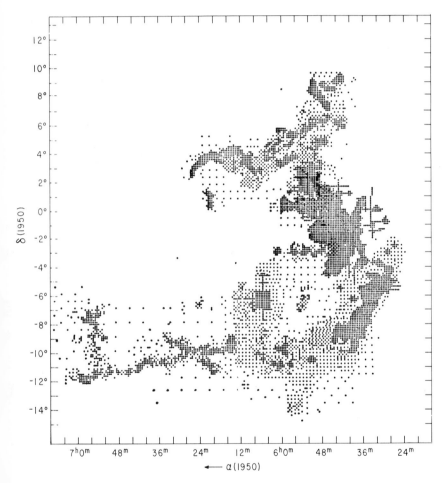

FIGURE 3. The location of observations made with the 4-ft telescope during the survey, each point being approximately the size of the antenna beam. The survey is not unbiased: In searching for new objects, observations were generally made on a 1° or 1/2° grid; once found, a cloud was sampled every beamwidth or every two beamwidths (*i.e.,* every 1/4° or 1/8°).

evidence for H II knots or reflection nebulae in the Palomar survey. An infrared survey of these filaments is highly desirable and it would be of interest to determine the magnetic field from stellar polarization in their vicinity. Similar objects have been detected with the Columbia telescope in other parts of the sky; they appear to constitute a fairly abundant class of molecular clouds that has been heretofore overlooked.

MON-ORI BRIDGE

According to Herbst and Racine, the Mon R2 complex of reflection nebulae and young stars is at a distance of 830 ± 50 pc[8] and, hence, is significantly further than the Orion Nebula; yet our CO map suggests the existence of a bridge connecting the Mon R2 and Ori A clouds. Rejecting the possibility that this is merely a chance alignment of unrelated objects, the alternatives would seem to be (1) that Mon R2 is closer than claimed or (2) that the bridge is being viewed nearly end on and is actually a molecular

TABLE 1

MASSES OF MOLECULAR CLOUDS IN THE ORION-MONOCEROS COMPLEX

Cloud	Mass ($10^5 M_\odot$)		
	CO	LTE	Virial
Ori A	1.1	1*	1.9
Ori B	0.86	0.6*	1.4
Mon R2	0.74	—	1.3
Bridge	0.22	—	0.73
North Filament	0.15	—	1.3
South Filament	0.25	—	2.1
Total	3.3		8.7

*Reference 7.

filament several hundred parsecs in length—that is, as long as or longer than the northern and southern filaments just discussed. The greater the distance of Mon R2 from the Orion Nebula, of course, the less plausible the latter alternative becomes; perhaps the most satisfactory resolution of this puzzle is that both are correct. In any case, our data clearly raise the possibility of a previously unsuspected relation between Mon R2 and the Orion Nebula and emphasize the need to carefully scrutinize the question of their relative distance.

MASSES

Preliminary mass estimates for the main molecular clouds in the Orion-Monoceros complex are given in TABLE 1. The CO mass in the second column is simply derived on the assumption that W_{CO}, the integrated intensity of the line of the normal isotopic species that we have surveyed, is roughly proportional to the H_2 column density when averaged over a cloud; in column 2, we have adopted $N_{H_2}/W_{CO} = 2 \times 10^{20}$ cm^{-2}

$(K \text{ km } s^{-1})^{-1}$, the value recently derived from a comparison of Columbia CO and COS-B gamma ray surveys of the first galactic quadrant.[9] The LTE mass is that derived by Chin from his ^{13}CO and CO surveys. The virial mass is derived by applying the virial theorem to the widths of the CO lines in the present survey: $M_{vir} = CR\Delta v^2/G$, where R is the cloud radius, Δv is the full velocity width at half intensity of the CO line averaged over the cloud, and C is a factor on the order of unity that depends on the geometry and density structure.

Space does not permit discussion of the relative merits of the CO and LTE masses, except to say that the two are probably of comparable accuracy and that neither can be trusted to an accuracy of better than 50%. The point that deserves emphasis is that, while the CO and LTE masses are in fairly good agreement, both are systematically less than the virial masses by a factor of about three. Our survey data therefore suggest that the molecular clouds in Orion and Monoceros are not gravitationally bound objects—unless surrounded by fairly massive and so far unobserved H I halos—and will disperse in a time on the order of $t \simeq 2R/\Delta v \simeq 2 \times 10^7$ y. With the Orion region as a cautionary example, masses of molecular clouds elsewhere in the Galaxy determined from the virial theorem should only be regarded as fairly loose upper limits.

RELATION TO ORION'S CLOAK

It is remarkable that, with the exception of Mon R2, the main areas of star formation in our survey appear to lie like blisters on the forward surface of a large blunt cone or wedge of molecular clouds pointing towards the center of Barnard's Loop and the older subassociations of OB stars in Orion (FIGURE 2)—a region where our negative CO results, 21-cm observations, and galaxy counts[10] indicate that the density is now quite low, but that once presumably contained the dense molecular gas from which the older OB stars formed. Indeed, Cowie et al. have shown from Copernicus uv spectra and other data that the older subassociations are the apparent center of a systematic high-velocity outflow of gas with a maximum velocity greater than 100 km s^{-1} and a diameter of about 120 pc—a feature they have dubbed Orion's Cloak.[11] The highest velocity outflow, attributed by Cowie et al. to a radiative shock at the outer edge of the cloak, may be the result of a single supernova about 3×10^5 y ago, but the far more massive outflow observed at intermediate velocities apparently requires multiple supernovae explosions (~ 10) or stellar winds or both operating over a period of about 3×10^6 y.

A large-scale sustained wind of this kind that has dispersed and ionized much of the molecular gas in its path can hardly fail to have a pronounced effect on that which remains, and it is plausible to assume that both the swept back, conical configuration of clouds that we observe and the patches of star formation apparently facing the wind, (including the Orion Nebula) are a direct result of Orion's Cloak. If this assumption is correct, self-propagating star formation in large molecular complexes may operate on a larger scale, and may be a more coherent process, than has generally been assumed, resulting in the wholesale dispersal of the complex in a fairly short time.

I am indebted to R. Maddalena for preparing FIGURES 1 and 3 and for calculating the cloud masses in TABLE 1, and I wish to thank him and Mark Morris for a series of illuminating discussions on the questions posed by the Orion region; the three of us are preparing a much fuller description of the work reported here.

REFERENCES

1. GORDON, C. P. 1970. Astron. J. **75:** 914.
2. HEILES, C. & E. B. JENKINS. 1976. Astron. Astrophys. **46:** 333.
3. MENON, T. K. 1958. Astrophys. J. **127:** 28.
4. WOERDEN, H. VAN. 1967. *In* Int. Astron. Union Symp. 66, Radio Astronomy and the Galactic System. H. van Woerden, Ed. Academic Press. New York and London.
5. TUCKER, K. D., M. L. KUTNER & P. THADDEUS. 1973. Astrophys. J. **186:** L13.
6. DICKMAN, R. L. 1978. Astrophys. J. Suppl. Ser. **37:** 407.
7. KUTNER, M. L., K. D. TUCKER, G. CHIN & P. THADDEUS. 1977. Astrophys. J. **215:** 521.
8. HERBST, W. & R. RACINE. 1976. Astron. J. **81:** 840.
9. LEBRUN, F. 1982. Private communication.
10. BAUD, B. & J. G. A. WOUTERLOOT. 1980. Astron. Astrophys. **90:** 297.
11. COWIE, L. L., A. SONGAILA & D. G. YORK. 1979. Astrophys. J. **230:** 469.

DISCUSSION OF THE PAPER

A. STARK (*Bell Telephone Laboratories, Holmdel, N.J.*): It seems that the velocity gradient in the Orion A cloud has a magnitude and direction corresponding to nonrotation in an inertial frame, which minimizes the centrifugal force. If the cloud were formed from chaotic material, and was barely bound, I would expect that the material that formed the cloud would also have this property.

THADDEUS: That may well be true, but the region is so complex that it is difficult to tell which is the most important effect.

R. WILSON (*Bell Telephone Laboratories, Holmdel, N.J.*): At each step of your calculation there seems to be a problem with the mass estimate. For example, there are several difficulties in interpreting the ^{13}CO excitation temperature. Do you really think that your mass estimates are good to within a factor of two or three?

THADDEUS: Well, it's like investigating the biography of a shifty politician. The more you look at the various events in the life of the man, the more you decide you don't know what's happening. That's true to a certain extent with these mass calculations. My own feeling is that they are order of magnitude estimates in many ways.

T. MOUSCHOVIAS (*University of Illinois, Urbana, Ill.*): Are there polarization observations available in the neighborhood of the retrograding cloud and, if so, what is the orientation of the polarization vectors with respect to the axis of rotation of the cloud?

THADDEUS: There is no information available.

ORION'S CLOAK AS A MODEL FOR SUPERSHELLS
OF GAS AROUND OB ASSOCIATIONS*

Lennox L. Cowie

Physics Department
Center for Space Research
and
Center for Theoretical Physics
Massachusetts Institute of Technology
Cambridge, Massachusetts 02139

If our current belief that type II supernovae form from massive stars is true,[24] and if the ratio of type II to type I supernovae is near unity, as the observations of external galaxies suggest,[1] then the most active supernova sites in the Galactic disk should be the OB associations. However, relatively few of the radio-selected sample of supernova remnants occur in the associations.[2] Perhaps this is not unreasonable, since it is clear that the dynamical evolution of supernova explosions in the associations should be very different from that in the field. The main reasons for this are the homogenizing effects of photoionization on the surrounding gas[3] and the evacuation of gas by previous explosions. As has been recently pointed out by several groups, the most attractive feature of this general point is the possibility of reconciling the higher supernova birthrates based on extragalactic counts[1] with the lower radio remnant birthrate.[2,4,5]

I now want to consider the Orion OB1 association. I will include λ Ori in Ori OB1, despite the fact that it is normally considered to be distinct. The reason for this will become obvious. As in everything else, one would hope to use Orion as a paradigm—in this case, for a class of event that I want to call the association supershell, which is the composite remnant of the supernovae in an association.

Orion OB1 was the association most heavily observed by the Copernicus satellite uv spectrometer and, even in the early surveys, very high signal-to-noise exposures were obtained of a number of the association stars. As was first pointed out by Cohn and York, the strong interstellar uv absorption lines in the Orion stars are very unusual.[6] In FIGURE 1, which uses observations made by D. G. York and myself, one can see the comparison between an Orion star (23 Orionis) and the normal star 54 Crucis. The peculiar feature is the negative velocity gas at -100 km s^{-1}. In the Copernicus surveys, gas at this high a velocity is unique to Orion.

In an attempt to understand this particular Orion phenomenon, York, Songaila and I mapped a large number of stars in Orion using Copernicus.[7] The results are fascinating for several reasons. First, the presence of the highest negative velocity features seem to be a hallmark of the Orion stars both in Ori OB1 and λ Ori (FIGURE 2). Second, there is no corresponding very high positive velocity feature. Third, there are sporadic intermediate velocity features at both positive and negative velocities in many of the stars (FIGURE 3). Finally, the column density and ionization structure of the highest velocity features are reasonably invariant from star to star, with a total

*This research was supported by grants from the National Aeronautics and Space Administration, nos. NAGW-308, NSG-7643, and NGL-22-009-638.

17

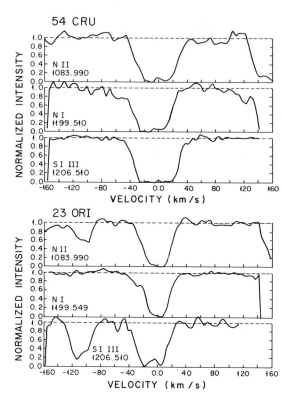

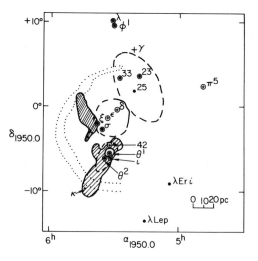

FIGURE 1. A comparison between an Orion star (23 Orionis) and a star with more normal interstellar absorption (54 Crucis). The absorption feature at -100 km s^{-1} in 23 Ori is seen in many of the Orion stars.

FIGURE 2. The spatial distribution of high-velocity gas in observed Orion stars. Stars with high-velocity components are circled. (From Cowie et al.[7])

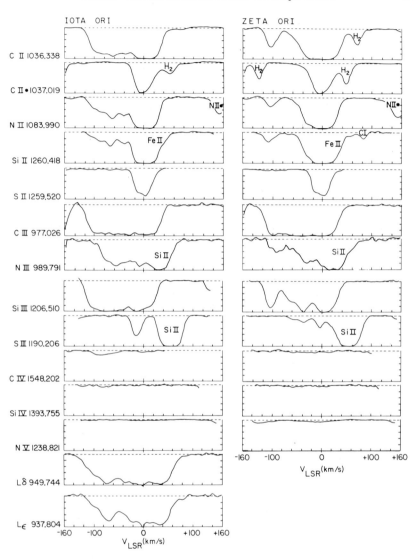

FIGURE 3. The stars ι and ζ Orionis show both the highest-velocity component and a considerable amount of intermediate velocity structure. (From Cowie *et al.*[7])

hydrogen column density of a few 10^{17} cm^{-2} distributed in species ranging from singly to triply ionized.

From these results, we inferred that the structure of the gas surrounding Orion was as shown in FIGURE 4. Surrounding the Ori OB1 and λ Ori regions is a radially expanding thin, uniform, low column density shell of fast-moving (100 km s^{-1}) gas. The absence of very high positive velocity features suggests that none of the association stars lie behind this shell, which has a radius of about 50–100 pc. Inside this shell is a slower-moving inhomogeneous region of higher density cloudy gas that produces the more sporadically distributed gas at velocities in the 30–100 km s^{-1} range. Some of the stars must lie within the rearward (positive velocity) side of this region—since they possess positive velocity features—though the great majority do not. The slower-moving regions contain both neutral and ionized gas.

We can also tie earlier observations into this picture. The optical observations of absorption lines in Orion by Adams[8] and, later, Hobbs[9] and the 21-cm studies of Menon[10] and Gordon[11] and Hα studies[12] showed this region to have an unusual amount of structure at the lower velocities (~20 km s^{-1}). This structure is presumably produced by the slower-moving inhomogeneous shell, with the higher column density gas seen by these observations having a slightly lower velocity than the lower column density gas seen in the uv. Ionization and reflection from the inner edge of the inhomogeneous shell must form Barnard's Loop.

In Reference 7, we interpreted the structure described above as follows. The homogeneity, ionization structure, and column density of the highest velocity feature

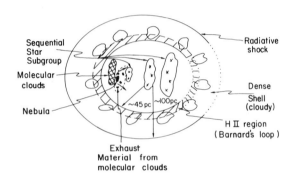

FIGURE 4. A schematic outline of the Orion structure. In the interior there is an evacuated cavity containing most of the association stars and the molecular cloud. At the edge of this lies a dense, inhomogeneous, relatively slowly moving (~20 km s^{-1}) shell. Finally, outside this, there lies a fast-moving (100 km s^{-1}) radiative shock sweeping up low-density (few 10^{-3} cm^{-3}) gas. (From Cowie *et al.*[7])

suggest that this feature is, in fact, a one hundred kilometers per second radiative shock. The column densities of individual ionization species of the various elements behind such a shock are predicted independently of the preshock density[13,14] and agree quantitatively with the observed values. The density of the gas being swept up by the feature can be inferred from the radius of the shell and the column density and is a few 10^{-3} cm^{-3}. Pressures in the feature can also be inferred from the fine-structure lines of C II and Si II. In the shock model, the fine-structure lines give an upper limit to the preshock density of a few 10^{-3} cm^{-3}. Finally, the characteristic age of the feature obtained by dividing the radius by the velocity is between half a million and a million years. This is relatively young compared with stellar lifetimes or with the age of the association and suggests that the progenitor produced a relatively instantaneous energy input and was probably the most recent supernova in Orion.

The kinematic age of the more inhomogeneous higher column density gas is significantly longer than that of the highest-velocity gas because of its lower velocity.

In this case, the age is comparable to the association age and the motions may be caused by the composite effects of stellar winds and supernovae within the association. While in an inhomogeneous medium the supernova energy propagates primarily in the diffuse substrate interspersed among the clouds forming the shell, the periodic supernova blast waves can transfer some momentum to the clouds and gradually generate a radial velocity.[15]

In this picture, we see the most recent supernova in the highest-velocity gas and the composite effects of the history of supernova formation in the denser slower-moving material.

We were struck by a number of very nice features about all this. First, there is distance and age information. We can see at once that λ Ori and Ori OB1 are at the same distance to within 100 pc. The size of the supershell is roughly fixed and provides a crude yardstick for absolute distance estimates. Second, we can measure a rough age for the association of 5–10 My from the age of the inhomogeneous shell. In comparing this age with spectroscopic ages, the complication is determining which subgroup of the association this age corresponds to. In Orion's case, the ages of the various subgroups range up to 12 My.[21] Finally, we can compare the interval between supernovae inferred from taking the inverse of the number of massive stars divided by their evolution times and comparing this with the time to the last supernova. The two agree quite well at about 0.5 My.

Unfortunately, Orion has proved to be, as usual, somewhat anomalous. Hu, Taylor, York, and I have carried out a survey of 13 additional associations to look for similar supershells and found only one similar structure—that around the Carina OB1 and OB2 associations.[16] Hu, Songalia and I have also looked at the optical absorption line and 21-cm structure around the Car OB1 and Car OB2 associations and found it to be almost identical to Orion.[17] It is rather an interesting point that the Carina supershell also contains two associations (Car OB1 and Car OB2) that are normally distinguished and are considered on standard distance estimates to be at about 500 pc apart,[18] but which are, in fact, at the same distance.

I want to close with some brief theoretical considerations of the evolution of the class of association supershells based on Bill Jeffrey's MIT senior thesis.[19] It is likely that, in the early period, 2–3 My after an association forms, the surrounding region is photoionized and homogenized and that the bubble blown by the association winds and supernovae is isotropic, homogeneous, and slowly expanding. This type of structure was first described by Bruhweiler et al.[20] Only later, as the earliest-type stars die and the photoionization rate drops, does the shell penetrate the Strömgren sphere and break up, allowing the energy from individual supernovae to break out and form fast shocks in the diffuse (around a few 10^{-3} cm^{-3}) component of the interstellar matter (ISM) (see, e.g., McKee and Ostriker for a description of the ISM[22]). In this picture, the later stages of the association supershell evolution are of the Orion-Carina form. In the earlier stages, the supershells look very much like rather homogeneous supernova remnants surrounding the associations. I want to emphasize that these early-stage supershells may indeed be identified as supernova remnants and appear as such in standard lists. The most obvious case may be the Monoceros loop that surrounds Mon OB2.[23] This very homogeneous radio-emitting shell is somewhat of an anomaly as a supernova remnant both because of its very large size (a radius of 60 pc), its very uniform structure, and its anomalously low radio surface brightness.[2] However, it fits very well into the early stages of the evolutionary sequence of the supershells.

This paper is heavily based on the work of Esther Hu, Bill Jeffrey, and Toni Sonagaila, and I would like to thank them for allowing me to discuss their results in advance of publication. The observations shown and discussed were carried out using Copernicus and the International Ultraviolet Explorer and at Cerro Tololo Interamerican Observatory and were made possible by the assistance of the staffs of all three facilities.

REFERENCES

1. TAMMANN, G. 1974. In Supernovae and Supernova Remnants. C. B. Cosmovici, Ed. D. Reidel. Dordrecht.
2. CLARK, D. H. & J. L. CASWELL. 1976. Mon. Not. R. Astron. Soc. **174:** 267.
3. ELMEGREEN, B. G. 1976. Astrophys. J. **205:** 405.
4. KAFATOS, M., S. SOFIA, F. C. BRUHWEILER & T. R. GULL. 1980. Astrophys. J. **242:** 294.
5. HIGDON, J. C. & R. E. LINGENFELTER. 1980. Astrophys. J. **239:** 867.
6. COHN, H. & D. G. YORK. 1977. Astrophys. J. **216:** 408.
7. COWIE, L. L., A. SONGAILA & D. G. YORK. 1979. Astrophys. J. **230:** 469.
8. ADAMS, W. S. 1949. Astrophys. J. **109:** 354.
9. HOBBS, L. 1969. Astrophys. J. **157:** 135.
10. MENON, T. K. 1958. Astrophys. J. **127:** 28.
11. GORDON, C. P. 1970. Astrophys. J. **75:** 914.
12. REYNOLDS, R. & P. OGDEN. 1979. Astrophys. J. 229, 942.
13. RAYMOND, J. C. 1979. Astrophys. J. Suppl. Ser. **39:** 1.
14. SHULL, J. M. & C. F. MCKEE. 1979. Astrophys. J. **227:** 131.
15. COWIE, L. L., E. M. HU & A. SONGAILA. 1982. In preparation.
16. COWIE, L. L., E. M. HU, W. TAYLOR & D. G. YORK. 1981. Astrophys. J. **250:** L25.
17. COWIE, L. L., C. F. MCKEE & J. P. OSTRIKER. 1981. Astrophys. J. **247:** 908.
18. HUMPHREYS, R. 1978. Astrophys. J. Suppl. Ser. **38:** 309.
19. JEFFREY, W. 1981. The Effects of Supernovae in OB Associations. Senior Thesis. MIT Physics Department.
20. BLAAUW, A. 1964. Annu. Rev. Astron. Astrophys. **2:** 219.
21. BRUHWEILER, F. C., T. R. GULL, M. KAFATAS & S. SOFIA. 1980. Astrophys. J. **238:** L27.
22. MCKEE, C. F. & J. P. OSTRIKER. 1977. Astrophys. J. **218:** 148.
23. WALLERSTEIN, G. & T. S. JACOBSEN. 1976. Astrophys. J. **207:** 53.
24. OSTRIKER, J. P., D. O. RICHSTONE & T. X. THUAN. 1974. Astrophys. J. **188:** L87.

DISCUSSION OF THE PAPER

C. HEILES (*University of California, Berkeley, Calif.*): You present a picture in which the volume density of the gas surrounding the Orion region is very small, perhaps 10^{-2} or 10^{-3} cm^{-3}. Yet the 21-cm observers find a large atomic hydrogen cloud with a radius of about 100 pc and a density of about 2.5 cm^{-3}. Can you reconcile these two points of view?

COWIE: The region is inhomogeneous. The bulk of the mass is contained in denser clouds or filaments, whereas most of the volume is occupied by hot, rarified gas.

P. SOLOMON (*State University of New York, Stony Brook, N.Y.*): The picture you present seems to account for the formation of a thin shell of about 100 $M_\odot$. But does it tell us anything about star formation?

COWIE: Although it may not be very informative about the molecular cloud or star formation, it does tell us about the origin of some of the inhomogeneity of the diffuse interstellar medium. It is an effect rather than a cause of star formation.

P. THADDEUS (*Goddard Institute for Space Studies, New York, N.Y.*): How did you arrive at your estimate of the supernovae rate in Orion?

COWIE: The rate can be estimated from the number and expected lifetimes of the OB stars in the association. It is in accord with the value required to explain the observed kinematics of the shell.

G. FIELD (*Harvard-Smithsonian Center for Astrophysics, Cambridge, Mass.*): You mentioned two timescales, 0.5 and 5 million years. The former is derived from the fast-moving hot gas. What is the latter? Can I conclude that there have been ten supernovae?

COWIE: The shorter age estimate is based on the radiative shock at the outside edge of the association. The longer age is based on the denser, slower-moving structure. The shorter-age phenomenon occurs frequently, so there have been ten or twenty supernovae in Orion.

A. YAHIL (*State University of New York, Stony Brook, N.Y.*): Woosley and Weaver argue that a minimum mass for a type II supernova is 10 $M_\odot$, whereas you mentioned 4 or 5 $M_\odot$. Does this difference cause any difficulty in your model?

COWIE: There is no problem because the expected rate is dominated by O9 stars.

M. L. KUTNER (*Rensselaer Polytechnic Institute, Troy, N.Y.*): The brightest part of Barnard's Loop is where it crosses the dark cloud L1630. Is this explainable in terms of your three-dimensional picture, or is it a coincidence?

COWIE: It is probably not a coincidence because the densest regions should be the brightest.

PHYSICAL CONDITIONS OF THE ORION NEBULA
DERIVED FROM OPTICAL AND ULTRAVIOLET DATA

Manuel Peimbert

Instituto de Astronomía
Universidad Nacional Autónoma de México
México 20 D.F., México

INTRODUCTION

The Orion Nebula has been the most observed H II region in the optical and uv regions due to its apparent size, low reddening, and high emission measure. Many models attempting to explain different aspects of the Orion Nebula have been presented in the literature. In this review, I will be mainly concerned with some aspects of its structure, dust content, temperature distribution, and abundances of the gaseous component.

STRUCTURE

By structure I mean mainly the distribution and motions of stars, gas, dust, and ionization fronts. We can distinguish three different domains centered on the Trapezium: $r < 2'$, $r < 10'$, and $r < 32'$.

The Region with $r < 2'$

This is the brightest region of the Orion Nebula and contains about half the emission. Ten to twenty years ago, most of the models of the Orion Nebula were spherically symmetrical; however, about eight years ago, Zuckerman[1] and Balick *et al.*[2] proposed a model of the Orion Nebula in which the molecular cloud is behind but very close to the Trapezium ($\sim$0.1–0.2 pc) and the direction to the observer is density bounded. With this model, they were able to explain the main features of the observed radial velocity pattern and the correlation of radial velocity with degree of ionization (*e.g.,* Kaler[3]), in the sense that the smaller the ionization degree the larger the redshift. At the H I–H II interface behind the Trapezium, the gas suffers a process of acceleration due to the large density gradient expected. The lines of low degrees of ionization (*e.g.,* O I, C II, N II, and S II) are formed in regions near the ionization front, where pressure gradients have been only partially effective in accelerating the gas, while the lines of high degrees of ionization (*e.g.,* He I, O III, and Ne III) arise in the flow closer to us, where the gas travels at the sound speed relative to the ionization front. Tenorio-Tagle has computed an evolutionary numerical model to reproduce this situation and has called it the champagne model.[4] Refinements on this basic model that introduce secondary flows and different geometrical considerations have been introduced by many authors (*e.g.,* References 5–7).

The main flow fails to explain the velocity pattern of many regions within $r < 2'$ (see References 8–14) and other causes have to be taken into account, such as the

24

disruption of partially ionized globules with neutral cores,[12,14] mass loss from Herbig-Haro objects,[15] and mass loss from θ^2C, θ^1D, and θ^2A Orionis.[16–18]

In addition to the complex velocity field pattern, there are two observations that cannot be explained by the champagne model of the core of the Orion Nebula. (1) The predicted pressure gradient is not observed (see TABLE 1, which presents the densities and temperatures from the [C II] 157-μm emission line,[19] the [C I] lines $\lambda\lambda$9850 and 8727,[20] and the optical lines[21–23]); if anything, the regions of higher ionization seem to be at higher densities. (2) θ^1C Ori exhibits interstellar absorption lines both of a low degree of ionization, such as those of C I, N I, O I, and Mg I, and of a high degree of ionization, such as those of S III, C IV, and Si IV, with radial velocity behaviors similar to those of the emission lines.[18] These observations imply that the material closer to the star, which has a higher degree of ionization and is farther away from us, is blueshifted with respect to the material closer to us, which has a lower degree of ionization—exactly the opposite of the champagne model prediction.

The Region with r < 12′

The Fabry-Perot observations of Deharveng show extended regions where the [N II] lines are split by 20–25 km s^{-1}.[24] The gas velocity distribution of these regions

TABLE 1

DENSITIES, TEMPERATURES, AND PRESSURES
IN THE CENTRAL REGIONS OF THE ORION NEBULA

Identity	log N_e	log N	log T	log P/k	v (km s^{-1})
[C II]	1.50	5.0	2.3	7.3 ± 0.5	+8
[S II]	3.65	3.95	3.9	7.85 ± 0.2	+0
[O II]	3.60	3.90	3.9	7.80 ± 0.2	+0
[Cl III]	3.70	4.00	3.9	7.90 ± 0.3	−3
[Ar IV]	4.00	4.30	3.9	8.20 ± 0.4	−3

(A, B, and C) cannot be explained by the model for the core. Deharveng has proposed a model with an incomplete shell in expansion to explain the [N II] velocity pattern.

We will concentrate on region A, where θ^2A Ori is located. There are four main arguments that support the idea that there is an H II region around θ^2 Ori disconnected from the main Trapezium H II region, just as M43 is located north of the main body of the Orion Nebula: (1) the different radial velocity field,[24] (2) from 10-μm observations, it has been suggested that the dust in the bar southeast of the Trapezium and northwest of θ^2A Ori is heated by the Trapezium stars and not by θ^2A Ori,[25] (3) from optical linear polarization observations, it is found that the bar is physically closer to θ^1 than to θ^2A, in spite of its projected position; consequently, θ^2 A and the regions it illuminates, particularly those to the east, are at a different distance,[26] and (4) the lower electron temperature and lower degree of ionization in the direction of θ^2, especially the large He^0/He$^+$ ratio, compared to those in the direction of θ^1 indicate that the radiation fields of θ^1C Ori and θ^2A Ori are uncoupled.[22] About 20% of all the ionization photons in the Orion Nebula are produced by θ^2 Ori A. The lack of substantial dust and CO emission to the southeast of the bar and the fact that the extinction is lower in the direction of θ^2A Ori than in the direction of θ^1 Ori support the idea that θ^2A Ori is a foreground and not a background object relative to the Trapezium. A schematic model is presented in FIGURE 1.

The Region with r < 32′

There are several lines of evidence that indicate the presence of dust particles in or near the Orion Nebula. The reddening increases toward the Trapezium cluster (see Reference 27 and references therein). A considerable fraction of the continuum emission in the optical[28-30] and uv[31-35] regions is due to dust-scattered light from θ^1 and θ^2 Ori. The continuum and line emission show strong polarization.[26,36] Gull has suggested, from a comparison of the spectra of the central and outer regions of the

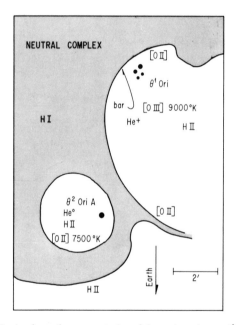

FIGURE 1. A schematic representation of the regions close to θ^1 and θ^2 Ori.

Orion Nebula, that a substantial fraction of the line emission from the outer regions is due to dust-scattered light.[37]

By adopting a model of the Orion Nebula in which the outer region, r > 12′, is just a reflection nebula, it is possible to explain both the presence of lines of high degrees of ionization to the edge of the Orion Nebula in all directions and, at the same time, the presence of regions of low degrees of ionization from a few minutes of arc from the center to the very edge. It is not possible to explain the observed ionization structure— that of helium, in particular—by radiative transfer models with spherical symmetry and without dust-scattered light. The contribution of the dust-scattered light to the emission line intensities in the inner 10' is generally small and, to a first approximation, can be neglected.

The reflection envelope, $r > 12'$, would also imply a more reasonable density distribution. Osterbrock and Flather, attempting to explain the difference between the root mean square densities, N_e (rms), and the local densities derived from the 3726/3729 O II line intensity ratios, suggested a model of the Orion Nebula with $r < 24'$ in which only one-thirtieth of the volume is filled with high density gas and the rest is empty. This model fits the high-frequency radio flux but predicts a larger low-frequency radio flux than is observed. To solve this discrepancy, Menon[39] and Simpson[40] adopted models in which the filling factor becomes smaller with distance from the center of the Orion Nebula, in practice reducing the contribution of the outer parts to the radio flux; no physical argument has been advanced to explain why the filling factor has to decrease so drastically with distance from a value of $1/5$ to a value of $1/200$. Alternatively, if we assume that the radius of the Orion Nebula is only 12', then, with a constant filling factor of $1/30$, it is possible to explain both the high- and the low-frequency radio emission. By adopting a slab model instead of a spherical one, it is possible to increase the filling factor; a filling factor of 1 is obtained with a slab 0.5 pc thick.

From the maps of the Orion Nebula by Strand and Teska[41] and two plates of 3 and 9 min exposure from the 3-m Lick telescope, Ezquerro estimated that there were 1833 stars in the Orion Nebula cluster with $m_v < +18$. This number, coupled with Serrano's initial mass function,[43,44] with minimum masses of 0.1 $M_\odot$ and 0.5 $M_\odot$, yielded a minimum mass for the cluster of 1060 $M_\odot$ and a probable mass of 2640 $M_\odot$, respectively. The total mass of the ionized gas in the Orion Nebula amounts to about 20 $M_\odot$,[40,45] which is a very small amount if we consider that the mass of the visible stellar cluster is a few thousand solar masses and that the age of some of the stars in the cluster that are reaching the main sequence is several million years. Therefore, in addition to the possibility that this gas comes from the molecular cloud,[1,2] two other possibilities should be considered: (1) that the gas is just the residue of the process of star formation and (2) that the gas is produced by mass loss from the cluster stars through stellar winds[16-18] and flare activity, since most of the cluster stars are flare or T Tauri.[46] Support for these two possibilities comes from the fact that the distribution of gas and that of stars are very similar.[47]

CHEMICAL ABUNDANCES

Temperature

Preliminary results based on optical and radio data indicated the presence of a temperature gradient in the Orion Nebula, with T_e increasing outwards.[48,49] Such a gradient, if present, would imply that a large fraction of the main coolants, such as CNO, is embedded in grains in the outer regions.[45] More recent observations do not indicate such a gradient.[6,22,50,51] These results imply that there are no substantial abundance gradients and, consequently, either that the fraction of heavy elements embedded in dust grains is negligible or that it is almost the same throughout the nebula.

Dust Content

As mentioned in *Structure,* above, dust is present in the Orion Nebula. McCall has summarized the arguments that indicate the presence of large dust grains in the Orion Nebula;[52] according to him, the small grains have been destroyed by evaporation throughout the nebula and the difference in O abundance between the Sun and the Orion Nebula is probably due to the large fraction of O embedded in the large dust grains.

The existence of a minimum in the dust-to-gas ratio in the center of the Orion Nebula (dust hole) seems to be well established;[28,30] this hole could be due to dust destruction or to an excess of gas created by stellar winds from the central stars.[49] In any case, the presence of the hole and the lack of abundance gradients across the nebula argue against the idea that a large fraction of the heavy elements is trapped in dust grains across the Orion Nebula.[53]

Gas Phase Abundances

There have been many chemical abundance determinations of the Orion Nebula based on optical and uv emission lines.[22,32,53–56] The results are presented in TABLE 2.

TABLE 2

CHEMICAL ABUNDANCES IN THE ORION NEBULA AND THE SUN*

Element	Orion ($t^2 = 0.00$)	Orion ($t^2 = 0.02$)	Sun
He	11.02	11.01	. . .
C	8.35	8.57	8.67
N	7.57	7.68	7.99
O	8.52	8.65	8.92
Ne	7.66	7.80	8.00
Mg	≤6.6::	≤6.7::	7.62
S	6.97	7.10	7.23
Cl	4.95	5.05	. . .
Ar	6.6	6.65	6.52
Fe	6.5::	6.6::	7.40

*Given in log $N(X)$ with H = 12.00.

From a discussion of the best electron temperature determinations,[55] we recommend the values with a mean square temperature variation, t^2, equal to 0.02.

In TABLE 2, we compare the abundances derived for the Orion Nebula with those derived for the Sun.[53,57–61] The elements can be divided into three groups: those expected to be mostly in gaseous form, *e.g.,* H, He, Ne, and Ar; those likely to be in gaseous form,[53,62,63] *e.g.,* C, N, and O; and those likely to be components of dust,[53,62] *e.g.,* Fe. The absolute accuracy of the abundance determinations of C, N, O, Ne, and S is ∼0.2 dex, that of Ar is ∼0.3 dex, and that of Mg and Fe is ∼0.5 dex.[53,64] Ar abundances in Orion and the Sun seem to be similar; however, the Ar abundance in the Sun is poorly known. On the other hand, the abundances of C, N, O, S, and Ne seem to be deficient by ∼0.2 dex, though I think that the difference is marginal. If real, it could imply either that about one-third to one-half of the C, N, O, and S is embedded in dust grains or that the Orion Nebula as a whole is about 0.2 dex deficient in heavy elements with respect to the Sun. Two more points should be raised. The relative ratios of C, N, O, and S with respect to Ne, which is not expected to be in dust grains, are similar to the solar ratios, implying that most of these elements are in gaseous form. The Orion Nebula is 500 pc further away from the galactic center, and the presence of abundance gradients in the galaxy[65] implies that an H II region at the solar distance would have abundances 0.05 dex higher than those of the Orion Nebula.

SUMMARY

The structure and chemical composition of the Orion Nebula were reviewed. The following five points were stressed. (1) The velocity and pressure patterns of the core region are more complex than those predicted by models in which the gas flow originates in the molecular cloud. (2) There is an H II region around θ^2 Ori that is disconnected from the main H II region produced by the Trapezium. (3) The emission lines from the outer regions ($r \gtrsim 12'$) are mainly produced by dust-scattered light, *i.e.,* the outer shell is a reflection nebula and not an H II region. (4) The gaseous abundances of C, N, O, Ne, and S relative to H are 0.2 $\pm$ 0.2 dex smaller than those in the Sun. (5) The fraction of C, N, O, and S embedded in dust grains is less than one-half.

REFERENCES

1. ZUCKERMAN, B. 1973. Astrophys. J. **183:** 863–69.
2. BALICK, B., R. H. GAMMON & R. M. HJELLMING. 1974. Publ. Astron. Soc. Pac. **86:** 616–34.
3. KALER, J. B. 1967. Astrophys. J. **148:** 925–26.
4. TENORIO-TAGLE, G. 1979. Astron. Astrophys. **71:** 59–65.
5. MEABURN, J. 1975. *In* H II Regions and Related Topics. T.L. Wilson and D. Downes, Eds.: 222–44. Springer-Verlag. New York.
6. PANKONIN, V., C. M. WALMSLEY & M. HARWIT. 1979. Astron. Astrophys. **75:** 34–43.
7. BALICK, B., T. R. GULL & M. G. SMITH. 1980. Publ. Astron. Soc. Pac. **92:** 22–31.
8. WILSON, O. C., G. MUNCH, E. M. FLATHER & M. F. COFFEN. 1959. Astrophys. J. Suppl. Ser. **4:** 199–256.
9. SCHEGLOV, P. V. 1968. *In* Int. Astron. Union Symp. 34, Planetary Nebulae. D.E. Osterbrock and C. R. O'Dell, Eds.: 270–72. D. Reidel. Dordrecht.
10. MEABURN, J. 1971. Astrophys. Space Sci. **13:** 110–27.
11. DOPITA, M. A., A. H. GIBBONS, J. MEABURN & K. T. TAYLOR. 1973. Astrophys. Lett. **13:** 55–59.
12. ELLIOT, K. H. & J. MEABURN. 1974. Astrophys. Space Sci. **28:** 351–64.
13. MÜNCH, G. & K. TAYLOR. 1974. Astrophys. J. **192:** L93–95.
14. LAQUES, P. & J. L. VIDAL. 1979. Astron. Astrophys. **73:** 97–106.
15. MÜNCH, G. 1977. Astrophys. J. **212:** L77–79.
16. SNOW, T. P., JR. & D. C. MORTON. 1976. Astrophys. J. Suppl. **32:** 429–65.
17. TAYLOR, K. & G. MÜNCH. 1978. Astron. Astrophys. **70:** 359–66.
18. FRANCO, J. & B. D. SAVAGE. 1982. Astrophys. J. **255:** 541–51.
19. RUSSELL, R. W., G. MELNICK, G. E. GULL & M. HARWIT. 1980. Astrophys. J. **240:** L99–103.
20. HIPPELEIN, H. & G. MÜNCH. 1978. Astron. Astrophys. **68:** L7–10.
21. ALLER, L. H. & W. LILLER. 1959. Astrophys. J. **130:** 45–56.
22. PEIMBERT, M. & S. TORRES-PEIMBERT. 1977. Mon. Not. R. Astron. Soc. **179:** 217–34.
23. SARAPH, H. E. & M. J. SEATON. 1970. Mon. Not. R. Astron. Soc. **148:** 367–81.
24. DEHARVENG, L. 1973. Astron. Astrophys. **29:** 341–46.

25. BECKLIN, E. E., S. BECKWITH, I. GATLEY, K. MATTHEWS, G. NEUGEBAUER, C. SARAZIN & M. W. WERNER. 1976. Astrophys. J. **207:** 770–79.
26. PALLISTER, W. S., H. G. PERKINS, S. M. SCARROTT, R. G. BINGHAM & J. D. H. PILKINGTON. 1977. Mon. Not. R. Astron. Soc. **178:** 93p–95p.
27. WARREN, W. H., JR. & J. E. HESSER. 1978. Astrophys. J. Suppl. Ser. **36:** 497–572.
28. O'DELL, C. R. & W. B. HUBBARD. 1965. Astrophys. J. **142:** 591–603.
29. PEIMBERT, M. & D. W. GOLDSMITH. 1972. Astron. Astrophys. **19:** 398–404.
30. SCHIFFER, F. H., III & J. S. MATHIS. 1974. Astrophys. J. **194:** 597–608.
31. BOHLIN, R. C. & T. P. STECHER. 1975. Bull. Am. Astron. Soc. **7:** 547–48.
32. PERINOTTO, M. & P. PATRIARCHI. 1980. Astrophys. J. **235:** L13–16.
33. MATHIS, J. S., M. PERINOTTO, P. PATRIARCHI & F. H. SCHIFFER III. 1981. Astrophys. J. **249:** 99–108.
34. CARRUTHERS, G. R. & H. M. HECKATHORN. 1981. Astrophys. Lett. **22:** 135.
35. BOHLIN, R. C., J. K. HILL, T. P. STECHER & A. N. WITT. 1982. Astrophys. J. **255:** 87–94.
36. HALL, R. 1974. In Planets, Stars and Nebula Studied with Photopolarimetry. T. Gehrels, Ed.: 881–87. University of Arizona Press. Tucson.
37. GULL, T. R. 1974. In Proceedings of Eighth Eslab Symposium. A. F. M. Moorwood, Ed.: 1–11. Frascati, Italy.
38. OSTERBROCK, D. E. & E. FLATHER. 1959. Astrophys. J. **129:** 26–43.
39. MENON, T. K. 1961. Publ. Natl. Radio Astron. Obs. **1:** 1–17.
40. SIMPSON, J. P. 1973. Publ. Astron. Soc. Pac. **85:** 479–92.
41. STRAND, K. A A. & T. M. TESKA. 1958. Ann. Deaborn Obs. **7:** 67.
42. EZQUERRO, R. M. 1974. Tesis de Licenciatura, Universidad Nacional Autonoma de México.
43. SERRANO, A. 1978. Ph.D. Thesis, University of Sussex.
44. SERRANO, A. & M. PEIMBERT. 1981. Rev. Mexicana Astron. Astrof. **5:** 109–24.
45. PEIMBERT, M. 1966. Astrophys. J. **145:** 75–78.
46. HARO, G. & E. CHAVIRA. 1969. Bol. Obs. Tonantzintla y Tacubaya. **5:** 59–78.
47. VANDERVOORT, P. O. 1964. Astrophys. J. **139:** 869–88.
48. BOHUSKI, T. J., R. J. DUFOUR & D. E. OSTERBROCK. 1974. Astrophys. J. **188:** 529–32.
49. PERRENOD, S. C., G. A. SHIELDS & J. CHAISSON. 1977. Astrophys. J. **216:** 427–32.
50. McCALL, M. L. 1979. Astrophys. J. **229:** 962–70.
51. KALER, J. B., D. J. LIEN & M. L. PECK. 1979. Astrophys. J. **234:** 909–11.
52. McCALL, M. L. 1981. Mon. Not. R. Astron. Soc. **194:** 485–502.
53. PEIMBERT, M. 1979. In Les Elements et leurs Isotopes dans l'Univers. A. Boury, N. Grevesse, and L. Remy-Battian, Eds.: 451–61. Université de Liege, Institut d'Astrophysique. Liege.
54. OLTHOF, H. & S. R. POTTASCH. 1975. Astron. Astrophys. **43:** 291–95.
55. TORRES-PEIMBERT, S., M. PEIMBERT & E. DALTABUIT. 1980. Astrophys. J. **283:** 133–39.
56. BOHLIN, R. C., J. K. HILL, T. P. STECHER & A. N. WITT. 1982. Astrophys. J. Submitted.
57. LAMBERT, D. L. 1978. Mon. Not. R. Astron. Soc. **182:** 249–72.
58. LAMBERT, D. L. & R. E. LUCK. 1978. Mon. Not. R. Astron. Soc. **183:** 79–100.
59. MEYER, J. P. 1979. In Les Elements et leurs Isotopes dans l'Univers. A. Boury, N. Grevesse, and L. Remy-Battian, Eds.: 465–76. Université de Liege, Institut d'Astrophysique. Liege.
60. MEYER, J. P. 1979. In Les Elements et leurs Isotopes dans l'Univers. A. Boury, N. Grevesse, and L. Remy-Battian, Eds.: 477–87. Université de Liege, Institut d'Astrophysique. Liege.
61. MEYER, J. P. 1979. In Les Elements et leurs Isotopes dans l'Univers. A. Boury, N. Grevesse, and L. Remy-Battian, Eds.: 489–501. Université de Liege, Institut d'Astrophysique. Liege.
62. BARLOW, M. J. 1978. Mon. Not. R. Astron. Soc. **183:** 367–395.
63. BARLOW, M. J. 1978. Mon. Not. R. Astron. Soc. **183:** 397–415.
64. PEIMBERT, M. 1981. In The Universe at Ultraviolet Wavelengths. R. D. Chapman, Ed.: 557–63. NASA. Greenbelt, Md.
65. PEIMBERT, M. 1979. In Int. Astron. Union Symp. 84, The Large Scale Characteristics of the Galaxy. W. B. Burton, Ed.: 307–15. Reidel. Dordrecht.

DISCUSSION OF THE PAPER

G. FIELD (*Harvard-Smithsonian Center for Astrophysics, Cambridge, Mass.*): If the 20 $M_\odot$ of gas in the Orion Nebula is attributed to outflow from the massive stars, one would expect a velocity broadening of $\sim$2000 km s^{-1} in the nebula. The observed value is only 30 km s^{-1}.

PEIMBERT: Only θ^1C, θ^1D, and θ^2A Ori seem to be losing mass by a stellar wind at a low rate and their mass-loss rate histories are not known. The C IV and Si IV lines are observed in absorption in the direction of θ^1C Ori with $V_{LSR} \simeq -40$ km s^{-1} and indicate the presence of decelerated material confined to regions very close to the Trapezium. These observations imply that the velocity pattern is not due solely to stellar winds from the very massive stars. There are some observations that indicate gas motions at velocities larger than 100 km s^{-1}, but they involve a very small fraction of the mass.

B. ZUCKERMAN (*University of Maryland, College Park, Md.*): I believe that the run of radial velocities observed in various ionic species in absorption against the Trapezium can be understood without any major modification to the model that I proposed in 1973. The highly ionized species are probably produced in a wind flowing out of the stars and will thus be blueshifted with respect to the lower excitation ions that may be associated with the neutral cloud that lies between us and the Orion Nebula. What is the extinction to θ^2 Ori?

PEIMBERT: The extinction is roughly $E_{B-V} = 0.4$ to θ^1 Ori and $E_{B-V} = 0.25$ to θ^2 A Ori.

CO AND SHOCKS RELATED TO THE EVOLUTION
OF THE ORION NEBULA

F. Peter Schloerb*

*Five College Radio Astronomy Observatory
Department of Physics and Astronomy
University of Massachusetts
Amherst, Massachusetts 01003*

Robert B. Loren†

*Millimeter Wave Observatory
McDonald Observatory
Electrical Engineering Research Laboratory
University of Texas, Austin
Austin, Texas 78712*

INTRODUCTION

Henry Draper's first photograph of the Orion Nebula was a milestone in the observational study of star formation and the interstellar medium. Modern astronomical techniques produce images that contain information complementary to that contained in Draper's optical photograph. In this paper, we discuss the features of the molecular emission from the giant molecular cloud that is associated with the nebula. Although it is invisible in optical photographs of the region, the molecular gas has played an important role in the evolution of the visible features. The processes that produce the optical features have, in turn, affected the structure and temperature of the molecular gas. Thus, we may study these properties *via* millimeter line observations to learn about the sequence of events that produced the present appearance of the nebula. Specifically, the study of the shocked regions of molecular gas that result from the expansion of the H II region allows us to test current models of its evolution.

Our interest in the relationship between an evolving H II region and a molecular cloud goes beyond the structure of the optical nebula. We would also like to know what this relationship has been in the past and what it is likely to be in the future. Since we believe that the Trapezium stars were formed from molecular gas that was initially part of the giant molecular cloud complex, we may also approach its study with an eye towards deducing the physical conditions under which stars form. This view is supported by recent discussions of the possible roles of O and B stars in the formation of new stars. Elmegreen and Lada suggest that the expansion of an evolving H II region provides the impetus for the formation of new stars in the shocked boundary layer of the molecular cloud.[1] Thus, the study of the shocked regions could provide important

*Supported by the Five College Radio Astronomy Observatory, which is supported by a grant from the National Science Foundation, No. AST 80-26702, and operated by permission of the Metropolitan District Commission of the Commonwealth of Massachusetts.

†Supported by a grant from the National Science Foundation, No. AST 79-20966, to the University of Texas at Austin.

32

clues about star formation. Finally, it is important to look further into the future of the cloud and ask how the formation of O and B stars affects the cloud's lifetime. Surveys of CO around old clusters[2] suggest that clouds are eventually disrupted by the formation of OB associations and, therefore, our study of the interaction between H II regions and molecular clouds may reveal the mechanism by which this disruption occurs.

MILLIMETER MOLECULAR LINE TRACERS OF SHOCKS

The sequence of events that occurs after an H II region reaches its initial Strömgren radius is well understood, although some details have yet to be established.[3] When the H II region has grown to its initial Strömgren radius, its density and that of the neutral cloud are the same, but their temperatures differ by two to three orders of magnitude. Thus, the H II region continues to expand, since its internal pressure exceeds that of the surrounding cloud by the same two to three orders of magnitude. This expansion is supersonic with respect to the sound speed of the neutral cloud, and a shock front (S-front) is generated that precedes the ionization front (I-front) into the surrounding cloud. Expansion of the H II region continues until either its internal pressure falls to that of the surrounding cloud or the S- and I-fronts encounter the cloud edge. In the latter case, the ionized gas is free to expand into the low-pressure intercloud medium and the pressure within the H II region is greatly reduced. The shock and ionization fronts continue to move into the cloud, however, due to the pressure caused by the momentum flux of ions across the I-front, and the structure of the transition region between the H II region and the molecular cloud will be fundamentally the same as if the H II region were confined completely.[1,4] We note that this latter possibility has surely occurred in the Orion Nebula—we can optically observe the ionized gas and its flow away from the neutral cloud.[5]

If the Orion Nebula is, in fact, evolving in this manner, it is logical to look behind the shock for evidence of the expansion of the H II region. Thus, we now consider the identifying characteristics of this postshock layer. One important feature of the postshock region is that it is accelerated by the shock relative to the ambient cloud. The velocity of the postshock layer (V_{psl}) can be shown to be[3]

$$V_{psl} = V_{shock}\left(1 - \frac{C_I^2}{V_{shock}^2}\right),$$

where V_{shock} is the shock velocity and C_I is the velocity of sound in the neutral cloud. Since C_I is typically much less than V_{shock}, the velocity of the postshock layer should be close that of the shock. Thus, depending on the geometry, a shock should produce a sudden velocity shift (of $\leq V_{shock}$) at its position. In the "classical" picture, one might expect that the velocity shift observed would be close to the speed of sound in the H II region (~ 15 km s^{-1}), since this is the value expected shortly after the H II region reaches its initial Strömgren radius. However, consideration of the molecular composition of the postshock gas during the expansion suggests that the velocities of molecular shocks should be less than this. Hill and Hollenbach find that a wave of H$_2$ dissociation (d-wave) precedes the I- and S-fronts into the molecular cloud; thus, the gas entering the shock in the initial phase of the expansion is largely atomic. Numerical models of the expansion of the H II region find that the d-wave is eventually overtaken by the

S-front and that a significant fraction of the material entering the shock will be molecular only after the H II region has expanded to about five to seven times its initial Strömgren radius.[6] By this point, the velocities of the S- and I-fronts will have fallen significantly below the H II region sound speed, so the velocity of the shocks probed by millimeter lines might be expected to be only a few kilometers per second. Unfortunately, since it is close to the velocity width of the ambient cloud, such a small velocity shift makes unambiguous detection of a shock difficult. Thus, sudden shifts observed in the velocity might be also interpreted in terms of inhomogeneities in the cloud.

A second potential tracer of shocked regions is the column density or volume density of molecules in the postshock layer. Since shocks compress the material that passes through them, the postshock density should be enhanced relative to that of the ambient cloud. For isothermal shocks, a reasonable approximation to the large-scale structure of shocks in molecular gas, the amount of compression is potentially quite large, since

$$\frac{\rho_2}{\rho_1} = \frac{V_{shock}^2}{C_I^2},$$

where ρ_2 is the postshock density and ρ_1 is the preshock density.[3] Thus, since the shock velocity (a few kilometers per second) may be many times the speed of sound in the molecular gas (a few tenths of a kilometer per second), the material might easily be compressed by a factor of 10^2. Tracers of high molecular density, such as HCN, which is only excited at densities of 10^4–10^5 cm^{-3}, might therefore be useful in delineating shocked regions. Column density tracers such as ^{13}CO may also serve this purpose, since the enhanced density will enhance the column density at the position of the postshock layer. Again, however, this potential tracer of shocked regions may be subject to other interpretations. In particular, one can always argue that clumps in clouds are due to initial inhomogeneities in the cloud rather than to the effects of shocks.

Although the above discussion noted that isothermal shocks are a good approximation to the large-scale structure of molecular shocks, the structure at smaller scales is not isothermal and one might, under some circumstances, expect to see a temperature enhancement in the postshock layer. Kwan and others have calculated the postshock temperatures for a shock moving into a molecular medium and find that they decrease rapidly from a few thousand degrees to ~200 K in only 10^2 y.[7] Even for a 10 km s^{-1} shock, this amounts to a size scale for high-temperature gas of only 10^{-3} pc, which would make the highest-temperature gas difficult to observe. However, shocks can play a role in heating the gas in the postshock layer, and this extra heating is potentially observable. Gas at 100–200 K will have a characteristic cooling time of $\tau_{cool} \simeq n_{H_2} kT/\Lambda$, where Λ is the cooling rate of molecular gas at temperature T and density n_{H_2}. Adopting the cooling coefficients of Goldsmith and Langer,[8] we find a cooling time of 10^5 y for gas at 100 K and densities of 10^5 cm^{-3}. Since this time is comparable to the length of time that the gas resides in the postshock layer, it is reasonable to assume that warm postshock molecular gas might exist. However, since this gas is close to a source of great luminosity, the OB association that formed the H II region in the first place, it is likely that warm gas will be present in the postshock layer anyway. Therefore, we must be careful to consider heating of the gas by stars before we can be sure that it is heated by a shock.

Finally, since the density and temperature of the postshock layer differ from those of the ambient molecular cloud, we might also expect the molecular abundance patterns in the postshock gas to be different. This would be especially true for the highest-temperature regions behind the shock front, where some reactions that are endothermic and slow under normal interstellar conditions might proceed at a significantly higher rate. It is conceivable, then, that there are chemical tracers of shocked regions as well as physical tracers. In the Orion plateau source, for example, the high abundance of some species is interpreted as the result of the "shock chemistry" that occurs in the high-temperature postshock region.[9,10] Hartquist *et al.* state that the most prominent chemical shock tracers are S- and Si-bearing molecules such as H_2S, SO and SiO,[10] but these are problematic, since the normal cloud chemistry of S and Si is not well understood. In addition, their usefulness as probes may be limited to shocks of a relatively narrow velocity range and to shocks that have an appreciable fraction of the incoming gas in the molecular form. High-velocity shocks (>10 km s^{-1}), which dissociate some fraction of the H_2 entering the shock, and shocks with an appreciable amount of H I entering the postshock layer may not show these tracers. In both instances, the H I present behind the shock would react to destroy the tracers and their precursors. On the other hand, low-velocity shocks (<4 km s^{-1}) do not raise the postshock temperatures by enough to activate the shock chemistry reactions.[10] Thus, though potentially useful for shocks of the right velocity, these particular shock tracers may not be the correct ones for the slow molecular shocks associated with the expansion of the H II region.

SHOCKED MOLECULAR GAS IN THE ORION MOLECULAR CLOUD

We now turn our attention to the Orion Molecular Cloud itself to look for evidence of a shocked layer at the boundaries of the H II region. Orion is, in many ways, an ideal cloud in which to search for the effects of shocks in the molecular gas since it is relatively close and can be observed at a wide variety of wavelengths. The latter point is especially important in the identification of a molecular feature with a shock. As noted in the previous section, many of the potential shock tracers of molecular regions do not offer unambiguous evidence of a shock. We anticipate, therefore, that evidence for shocks in molecular clouds will be largely circumstantial and that the plausibility of a shock origin will be strengthened by an association of several possible shock tracers. Particularly useful are optical tracers of S- and I-fronts, such as [S II],[11] which can help delineate possible shocked regions. Exceptionally good monochromatic photographs of [S II] in the Orion Nebula, which show enhancements in bright filaments on the eastern side of the nebula and in the bright bar feature to the southeast of the Trapezium stars, are given by Gull.[12] (These features are shown in FIGURE 1 in a sketch map of the region based on the [S II] photograph.) In both cases, millimeter shock tracers are associated with these optical features and there is good reason to believe, therefore, that we are probing the shocked shells of molecular gas surrounding the H II region.

Much of our discussion of the shocked molecular gas in these regions is based on observations of CO obtained from two surveys of the Orion Molecular Cloud. The first survey, by Loren, of CO and ^{13}CO was obtained with a 2.3' beam size over a 60' × 80' region with a fully sampled 2' grid spacing using the Texas Millimeter Wave

Observatory (MWO) 4.9-m antenna.[13,24] The $T_A^*(CO)$ and $T_A^*(^{13}CO)$ distribution from this work is shown in FIGURE 2, along with the column density of ^{13}CO ($N^{13}CO$), which was derived from them under the assumptions that both CO and ^{13}CO are thermalized and that CO is optically thick. The second survey complements the first, since it covers a more limited region with higher resolution. Schloerb *et al.* mapped CO, ^{13}CO, and HCN over a 15′ × 30′ region with a 50″ beam size on a fully sampled 45″ grid using the Five College Radio Astronomy Observatory (FCRAO) 14-m antenna. Maps of the brightness temperature of the emission from these species are shown in FIGURE 3. For both surveys, the spectral resolution was 250 kHz and calibration was obtained by reference to an ambient temperature chopper wheel.

Shocked Molecular Gas at the Perimeter of the Optical H II Region

As discussed in Reference 13, maps like those presented in FIGURES 2 and 3 present only a part of the information contained in the millimeter-wave spectrum, since the considerable velocity structure that is apparent in the lines near the center of Orion is not represented. Thus, early maps of the region, which characterized the molecular emission in terms of peak or velocity-integrated intensities, do not show many of the shock features clearly. One of the unusual features that appears in maps of individual

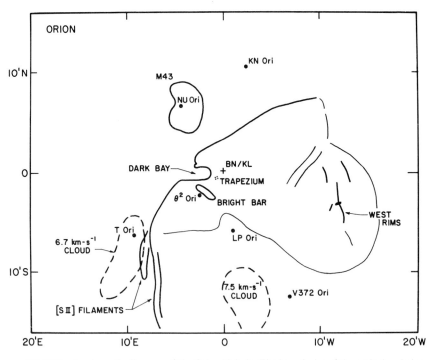

FIGURE 1. A schematic diagram of the Orion Nebula. The boundaries of the optical emission are based on Gull's monochromatic [S II] photograph.[12]

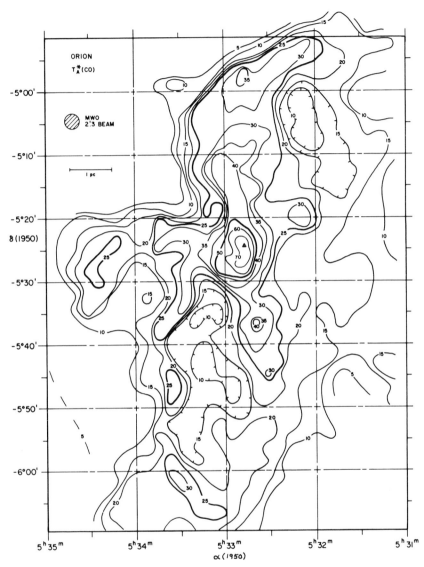

FIGURE 2a. A map of $T_A^*(CO)$ in the molecular cloud surrounding the Orion Nebula (L1641) obtained with the MWO 4.9-m antenna. The position of the KL nebula is denoted by a ▲.

velocities is a ring of high-velocity gas that occurs along the eastern, southern, and southwestern perimeter of the optical emission.[13] The CO line profiles in this ring are split about the nominal core velocity of 8.5 km s^{-1}; we show several profiles from Loren's map located on an isophote map of Hβ in FIGURE 4.[16] The velocity splitting becomes apparent at the eastern boundary of the H II region and becomes very prominent at the southern and southeastern boundaries. Since the splitting is also observable in ^{13}CO lines, it is not likely to be due to the effects of self-absorption. Maps of CO emission at particular velocities show that the velocity splitting is confined to the perimeter of the H II region. In FIGURES 5a and 5b, we show Loren's maps at 6.1 and 12.6 km s^{-1} to illustrate this point.[13] The ring at 12.6 km s^{-1} (FIGURE 5b) is complete from the east side of the nebula around to the southwest. The low-velocity component

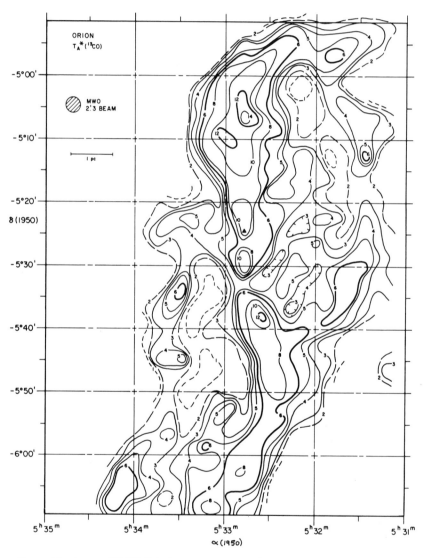

FIGURE 2b. A map of $T_A^*(^{13}CO)$ in the molecular cloud surrounding the Orion Nebula (L1641) obtained with the MWO 4.9-m antenna. The position of the KL nebula is denoted by a ▲.

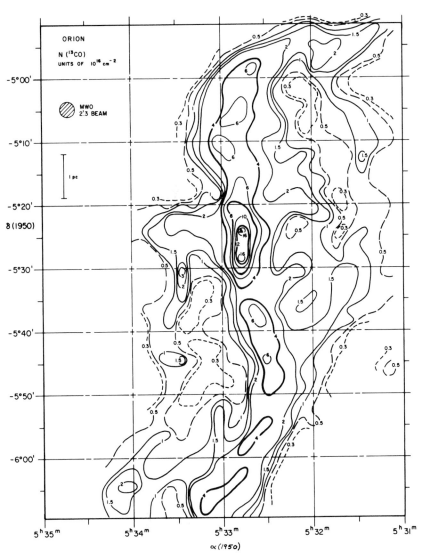

FIGURE 2c. A map of the ^{13}CO column density in the molecular cloud surrounding the Orion Nebula (L1641) obtained with the MWO 4.9-m antenna. The position of the KL nebula is denoted by a ▲.

(FIGURE 5a) is less complete and occurs in one segment within the dark lane between M42 and M43 and in another segment at the southwest perimeter of the nebula near the pre-main-sequence (PMS) star T Ori. In both cases, the low-velocity component correlates well with optical obscuration (particularly in the dark lane), while the high-velocity gas is present whether there is obscuration or not. This association suggests that the low-velocity material occurs in the foreground portions of the molecular cloud, while the high-velocity component is more closely connected to the background gas. Apparently, the high-velocity ring of molecular gas is an expanding shell located at the perimeter of the H II region.

The high-velocity ring has many of the desired shock tracers outlined in the previous section; there are, therefore, many good reasons to associate it with a shock propagating into the cloud. First and foremost, it occurs at the perimeter of the H II region, where we anticipate that shocked gas should exist. Second, it is correlated with optical tracers of S- and I-fronts along the southeastern perimeter of the nebula. Third, the expanding shell has been accelerated radially by at least 2–3 km s^{-1}, in good agreement with the expected magnitude of the acceleration due to shocks entering molecular gas. The thickness of the shell is in reasonable agreement with that expected for the postshock layer,[17] since it is about 20% of the radius of the optical nebula. Finally, the expansion time for the shell, $\tau_{exp} \simeq 10^{18}$ cm$/10^5$ cm s$^{-1} \simeq 10^6$ y, is

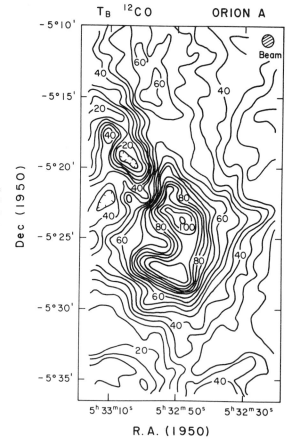

FIGURE 3a. A map of the brightness temperature of CO in the central portion of the Orion Molecular Cloud obtained with the FCRAO 14-m antenna.

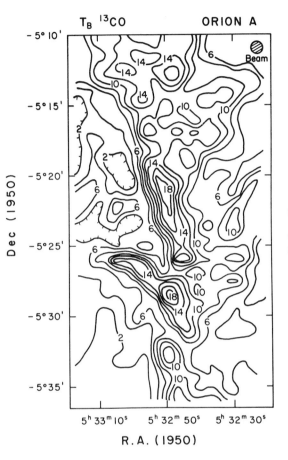

FIGURE 3b. A map of the brightness temperature of ^{13}CO in the central portion of the Orion Molecular Cloud obtained with the FCRAO 14-m antenna.

comparable to the age of the Trapezium stars. Thus, the high-velocity ring appears to fit well into the evolutionary picture of an H II region, and we note that similar expanding shells probably exist in other regions (for example, W80[4]).

Shocked Molecular Gas near the Bright Bar

A second molecular region of note was discovered in the high-resolution FCRAO maps of the portion of the molecular cloud immediately surrounding the brightest and densest part of the H II region. The morphology of the molecular emission in this central region follows that of the H II region closely and hints at a close relationship between the two. This is especially true in the vicinity of the "bright bar," which lies about 2′ southeast of the Trapezium stars. The bright bar has appeared in observations at all wavelengths, and has now been observed with a molecular feature as well. It is seen clearly enhanced in the [S II] photograph; it appears as a narrow ridge of hot dust emission in maps at 10 μm,[18] in far-ir maps,[19] and in radio continuum maps.[20] Since the bright bar is probably an ionization front seen edge on,[5] its association with the molecular bar seen in the maps of FIGURE 3 makes it a logical place to look for shocked molecular gas. In fact, since models of the nebula generally place the highest-density ionized gas in a cavity in the cloud surrounding the Trapezium,[5,21] the whole region around this high-density H II region might be a good place for such a search.

The maps shown in FIGURE 3 and the cross sections through the bar, the Trapezium stars, and the Kleinmann-Low (KL) nebula shown in FIGURE 6 demonstrate that the molecular bar lies parallel to the optical bar and is slightly southeast of it, away from the Trapezium. Since the molecular bar appears as a feature in CO, ^{13}CO, and HCN, we may infer that it is a region of enhanced temperature, column density, and space density. Each of these features was proposed as a possible tracer of shocked boundary layers in the previous section of this paper, and their association with a known ionization front makes a shock interpretation for the origin of the molecular bar tempting indeed. This interpretation is further supported by the velocities of the gas in

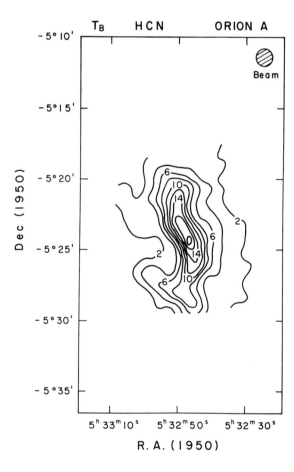

FIGURE 3c. A map of the brightness temperature of HCN in the central portion of the Orion Molecular Cloud obtained with the FCRAO 14-m antenna.

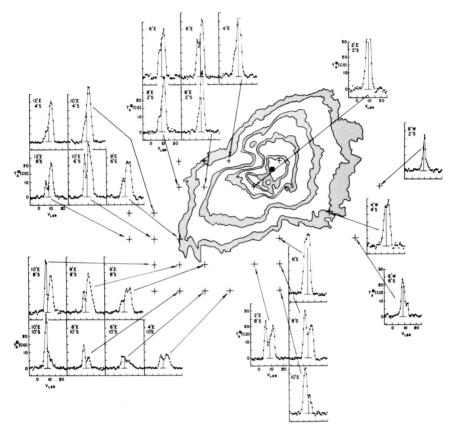

FIGURE 4. MWO CO line profiles at various positions on the perimeter of the H II region. The Hβ isophotes are taken from Dopita *et al.*[16]

the bar and surrounding the H II cavity. In FIGURE 7, we show maps of the CO emission in three 1 km s^{-1} resolution bins centered on 6.5, 8.5, and 10.5 km s^{-1}. The maps clearly show that the velocity of the molecular bar is different from that of the ambient molecular cloud, and also that it is part of a larger systematic variation that occurs along the horseshoe of hot gas traced by the peak CO intensity map in FIGURE 3a. It is interesting to note that the velocity of C II emission, a tracer of the ionized-neutral boundary, follows some of the same tendencies in its velocity structure.[22] If the molecular bar were, in fact, the shocked boundary layer at the edge of the ionized blister surrounding the Trapezium stars, we would anticipate that it would be accelerated outward from the stars at a velocity of up to a few kilometers per second. Depending on the geometry, then, systematic velocity gradients around the H II cavity, such as those in FIGURE 7, would be readily understandable in terms of such a model. Therefore, we feel that it is likely that the FCRAO maps of the central portion of the Orion Nebula are probing the shocked boundary layer produced by the expansion of the H II region generated by the Trapezium stars.

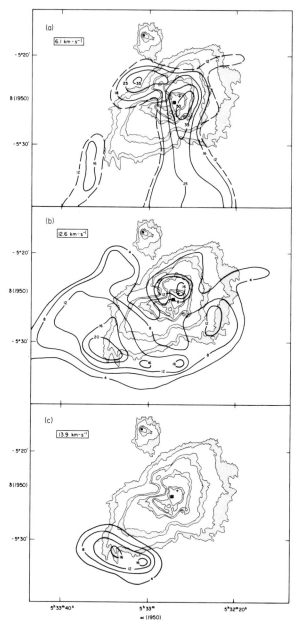

FIGURE 5. MWO CO maps at velocities of 6.1, 12.6, and 13.9 km s^{-1} superposed on the Hβ isophote map.

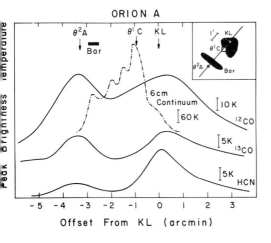

FIGURE 6. Spatial cross sections of radio continuum emission and molecular emission (CO, ^{13}CO, and HCN) through the KL nebula, the Trapezium stars, and the bright bar.

Uniqueness of the Shock Interpretation

The presence of the shock tracers suggested in *Millimeter Molecular Line Tracers of Shocks* is a necessary condition for shocks, but it is worthwhile reviewing whether this is a sufficient condition to prove that the shocks exist. Although we have found three of the four proposed shock tracers in the molecular bar (variations in chemical abundance have yet to be tested), it is difficult to rule out other plausible explanations. For example, it is particularly difficult to say that the shock has heated the gas since the Trapezium stars provide sufficient energy to heat the dust in the bar to the temperatures that are observed in CO.[15] A similar conclusion could be reached for the other high-temperature gas in the central region and, therefore, shock heating is not

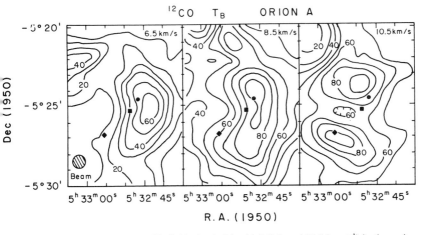

FIGURE 7. FCRAO CO maps of individual velocities (6.5, 8.5, and 10.5 km s^{-1}) in the region surrounding the high-density H II cavity near the Trapezium stars. The positions of KL (●), θ^1C (■), and θ^2A (♦) are shown.

required to explain the observed temperatures. The same is probably true of the $T_A^*(CO)$ peak in the vicinity of the PMS star T Ori at the edge of the H II region in the high-velocity ring. In this case, too, the high temperatures near T Ori could easily be explained by a close coupling of the gas to dust heated by the Trapezium stars. Since the column density is a local maximum at this position (see FIGURE 2), the density and grain-gas coupling is probably also greater than in surrounding regions, so this region would appear as a hot spot whether shock heating were important or not. Thus, we anticipate that the high temperatures observed in CO $(J = 1 - 0)$ are not good diagnostics of shocks and that they alone do not provide sufficient proof of the existence of shocks. On the other hand, higher J transitions might be a more useful diagnostic, since they are the dominant coolant of the postshock gas below a couple of hundred degrees.[7]

In a similar vein, one wonders whether the column density features that are observed result from a shock or from the expansion of the H II region into an initially inhomogeneous medium. We believe that a reasonable case could be made that the association between the edges of the H II region and the high molecular densities is the result of inhomogeneities in the ambient molecular cloud. We note that the greatest extent of the H II region (to the west and northwest) is in the direction of the lowest molecular column density in the maps of FIGURE 2. Since no molecular column density has accumulated in these directions, it is possible that the observed high column densities to the east, southeast, and south were initially present in the cloud and halted the advance of the expanding H II region in those directions. Thus, the observed high column density around the perimeter of the H II region could have been present initially and this shock tracer may also have its limits.

Our remaining shock tracer shows more promise of providing sufficient evidence of shocks for their identification, although it is conceivable that the velocity shifts observed in our proposed shock regions could be due to clumps in the cloud with slightly different velocities. However, it is difficult to imagine how such a process could have led to the systematic velocity variations that are observed. For example, the proposed expanding shell at the perimeter of the H II region would have to be reinterpreted in terms of two clouds of different velocities that have confined the expansion of the H II region. The arcuate shape of the 12.6 km s^{-1} cloud (FIGURE 5) would then have to be fortuitous in such a model and, similarly, the association of these velocity features with optical tracers of S- and I-fronts would have to be fortuitous. Therefore, of all the shock tracers we have discussed, the shift in velocity due to the passage of the shock and its occurrence near the perimeter of an H II region in association with other shock tracers probably provides the best evidence for the shock interpretation.

SUMMARY

We have discussed the effects of the expansion of an H II region on a molecular cloud and sought them in the Orion Molecular Cloud surrounding the Orion Nebula. Two regions in the molecular cloud show evidence for shocks associated with this expansion. A partial ring of molecular gas surrounding the eastern, southeastern, and southern perimeters of the H II region is probably an expanding shell driven outward by the expansion of the H II region. A similar phenomenon seems to be occurring in the region around the densest part of the H II region near the bright-bar I-front southeast of the Trapezium. Thus, the present appearance of the molecular gas does support the traditional concepts of the evolution of the Orion Nebula. It is clear, however, that this evolution has been far more complicated than the simple "classical" picture of the expansion of a uniform H II region into a uniform neutral cloud. Inhomogeneities, which are observable elsewhere in the cloud, have allowed the H II region to break through the edge of the neutral cloud and expand freely in directions where there is little molecular material. It is little wonder, then, that evidence for shocks at the H II region–molecular cloud boundary is difficult to come by and somewhat ambiguous.

Finally, the spatial coincidence of the KL nebula and the shocked boundary layer surrounding the high-density H II cavity near the Trapezium makes it tempting to speculate about the possible relationship between the shocks and star formation. Elmegreen and Lada have calculated that, at some point in its evolution, the shocked boundary layer should become gravitationally unstable and collapse to form new stars.[1] The clumpy appearance of the ^{13}CO map of this region obtained at FCRAO (FIGURE 3b) would seem to support the idea that this layer has, in fact, become unstable and is collapsing into a few small clumps. Furthermore, the OB associations of different ages in Orion have often been sited as evidence for some sort of successive star formation.[23,24] However, the application of the stability criterion for a shocked layer[1] to this situation suggests that it may actually be stable. The shocked layer will be unstable,[1] provided that its mass column density, σ, exceeds a certain critical value, σ_m:

$$\sigma_m = 2.3 \left(\frac{P_1}{\pi G}\right)^{1/2},$$

where P_1 is the pressure at the ionization front and G is the gravitational constant. For H II region parameters of $n_{H\,II} \simeq 10^4$ and $T_{H\,II} \simeq 10^4$, we find that $\sigma_m \simeq 1$ g cm^{-2}. Achieving this mass column density in a layer of the observed thickness ($\leq 4 \times 10^{17}$ cm) would require a density of greater than 10^6 cm^{-3}, which corresponds to a column density about ten times greater than that of the entire molecular cloud. Thus, "reasonable" physical parameters for the layer would suggest that it is stable against collapse and not in the process of forming stars. Further observations of this region may help to determine whether Elmegreen and Lada's mechanism for sequential star formation is correct or whether some other mechanism is necessary to explain the observed sequence of star formation in Orion.

ACKNOWLEDGMENTS

We are grateful to N. Scoville, G. Wynn-Williams, and J. Kwan for helpful discussions and useful suggestions concerning the manuscript. We also benefited from a discussion with P. Solomon following the presentation of this paper at the Draper Symposium. This is contribution number 504 of the Five College Astronomy Department.

REFERENCES

1. ELMEGREEN, B. G. & C. J. LADA. 1977. Astrophys. J. **214**: 725.
2. BASH, F. N., E. M. GREEN & W. L. PETERS. 1977. Astrophys. J. **217**: 464.
3. SPITZER, L. 1978. Physical Processes in the Interstellar Medium. Wiley and Sons. New York.
4. BALLY, J. & N. Z. SCOVILLE. 1980. Astrophys. J. **239**: 121.
5. BALICK, B., R. H. GAMMON & R. M. HJELLMING. 1974. Publ. Astron. Soc. Pac. **86**: 616.
6. HILL, J. K. & D. J. HOLLENBACH. 1978. Astrophys. J. **225**: 390.
7. KWAN, J. 1977. Astrophys. J. **216**: 713.
8. GOLDSMITH, P. F. & W. D. LANGER. 1978. Astrophys. J. **222**: 881.
9. LADA, C. J., M. OPPENHEIMER & T. W. HARTQUIST. 1978. Astrophys. J. **226**: L153.
10. HARTQUIST, T. W., M. OPPENHEIMER & A. DALGARNO. 1980. Astrophys. J. **236**: 182.
11. DOPITA, M. A. 1978. Astrophys. J. Suppl. Ser. **33**: 437.
12. GULL, T. R. 1974. *In* Proc. ESLAB Symp. on H II Regions and the Galactic Center. A. F. M. Moorwood, Ed.:1. ESRO-8P103. Organ. Eur. Rech. Spat. Neuilly.
13. LOREN, R. B. 1979. Astrophys. J. **234**: L207.
14. LOREN, R. B. 1982. In preparation.
15. SCHLOERB, F. P., P. F. GOLDSMITH & N. Z. SCOVILLE. 1982. *In* Regions of Recent Star Formation. P. Dewdney and R. Roger, Eds.:439. D. Reidel. Dordrecht.
16. DOPITA, M. A., S. ISOBE & J. MEABURN. 1975. Astrophys. Space Sci. **34**: 91.
17. LASKER, B. M. 1966. Astrophys. J. **143**: 700.
18. BECKLIN, E. E., S. BECKWITH, I. GATLEY, K. MATTHEWS, G. NEUGEBAUER, C. SARAZIN & M. W. WERNER. 1976. Astrophys. J. **207**: 770.
19. WERNER, M. W., I. GATLEY, D. A. HARPER, E. E. BECKLIN, R. F. LOWENSTEIN, M. TELESCO & H. A. THRONSON. 1976. Astrophys. J. **204**: 420.
20. MARTIN, A. H. M. & S. F. GULL. 1976. Mon. Not. R. Astron. Soc. **175**: 235.
21. ZUCKERMAN, B. 1973. Astrophys. J. **183**: 863.
22. JAFFE, D. T. & V. PANKONIN. 1978. Astrophys. J. **226**: 869.
23. BLAAUW, A. 1964. Annu. Rev. Astron. Astrophys. **2**: 213.
24. KUTNER, M. L., K. D. TUCKER, G. CHIN & P. THADDEUS. 1977. Astrophys. J. **215**: 521.

DISCUSSION OF THE PAPER

J. GIULIANI (*Princeton University, Princeton, N.J.*): Do you have an estimate for the physical thickness of the shock boundary layer?

SCHLOERB: Yes, an upper limit of 4×10^{17} cm, given by the telescope beam size.

P. SOLOMON (*State University of New York, Stony Brook, N.Y.*): Your requirement of 1 g cm^{-2} for collapse along the shock front is equivalent to 1000 magnitudes of visual extinction. This is much greater than anyone would claim for any part of the molecular cloud. The model then seems to fail.

M. ELITZUR (*University of Kentucky, Lexington, Ky.*): In the hot postshock region, CO will emit more strongly in the vibrational transitions than in the rotational transitions. This could provide a good shock tracer if one could find a direction without too much extinction.

ATOMIC CARBON IN ORION*

T. G. Phillips

Department of Physics
California Institute of Technology
Pasadena, California 91125

INTRODUCTION

The ground state fine-structure transitions of atomic carbon (C I) have often been considered major cooling lines for the material of the interstellar medium, where the temperature is in the range 15–100 K.[1] The other major cooling lines are likely to be the ground state fine-structure transition of C II and the rotational transitions of carbon monoxide (CO).

Emission from the ground state C I $^3P_1$–$^3P_0$ transition was first detected from the Orion Molecular Cloud-1 (OMC1) dense interstellar cloud in the direction of the Becklin-Neugebauer–Kleinmann-Low (BN-KL) infrared sources. It was detected by means of a heterodyne receiver aboard the NASA Kuiper Airborne Observatory.[2] A vital component in that detection was the laboratory measurement of the line frequency, first by careful uv spectroscopy (*e.g.,* 16.4 cm^{-1})[3] and, more recently, to high precision by Saykally and Evenson (492.163 GHz).[4]

There is considerable information available from the Copernicus satellite about the carbon abundance in regions of the interstellar medium surrounding the dense clouds where densities are low ($n_H \simeq 10^2$). It is primarily atomic and ionized, *i.e.,* C II $\simeq 90\%$, C I $\simeq 10\%$.[5] According to recent theories, *e.g.,* Reference 6, there should be regions at the edges of dense molecular regions, though within the uv penetration depth, where carbon is primarily in the form of C I. Deep in the interior of the dense clouds it is expected to be almost totally in CO. The investigations reported here will be concerned with C I $^3P_1$–$^3P_0$ emission from the interior and edges of the dense Orion cloud.

Data from the Orion Molecular Cloud[2,7] seem to show that C I is to be found throughout the cloud with some enhancement relative to CO at the edges. Questions arise as to how the models of the physics or chemistry of the dense clouds should be modified to account for these observations. Possibilities include the introduction of time-dependent chemistry, greater uv penetration, internal uv or ionization sources, shocks, and churning of the cloud material. Investigation of the Orion cloud has allowed some testing of these ideas. In particular, due to the highly developed status of the physical model of the Orion cloud, it has been possible to find some indications of cloud edge effects and to obtain a negative result for shocks.

OBSERVATIONS OF C I EMISSION

The most appropriate way to discuss the C I emission data is by comparison with CO. This is because the detailed balancing of collisional excitation and de-excitation,

*The California Institute of Technology observing program is supported by a grant from the National Aeronautics and Space Administration, no. NAG 2-1.

49

through spontaneous emission, are very similar for the C I $^3P_1-^3P_0$ and CO ($J = 1-0$) transitions. Also, the spatial and velocity distributions turn out to be observationally very similar.[7]

OMC1

The first detection of the C I $^3P_1-^3P_0$ was towards BN-KL in the Orion Molecular Cloud-1. FIGURE 1 shows the spectrum of C I, which was taken with an InSb heterodyne bolometer receiver at the Nasmyth focus on the NASA Kuiper Airborne Observatory 91.5-cm telescope. The C141 aircraft carries the observatory to altitudes of about 12 km, so avoiding the attentuating effects of the Earth's atmosphere at 610 μm.

The emission line for C I is at approximately the same velocity as that for CO and has a linewidth of 5.0 km s,$^{-1}$ compared with 5.9 km s^{-1} for ^{12}CO ($J = 1-0$) and 3.9 km s^{-1} for ^{13}CO(1-0). The corrected antenna temperature (T_A^*) for C I is about 11 K, corresponding to a brightness temperature of about 21 K (there is a large correction for the Planck distribution, since the transition corresponds to a temperature $h\nu/k \simeq 25$ K). Both the linewidths and line temperatures for C I (corrected for Planck effects) lie between those for ^{12}CO and ^{13}CO, indicating that the C I radiation has a large optical depth, since that for ^{12}CO is known to be on the order of 100 and that for ^{13}CO about 3. Phillips and Huggins deduced from these data that the C I/CO column density ratio is about unity in the direction of BN-KL.[7]

The relationship between C I and CO can be further followed by comparing the spatial distribution across the Orion cloud. FIGURE 2 is a strip map in right ascension

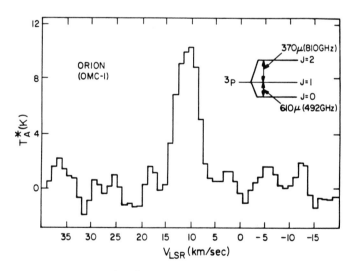

FIGURE 1. The spectrum of C I $^3P_1-^3P_0$ from the BN-KL region of the Orion Molecular Cloud. Shown inset is the fine structure splitting for the ground state of C I.

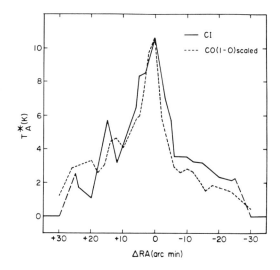

FIGURE 2. A right ascension strip map of the peak C I line temperature (solid line) across OMC1, centered on α_{1950} = 05:32:47; δ_{1950} = −05:24:30. For comparison, the peak CO(J = 1−0) line temperature is shown (dashed line), scaled to match the C I intensity at the cloud center.

centered on the BN-KL region and extending somewhat more than 0.5° east and west. This particular figure compares T_A^* for $^{12}CO(1–0)$ and C I and indicates that the distributions could be similar. However, if the Planck correction is applied to the C I data, it is found that the central temperature, $T_B \simeq 21$ K, is much less than the CO temperature of about 75 K, whereas the CO and C I Planck-corrected temperatures are similar in the outer regions of the cloud.

The Plateau Region

It is well known that the central BN-KL region of OMC1 contains an expanding gas flow region, which is observed by means of H_2 emission lines from shocked gas,[8] by means of high J CO emission from cooler gas,[9] by low J CO emission from the region known as the plateau,[10,11] and by water line maser emission.[12] It is interesting to inquire whether C I emission can be observed from this unusual region, whose chemistry can be rather different from that of the general dense interstellar cloud material.[10]

A careful examination of emission in the C I line was carried out with the NASA 3-m telescope (IRTF) at Mauna Kea, Hawaii. The diffraction-limited beamwidth is somewhat smaller for that telescope (~45 arcsec) than for the airborne telescope (~2.5 arcmin). The smaller beam is suitable for the plateau source, whose size is about 40 arcsec.[11] The effects of the Earth's atmosphere are much stronger from Mauna Kea than from the aircraft, but they affect the contributions from the molecular cloud and the plateau equally. By integrating for a considerable time, a high signal-to-noise ratio was obtained on the narrow molecular cloud line seen in FIGURE 1, but no signal was observed with the large (~50 km s⁻¹) linewidth characteristic of the plateau, or shocked gas region. It was deduced by Beichman et al. that the C I/CO column density ratio was less than 0.1 in the plateau.[13]

The Orion NGC 1977 Bright Rim Region

NGC 1977 is an H II region in the Orion cloud about 0.5° north of the BN-KL region. The H II region has a sharp edge adjacent to the northern extension of the Orion Molecular Cloud, which is known as OMC2. Strip maps of $^{13}CO(1-0)$ and C I emission are available for the north-south crossing of this edge, or bright rim region. FIGURE 3 shows a comparison of C I and ^{13}CO spectra for that strip. It is clear that C I is enhanced relative to CO in the immediate vicinity of the bright rim.

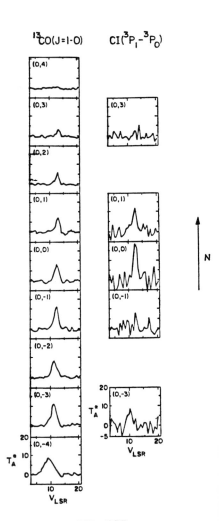

FIGURE 3. $^{13}CO(J = 1-0)$ and C $I(^3P_1-^3P_0)$ data near the rim in NGC 1977. The position of each panel is given in arcmin (α, δ) relative to α_{1950} = 05:32:47; δ_{1950} = −04:58:11.

NGC 1977

DISCUSSION

The conclusions that we can draw from the studies of C I emission in Orion are concerned with the abundance of C I relative to CO.

Comparison of the two species at the core of OMC1 indicates that the column densities are about the same. This conclusion is drawn from studies of both the line strengths and linewidths, from which the column density for CO was found to be about 5×10^{19} cm^{-2} and that for C I to be 3.5×10^{19} cm^{-2}.[7] However, from studies of the maps across the OMC1 cloud, it seems that there is a relative enhancement of C I away from the hot, dense central core. We do not yet have accurate abundance ratios for the cooler, less dense regions of the cloud, but we suspect that C I is dominant.

As mentioned in the introduction, recent theories of the chemistry of the dense interstellar medium[6,14,15] suggest that the clouds should be predominantly CO in the cores, but have a region near the surface, comparable to the uv penetration depth, in which C I is dominant with a column density of about 10^{17} cm^{-2}. This is to be compared with observed values of 10^{19} cm^{-2}. A revised model of the chemistry of the clouds by Graedel *et al.*, allows for a variable content of metals, leading to a variable electron abundance.[16] They are able to show, for a relatively restricted H$_2$ density range and for high values of the electron abundance ($\chi_e \simeq 10^{-5}$), that much of the carbon should be in C I form, even deep inside the cloud. This mechanism might operate in the core (BN-KL) region, but does not seem appropriate for the less active regions of the cloud, which could be similar to dark clouds in their physical construction. Dark clouds are known to have low electron abundances ($\chi_e \leq 10^{-8}$).[9] The problem for the outer regions of the Orion cloud seems acute.

The NGC 1977 bright rim map has shown that, at a well-defined cloud edge close to an ionized region, the C I/CO ratio is enhanced.[17] This is in agreement with the qualitative trend of the chemical arguments. We seem to have a situation in which the C I/CO ratio is qualitatively in agreement with chemical models, but the C I column density is too high.

There are various mechanisms by which C I can be formed at the expense of CO in the bulk of the cloud. For instance, if the cloud contains shocked regions[18] or is filled with star-formation regions, there may be enough activity to keep some of the carbon ionized. In this respect, the time-dependent chemical models of Langer, Iglesias, and Gerola and Glassgold, are important.[14,19,20] They show that, for about 10^6 y after such ionization, a molecular cloud will have C I as its dominant carbon-bearing species. Unfortunately, the attempt to test the relevance of shock activity had a negative result. The well-defined shocked region of the Orion plateau (surrounding BN-KL) was shown to be devoid of C I and, in fact, is one of the few regions in which we have demonstrated that CO is the dominant species.[13]

A physical process by which the time-dependent chemistry could be invoked might be a churning of small cloudlets within the bulk cloud, resulting in a bringing to the surface of any given cloudlet every 10^6 y or less. Of course, it has not been conclusively demonstrated that molecular clouds do have physical lifetimes greater than a few times 10^6 y. If cloud lifetimes are short, then the time-dependent chemical models would find strong support from the C I measurements.

CONCLUSIONS

Investigations of the Orion Molecular Cloud region have allowed the detection of interstellar C I $^3P_1-^3P_0$ emission and have provided much information concerning the relative abundance of C I and CO.

Various problems have arisen concerning the excessive abundance of C I observed as compared with current chemical models. Possible explanations for that excess invoke the presence of internal uv, shocks, or high electron abundances. Also, the chemical models indicate that time-dependent effects may be important if cloud lifetimes are less than 10^6 y, or if a chemical or physical process can periodically rejuvenate components of the cloud. At the moment, the evidence is probably against internal shocks as a relevant mechanism, on the basis of the lack of C I in the Orion plateau.

ACKNOWLEDGMENTS

My collaborators, Patrick Huggins, Jocelyn Keene, Charles Beichman, Ron Miller, Tom Kuiper, Alwyn Wootten, and Margaret Frerking, have all contributed to the studies of C I. William Langer has pointed out several important aspects of the carbon-chemistry problem.

REFERENCES

1. PENSTON, M. V. 1970. Astrophys. J. **166:** 771.
2. PHILLIPS, T. G., P. J. HUGGINS, T. B. H. KUIPER & R. E. MILLER. 1980. Astrophys. J. **238:** L103.
3. HERZBERG, G. 1958. Proc. R. Soc. London Ser. A **248:** 309.
4. SAYKALLY, R. J. & K. EVENSON. 1980. Astrophys. J. **238:** L107.
5. JENKINS, E. B. & E. J. SHAYA. 1979. Astrophys. J. **231:** 55.
6. LANGER, W. D. 1976. Astrophys. J. **206:** 699.
7. PHILLIPS, T. G. & P. J. HUGGINS. 1981. Astrophys. J. **251:** 533.
8. BECKWITH, S., S. E. PERSSON & G. NEUGEBAUER. 1979. Astrophys. J. **227:** 436.
9. WATSON, W. D., L. E. SNYDER & J. M. HOLLIS. 1978. Astrophys. J. **222:** L145.
10. ZUCKERMAN, G. & P. PALMER. 1975. Astrophys. J. **199:** L35.
11. KNAPP, G. R., T. G. PHILLIPS, P. J. HUGGINS & R. O. REDMAN. 1981. Astrophys. J. **250:** 175.
12. GENZEL, R., M. J. REID, J. M. MORAN & D. DOWNES. 1981. Astrophys. J. **240:** 884.
13. BEICHMAN, C. A., T. G. PHILLIPS, H. W. WOOTEN & M. A. FRERKING. 1982. *In* Proc. Symp. Neutral Clouds near H II Regions—Dynamics and Photochemistry. D. Reidel. Dordrecht. In press.
14. IGLESIAS, E. 1977. Astrophys. J. **218:** 697.
15. PRASAD, S. S. & W. T. HUNTRESS, JR. 1980. Astrophys. J. **239:** 151.
16. GRAEDEL, T. E., M. A. FRERKING & W. D. LANGER. 1982. Astrophys. J. In press.
17. WOOTTEN, H. A., T. G. PHILLIPS, C. A. BEICHMAN & M. A. FRERKING. 1982. Astrophys. J. In press.
18. NORMAN, C. & J. SILK. 1980. Astrophys. J. **238:** 158.
19. LANGER, W. D. 1976. Astrophys. J. **210:** 328.
20. GEROLA, H. & A. E. GLASSGOLD. 1978. Astrophys. J. Suppl. Ser. **37:** 1.

DISCUSSION OF THE PAPER

G. CARRUTHERS (*Naval Research Laboratory, Washington, D.C.*): In the plot of relative abundance of C^+, C, and CO *versus* optical depth in the cloud, what is the cause of the "dip" in the CO curve?

W. LANGER (*Plasma Physics Laboratory, Princeton, N.J.*): The dip is due to He^+ destruction of CO just before the H_3^+ dense-cloud chemistry becomes effective.

P. THADDEUS (*Goddard Institute for Space Studies, New York, N.Y.*): My impression is that molecular clouds may be dominated by diffuse, filamentary structures and, therefore, edge effects are generally very important.

PHILLIPS: There is still a problem in explaining the large amount of C I observed in molecular clouds, unless one invokes internal sources of uv radiation.

THE ORION NEBULA
LARGE-SCALE DISTRIBUTION OF FAR-INFRARED AND SUBMILLIMETER LINE EMISSION*

Martin Harwit

Department of Astronomy
Cornell University
Ithaca, New York 14853

INTRODUCTION

During the past six years, our group at Cornell University has been investigating radiative cooling in interstellar clouds through the emission of far-infrared and submillimeter radiation. Technical difficulties at these wavelengths are rapidly being overcome and an increasing number of research groups are discovering new kinds of gaseous domains previously unrecognized in interstellar space.

Except in the very densest clouds, thermal contract between the gaseous constituents and dust is tenuous. Gas that has been heated either radiatively or through shocks cools primarily through direct emission of line radiation. The thermal continuum emission by dust grains is unrelated to the gas temperature and usually represents direct re-emission of absorbed starlight. Gas and dust temperatures may be quite different, and the study of these two constituents may show independent thermal histories.

Line radiation from gas complexes arises through two radiative mechanisms, fine-structure transitions for atoms and atomic ions and rotational transitions for molecules and molecular ions. Transitions in molecular ions have not yet been detected but should become observable with improved instruments.

The first far-infrared line to be detected was the 88-μm [O III] $^3P_1 \rightarrow {}^3P_0$ fine structure transition, initially observed in M17.[1] A year later, it was also detected in M42.[2] Although the total radiation observed in this and subsequently detected lines is a small percentage of the total far-infrared luminosity of these regions, it represents a substantial contribution to the cooling of the ionized component. The far greater continuum flux goes toward cooling the grains, but remains decoupled from the gas. Watson and Storey have recently published a log of far-infrared and submillimeter fine-structure transitions reported by various observers for M42 and other gaseous nebulae.[3]

More recently, a number of emission lines of neutral oxygen, singly ionized carbon, and highly excited CO have been detected; these appear to be associated with ionization, shock, and, perhaps, dissociation fronts where temperatures are adequate for collisional excitation of higher fine-structure or rotational levels but may not be sufficiently high to excite states that emit visible light in downward transitions. Interstellar gas at temperatures between 20 and 2000 K, a range previously all but

*This research was supported by grants from the National Aeronautics and Space Administration, nos. NGR 33-010-146 and NSG 2347.

0077–8923/82/0395–0056 $1.75/1 © 1982, NYAS

inaccessible to direct observation, is being revealed for the first time through observations in this wavelength domain.

A rough comparison of different line and continuum intensities is provided in TABLE 1.

THEORETICAL EFFORTS

The suggestion that ionized hydrogen regions might be substantially cooled through far-infrared fine-structure line emission appears to have originated with

TABLE 1

A ROUGH COMPARISON OF SOME LINE AND CONTINUUM INTENSITIES
OBSERVED IN ORION

Radiation	Flux ($W cm^{-2}$)	Remarks
Hα	$\sim 6 \times 10^{-15}$	From Schmitter's maps[37]
All lines (3188–7330 Å)	$\sim 4 \times 10^{-14}$	From Johnson's compilation[38] and Schmitter's absolute values[37]
Trapezium dust continuum 9–38 μm	$\sim 7 \times 10^{-14}$	From Forrest et al.,[39] fitted to Gehrz et al.[40]
Total dust continuum 20–100 μm	$\sim 4 \times 10^{-12}$	Kleinmann-Low region, plus Trapezium ($\sim 4' \times 6'$ beam)[41,42]
[S III] 18-μm line	$\sim 2.4 \times 10^{-15}$	2.7′ beam centered on Trapezium[43]
[O I] 63-μm line	$\sim 8 \times 10^{-15}$	(4′ × 7′ beam)[16]
[O III] 51- and 88-μm lines	$\sim 8 \times 10^{-15}$	(4′ × 7′ beam)[44,45]
H$_2$ total emission	$\sim 10^{-14}$	Reference 46
CO in shocked region	$\sim 2 \times 10^{-15}$	Estimated from Storey et al.[21] and unpublished Cornell results for a ~ 2 square arc min emitting region
Radio free-free emission down to 3 μm	10^{-14}	4′ of arc beam[9,47]

Burbidge et al.[4] Gould listed some of the principal lines that might be sought at these wavelengths[5] and, while Delmer et al. calculated the far-infrared emission expected from planetary nebulae[6] and Pottasch modeled the radiation from neutral interstellar clouds,[7] Osterbock provided the first estimate of line emission from H I and H II regions, with rough estimates of line emission from the Orion Nebula.[8] These were put on a more secure footing by Petrosian, whose computations for line intensities in the far-infrared were remarkably accurate.[9] What is remarkable about these efforts is that none of these lines had ever been detected in astronomical observations until 1969, when Gillett and Stein first observed the Ne$^+$ 12.8-μm, $^2P_{1/2} \rightarrow {}^2P_{3/2}$ transition in the planetary nebula IC418, that the first far-infrared/submillimeter lines were only to be detected in the latter half of the 1970's, and that most of these lines had never been seen even in the laboratory.

It is still worth noting that more accurate fine-structure line intensities expected from specific sources were computed by Simpson,[10] that the question of interstellar cooling was reviewed quite generally by Dalgarno and McCray in 1972,[11] and that the first successfully observed molecular transitions—transitions between highly excited rotational levels of CO—were discussed by Kwan[12] and Hollenbach and McKee[13] well before their actual detection.[14]

The Cup-Shaped Cavity of Clumpy Ionized Gas

The flow of ionized hydrogen in Orion has been traced by Pankonin *et al.*, who found that the ionized gas is streaming toward us from a cavity that the ionizing stars have hollowed out in their motion away from the Earth into the molecular cloud.[15] This view follows from a general picture that has been built up by a large number of observers over the past decade, which shows that θ^1C Ori is moving away from us with a line-of-sight, local-standard-of-rest velocity $V_{LSR} = +11$ km s^{-1} into neutral gas moving away at a smaller velocity $V_{LSR} = +8$ km s^{-1}, while the ionized gas, as traced by the Doppler shift in H 76α lines, ranges from $V_{LSR} = 0$ to $V_{LSR} = -5$ km s^{-1}. The higher outflow velocities toward us are found at the edges of the ionized region. The low velocities at the center suggest that the ionizing stars plowing into the cloud have hollowed out a cup from which gas now streams toward us, is partially blocked by the presence of the stars at the center, and is forced to squeeze out at higher velocities around the edge of the cup—effectively like liquid splashing out of a cup at the edges when one blows into the cup at the center.

This is specifically mentioned here, since the conventional, often-cited "blister model" has no such cup-shaped cavity and would lead one to believe that the ionized gas leaving a plane surface beyond the ionizing stars would stream toward the observer fastest near the center and move primarily transverse to the line of sight at the edges. That clearly is not the case in Orion.

That the gas is clumpy in the ionized region is clearly shown by the far-infrared observations of doubly ionized oxygen.[16] However, somewhat surprisingly, this clumping appears to be substantially lower than originally postulated by Osterbrock and Flather.[17] The infrared observations involve the comparison of three measurements, the free-free radio emission from the ionized region, the 51.8-μm fine-structure transition in [O III], and the 88.35-μm transition of the same ion. The ratio of the two far-infrared line strengths is only mildly dependent on temperature, but, at densities between 10^2 and 10^4 cm^{-3}, it is a strong function of density. Collisional de-excitations rapidly diminish the 88-μm line strength relative to the 51-μm emission in this density range. For electron densities $n_e = 10^2$ cm^{-3}, the two lines are almost equally strong, while, at densities $n_e = 10^4$ cm^{-3}, the 51.8-μm line is eight times brighter than the 88-μm line. From this ratio, then, a value of n_e integrated along the line of sight can be determined. At the same time, the radio free-free emission provides us with a value of n_e^2 integrated along the line of sight. The respective densities obtained from the far-infrared line ratios and from the radio emission measure[18] are $n_e \cong 7080$, with a factor of two uncertainty, and $n_e \sim 2200$ cm^{-3}. These suggest a filling factor of ~3, again with a factor of two uncertainty for M42. In contrast, Osterbrock and Flather suggest that only 1/30 of the volume might be filled, though at densities 30 times the mean density.

The differences between these estimates have not yet been fully reconciled.

THE EDGES OF THE IONIZED REGION

The 63-μm [O I] fine-structure $^3P_1 \rightarrow {}^3P_2$ line was first observed from the NASA Lear jet with a beam covering a 4' x 6' field of view in Orion, which was found to emit $\sim 8 \times 10^{-15}$ W cm^{-2} at the Earth.[16] Storey *et al.*, who examined this region with a beam approximately 75 arcsec in diameter from the Kuiper Airborne Observatory, found the emission to be patchy.[19] Radiation was observed from two regions associated with ionization fronts, one of them the region of the ionization bar in which the 63-μm emission appeared to be particularly strong. They found no emission from the Kleinmann-Low nebula. The peak emissions found in the five small regions examined are much lower than the average value predicted by the Cornell observations on the 4' x 6' field of view.

The excitation temperature for the $^3P_1$ state is 228 K and the observed emission can be plausibly associated with either ionization or shock fronts at the interface between the ionized and neutral region. The natural next step to undertake consists of a search for the 145.5-μm transition $^3P_0 \rightarrow {}^3P_1$, which has an excitation temperature of ~ 330 K and a line strength with a substantially different density dependence. The ratio of these lines could then provide at least partial information on the density and temperature of the emitting region. Our group at Cornell has made these measurements and obtained upper limits for the 145.5-μm flux. The interpretation of the results is complicated, however, by the proximity of the line to a telluric ozone absorption feature and no definite upper limit can yet be reported.

Information on the boundary between the ionized and neutral region in Orion can also be obtained from a comparison of the 157-μm [C II] emission and the [C II] radio recombination lines. It will be necessary, however, to observe the far-infrared line with a relatively narrow beam, because, as mentioned below, in other nebulae, the infrared-emitting region appears to be far more extended than the radio-emitting region. These observations may turn out to be difficult because of the extent of the infrared emission and the difficulty of chopping against a region sufficiently far away to be dim in the 157-μm line.

AN IONIZED CARBON HALO?

In recently published observations, we have found that the [C II] emission from M17 is enormously extended.[20] It stretches at least 0.25° across the sky and has a luminosity of at least 2000 $L_\odot$. Observations carried out from the Lear jet also suggest that the emission in this line from NGC 2024 is extended. The [C II] fine-structure emission is far more extended than the radio-recombination emission from singly ionized carbon. This is explained by the fact that the fine-structure levels are readily depopulated by collisions. High densities, such as those found in most recombination-line regions do not, therefore, produce unusually bright 157-μm emission. The radiation emitted per atom is constant once the density substantially exceeds the critical density, which is $n_H \simeq 3 \times 10^3$ cm^{-3}. The distribution of the 157-μm radiation has not yet been mapped for M42. Its total intensity, however, is $\sim 10^{-15}$ W cm^{-2}, which corresponds to $\sim 80 \, L_\odot$ at a distance of 500 pc.

FAR-INFRARED/SUBMILLIMETER OBSERVATIONS OF HIGHLY EXCITED MOLECULES

Within the last two years, the Berkeley group has discovered radiation from highly excited CO molecules in the Kleinmann-Low nebula. Rotational transitions from upper levels in the range from $J = 21$ to $J = 30$ were studied.[14,21] This radiation is believed to emanate from shocked molecular hydrogen regions and gases left in the wake of shocks. The lines observed lie in the 87–124 μm wavelength range. Details of these measurements are presented by the Berkeley group elsewhere in these proceedings.

By observing at longer wavelengths, our group has recently also observed the $J = 17 \rightarrow 16$ transition at 153 μm.[22] The interesting feature of this transition is the virtual independence of its strength from temperature and density conditions over a very wide range of temperatures above ~ 750 K and molecular hydrogen densities $\gtrsim 10^6$ cm^{-3}. This enables one to define the mass of CO in the Kleinmann-Low nebula at temperatures $\gtrsim 750$ K, at least within a factor of two, to be 8×10^{30} g. By assuming cosmic abundance and that one-quarter of the carbon is present in the form of CO, as justified by Storey et al., one finds $\sim 1.5 \, M_\odot$ molecular hydrogen to have been heated to these temperatures within the nebula.

Since direct observations of the hydrogen is complicated by enormous near-infrared opacities—opacities believed to play a negligible role at 153 μm—this indirect estimate of the total mass of material involved in shocks is significant.

In passing, it is interesting to note that the number of CO molecules in a beam one arcmin square is $N_{CO} \simeq 10^{53}$. If there were approximately the same number of oxygen atoms in the beam and the temperature was 10^3 K, with a hydrogen density $n_{H_2} \simeq 10^6$ cm^{-3}, the expected 63-μm flux would be $\sim N_O A h\nu/3$, where $A = 9 \times 10^{-5}$ s^{-1} is the Einstein coefficient, the energy per emitted photon is $h\nu = 3 \times 10^{14}$ erg, and the factor $1/3$ accounts for the weight of different states. This assumes that the density of H_2 substantially exceeds the critical density. The expected flux is then on the order of 3×10^{-16} W cm^{-2}, well below the total observed, but well above the upper limit for the Kleinmann-Low nebula found by Storey et al., 5×10^{-17} W cm^{-2}.[19]

The Berkeley group has also made a tentative detection of the $^2\Pi_{3/2}$, $J = 5/2 \rightarrow 3/2$, transition of OH approximately 30 arcsec north of the Kleinmann-Low nebula. They calculate a column density of 3.6×10^{15} cm^{-2} molecules of shocked OH. At a distance of 500 pc, the number of molecules in a 1′ beam then becomes 7×10^{50} cm^{-3}. This is still very low compared to the amount of CO present. The other form in which oxygen might be found is H_2O. Phillips et al., however, estimate typical water vapor concentrations in the Kleinmann-Low nebula to be $n_{H_2O}/n_{H_2} \simeq 10^{-5}$.[24] This leaves as a puzzle the whereabouts of oxygen, unless carbon is more abundant than oxygen in the gas in the Kleinmann-Low nebula.

TURBULENT STATE OF THE MOLECULAR CLOUD

The existence of heated masses of shocked molecular hydrogen, taken together with far-infrared polarization measurements, leads to a picture of a molecular cloud turbulent in its interior and calm at the surface facing us.

The relevant data are these. Observations in the 1- to 13-μm region[25-28] reveal strong ($\gtrsim 15\%$) polarization in the dust absorption features at 1.6, 3, and 10 μm, but substantially reduced polarization between these features. The direction of polarization is remarkably uniform over a region that covers the Kleinmann-Low nebula stretching across at least 20 arcsec, $\sim 10^{17}$ cm at the distance of the nebula. At far-infrared wavelengths, however, our group found remarkably little polarization, if any.[29] At 60 and 100 μm, we found a polarization that could, at most, be 1%, but was also consistent with zero polarization.

These data are most easily interpreted as follows. At the surface of the molecular cloud facing us, the gas is relatively calm and can maintain a coherent magnetic field direction over its entire surface. The dust in this layer is thoroughly oriented by this magnetic field and preferentially transmits radiation polarized along the direction of the magnetic field lines. Further in the interior of the cloud, grains no longer appear to be systematically oriented. The near-infrared radiation at wavelengths distant from the absorption features is no longer strongly polarized; at the longest wavelengths, where the nebula is optically thin[30] and where we might expect grains to radiate preferentially along their long axes in a sense largely perpendicular to the magnetic field direction, we find no polarization at all.

Zero polarization might be consistent with local thermal equilibrium between dust and gas,[31] since systematic dust orientation depends on the presence of a thermal driving force. This equilibrium is, however, made quite unlikely by the pervasive presence of high-velocity clouds, which suggest rapidly varying gas temperatures. A more likely interpretation offered by the polarization studies and the observation of hot high-velocity clouds is that the entire Kleinmann-Low region is well churned up, so that the magnetic field directions are randomized or the dust grains' orientation is locally altered throughout the nebula by the gas flow mechanism originally proposed by Gold.[32] Johnson *et al.* have recently proposed a variant of flow-induced polarization.[33]

THE NEED FOR FURTHER LABORATORY AND THEORETICAL ANALYSES

Most of the lines discussed here have never been observed in the laboratory and were first detected in astronomical searches. Several of the lines have been difficult to find; the wavelengths of the transitions were not known with sufficient accuracy. The [C II] transition now observed at ~ 157 μm was referred to as the 156-μm line. The $^3P_0 \rightarrow {}^3P_1$ [O I] transition, which for many years was referred to as the 147-μm line, has now been found at 145.5 μm in laboratory studies. Accuracy in laboratory and theoretical wavelength determinations is important for two reasons. First, the spectral region to be scanned in an astronomical observation can be narrowed down and properly scrutinized to find a weak feature in the presence of strong continuum radiation or noise. Second, if we can be sure that the line to be sought does not lie behind an atmospheric spectral feature, even failure to detect line emission can provide information of astrophysical importance. For the above-mentioned [O I] transition, proximity to an atmospheric ozone feature makes observations quite sensitive to the line-of-sight velocity of the gas cloud observed. Certain regions will only exhibit this line at times of the year when the line shift places the [O I] feature well outside the ozone absorption line. Uncertainties in line positions therefore translate into uncertainties about when to observe a source and how to calibrate the strength of an observed feature.

The correct interpretation of observations also depends on the availability of accurate values for collision cross sections and radiative transition probabilities. Collisional excitation through collisions with hydrogen molecules[34] are particularly in demand, but we also lack many of the other cross sections and Einstein A coefficients needed for a complete analysis. For atomic oxygen alone, cross sections for excitation by collisions, as well as charge exchange reactions with protons,[35,36] need to be better understood.

REFERENCES

1. WARD, D. B., B. DENNISON, G. GULL & M. HARWIT. 1975. Astrophys. 202: L31.
2. BALUTEAU, J.-P., E. BUSSOLETTI, M. ANDEREGG, A. F. M. MOORWOOD & N. CORON. 1976. Astrophys. J. 210: L45.
3. WATSON, D. M. & J. W. V. STOREY. 1980. Int. J. Infrared Millimeter Waves 1: 609.
4. BURBIDGE, G. R., R. J. GOULD & S. R. POTTASCH. 1963. Astrophys. J. 138: 945.
5. GOULD, R. J. 1963. Astrophys. J. 138: 1308.
6. DELMER, T. N., R. J. GOULD & W. RAMSAY. 1967. Astrophys. J. 149: 495.
7. POTTASCH, S. R. 1968. Bull. Astron. Inst. Neth. 19: 469.
8. OSTERBROCK, D. E. 1969. Philos. Trans. R. Soc. London Ser. A 264: 241.
9. PETROSIAN, V. 1970. Astrophys. J. 159: 833.
10. SIMPSON, J. P. 1975. Astron. Astrophys. 39: 43.
11. DALGARNO, A. & R. A. MCCRAY. 1972. Annu. Rev. Astron. Astrophys. 10: 375.
12. KWAN, J. Y. 1977. Astrophys. J. 216: 713.
13. HOLLENBACH, D. J. & C. F. MCKEE. 1979. Astrophys. J. Suppl. Ser. 41: 555.
14. WATSON, D. M., J. W. V. STOREY, C. H. TOWNES, E. E. HALLER & W. L. HANSEN. 1980. Astrophys. J. 239: L129.
15. PANKONIN, V., C. M. WALMSLEY & M. HARWIT. 1979. Astron. Astrophys. 75: 34.
16. MELNICK, G., G. E. GULL & M. HARWIT. 1979. Astrophys. J. 227: L29.
17. OSTERBROCK, D. E. & E. FLATHER. 1959. Astrophys J. 129: 26.
18. SCHRAML, J. & P. G. MEZGER. 1969. Astrophys. J. 156: 269.
19. STOREY, J. W. V., D. M. WATSON & C. H. TOWNES. 1979. Astrophys. J. 233: 109.
20. RUSSELL, R. W., G. MELNICK, S. D. SMYERS. N. T. KURTZ, T. R. GOSNELL & M. HARWIT. 1981. Astrophys. J. 250: L35.
21. STOREY, J. W. V., D. M. WATSON, C. H. TOWNES, E. E. HALLER & W. L. HANSEN. 1981. Astrophys. J. 247: 136.
22. STACEY, G., N. KURTZ, S. SMYERS, M. HARWIT, R. RUSSELL & G. MELNICK. 1982. Submitted.
23. STOREY, J. W. V., D. M. WATSON, C. H. TOWNES. 1981. Astrophys. J. 244: L27.
24. PHILLIPS, T. G., N. Z. SCOVILLE, J. KWAN, P. J. HUGGINS & P. G. WANNIER. 1978. Astrophys. J. 222: L59.
25. LOER, S. J., D. A. ALLEN & H. M. DYCK. 1973. Astrophys. J. 183: L97.
26. BREGER, M. & J. HARDOP. 1973. Astrophys. J. 183: L77.
27. DYCK, H. M., R. W. CAPPS, W. J. FORREST & F. C. GILLETT. 1973. Astrophys. J. 183: L99.
28. DYCK, H. M. & C. A. BEICHMAN. 1974. Astrophys. J. 194: 57.

29. GULL, G. E., R. W. RUSSELL, G. MELNICK & M. HARWIT. 1980. Astron. J. **85:** 1379.
30. WERNER, M. W., I. GATLEY, D. A. HARPER, E. E. BECKLIN, R. F. LOWENSTEIN, C. M. TELESCO & H. A. THRONSON. 1976. Astrophys. J. **204:** 420.
31. DENNISON, B. 1977. Astrophys. J. **215:** 529.
32. GOLD, T. 1952. Mon. Not. R. Astron. Soc. **112:** 215.
33. JOHNSON, P. E., G. H. RIEKE, M. J. LEBOWSKY & J. C. KEMP. 1981. Astrophys. J. **245:** 871.
34. FLOWER, D. R., J.-M. LAUNAY & E. ROUEFF. 1977. 21st International Astrophysics Colloquium at Liège, Belgium, Les Spectres des Molecules Simples au Laboratoire et en Astrophysique.
35. BAHCALL, J. N. & R. A. WOOLF. 1968. Astrophys. J. **152:** 701.
36. BLACK, J. H. & A. DALGARNO. 1977. Astrophys. J. Suppl. Ser. **34:** 405.
37. SCHMITTER, E. F. 1971. Astron. J. **76:** 571.
38. JOHNSON, H. M. 1968. Nebulae and Interstellar Matter. B. M. Middlehurst and L. H. Aller, Eds: 65. University of Chicago Press. Chicago.
39. FORREST, W. J., J. R. HOUCK & R. A. REED. 1976. Astrophys. J. **208:** L133.
40. GEHRZ, R. D., J. A. HACKWELL & J. R. SMITH. 1975. Astrophys. J. **202:** L33.
41. WARD, D. B., B. DENNISON, G. E. GULL & M. HARWIT. 1976. Astrophys. J. **205:** L75.
42. HOUCK, J. R., D. F. SCHAACK & R. A. REED. 1974. Astrophys. J. **193:** L139.
43. MCCARTHY, J. F., W. J. FORREST & J. R. HOUCK. 1979. Astrophys. J. **231:** 711.
44. DAIN, F. W., G. E. GULL, G. MELNICK, M. HARWIT & D. B. WARD. 1978. Astrophys. J. **221:** L17.
45. MELNICK, G., G. E. GULL & M. HARWIT. 1979. Astrophys. J. **227:** L35.
46. BECKWITH, S. 1981. Int. Astron. Union Symp. 96, Infrared Astronomy. C. G. Wynn-Williams and D. P. Cruishank, Eds: 167. D. Reidel. Dordrecht.
47. GOUDIS, C. 1975. Astrophys. Space Sci. **36:** 105.

DISCUSSION OF THE PAPER

P. GOLDSMITH (*University of Massachusetts, Amherst, Mass.*): Have you resolved the CO $J = 17 \rightarrow 16$ transition and, if so, what is the linewidth?

HARWIT: No. We are currently using classical optical techniques. Such lines will be resolvable when heterodyne techniques become available in this wavelength band.

B. DRAINE (*Institute for Advanced Study, Princeton, N.J.*): What is the intensity of the $17 \rightarrow 16$ CO line?

HARWIT: 7×10^{-17} W cm^{-2} in a $1'$ field of view.

STARS OF LOW TO INTERMEDIATE MASS
IN THE ORION NEBULA

G. H. Herbig

Lick Observatory
Board of Studies in Astronomy and Astrophysics
University of California, Santa Cruz
Santa Cruz, California 95064

We are all familiar with the profusion of hot stars in Orion and with the development of the idea (going back to Ambartsumian and Blaauw) that the Orion OB associations represent several generations of massive star formation from the Orion Molecular Cloud. The idea that lower-mass stars have formed as well did not develop in such a persuasive fashion, but some of the raw evidence has been with us for over thirty years, since the discovery of hundreds of Hα emission (T Tauri) stars (mostly by Haro and his coworkers) in the same areas where early-type stars are found.

The distribution of these faint, presumably low-mass stars and their massive OB counterparts in the older Orion associations and around the periphery of the Orion Nebula does not differ in any conspicuous way from that seen in some other OB associations. On the other hand, it appears that a unique situation is visible in the core of the youngest Orion association, Orion Id, around the bright multiple star θ^1 Orionis at the center of the Orion Nebula. It is upon that region, and what appears to be taking place there, that I intend to concentrate.

The history of recorded observations of that particular area is an interesting one, going back to the first drawing of the Trapezium stars made by Huyghens in 1656.* The historic 1882 photograph taken by Henry Draper in blue-violet light (FIGURE 1) shows the H II region most impressively, but even the brighter stars are almost drowned in the background. Modern photographs in the same wavelength region (FIGURE 2) improve the situation only somewhat through their better angular resolution. The problem is apparent from the energy distribution of the nebula (FIGURE 3): a strong continuum due to scattered light from the Trapezium stars plus atomic sources, and the intense emission lines. The conventional blue-violet photographic passband includes a major contribution from both sources. If only the continuum between the lines contributed, the mean surface brightness over the blue-violet region would drop by a factor of about three. If the fainter stars in the nebula are red, then one ought to gain by going to longer wavelengths where the continuum flux rises only moderately and the emission lines are more widely spaced. Although there is some improvement in the detectability of faint stars by going to the conventional red (FIGURE 4), the gain is not dramatic because the powerful Hα line lies in that passband. Clearly, one must work at long wavelengths, but with the major emission lines screened out. TABLE 1 shows numerically the considerations that go into these passband choices.

The first photographic venture into the deep red beyond Hα was by R. J. Trumpler

---

*E. S. Holden wrote a detailed memoir that describes the early visual observations of the nebula and the stars in it and that ends with the opening of the photographic era in *Washington Observations* for 1878, Appendix I.

64

0077–8923/82/0395–0064 $1.75/1 © 1982, NYAS

FIGURE 1. The Orion Nebula, photographed by Henry Draper 1882 March 14 with the 11-in. refractor and an exposure of 137 min on "gelatino-bromide plates." Reproduced from Figure 40 of *Astronomical and Meteorological Observations Made During the Year 1878 at the U. S. Naval Observatory (1882)*.

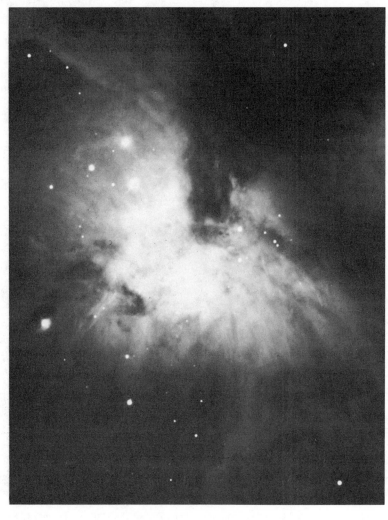

FIGURE 2. The Orion Nebula photographed in the blue-violet passband 390–480 nm, approximately the same as that of FIGURE 1, but with the Lick 120-in. reflector.

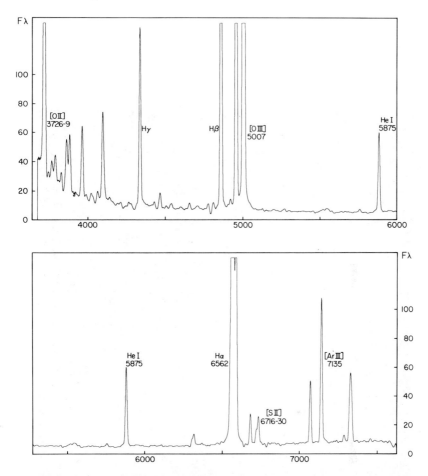

FIGURE 3. The energy distribution of a bright area in the Orion Nebula near the Trapezium. The original scans were obtained by D. E. Osterbrock, to whom the author is indebted for permission to reproduce this plot, with the 120-in. Cassegrain IDS scanner. The ordinate is 10^{16} F_λ, the unit being erg cm^{-2} s^{-1} nm^{-1} (arcsec)$^{-2}$. The original resolution of about 0.7 nm has been degraded by smoothing.

at Lick Observatory just fifty years ago. His historic photograph is shown as FIGURE 5. Trumpler named the curious concentration of faint stars around the bright early-type multiple system θ^1 Ori the Trapezium Cluster.[1] It is shown at a better scale on modern 120-inch plates (FIGURES 6 and 7). Several years later, Baade and Minkowski at Mount Wilson rediscovered and rediscussed the cluster,[2] and the photographs they published at that time made it famous.† There is, however, for our purpose, another

†W. Huggins reported that Lord Rosse had " . . . in some parts of the nebula . . . observed a large number of exceedingly minute red stars. . . ."[3] No mention of this observation has been found in Rosse's published papers. On that evidence alone it is hardly possible to say whether this represented a prediscovery detection of the Trapezium Cluster.

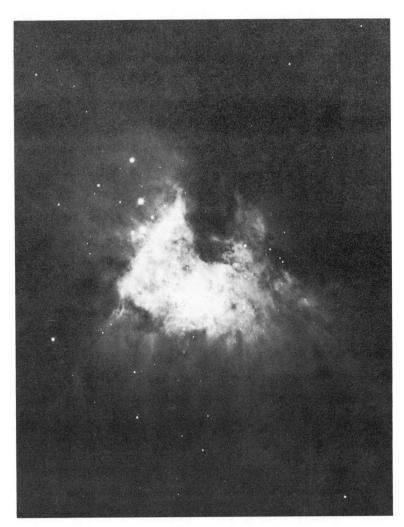

FIGURE 4. The Orion Nebula in the red passband 590–680 nm; 120-in. reflector.

TABLE 1

SURFACE BRIGHTNESS OF ORION NEBULA AS IT AFFECTS THE DETECTABILITY OF FAINT STARS

Corresponding Photograph	λ (nm)		Surface Brightness of 1 arcsec² of Nebula near Trapezium ($10^{15}\,F_\lambda$)	Nebular Equivalent Magnitude	Magnitudes of (Hypothetical) Dwarf Stars of $E(B-V)=0.4$, All Having $B=17.2$		
					AOV	KOV	MOV
—	440	Continuum only	9	$B=17.2$	$B=17.2$	17.2	17.2
FIGURES 1 and 2	380–501	Continuum plus lines	39	$B=15.7$			
—	550	Continuum only	5.6	$V=17.1$	$V=16.8$	16.0	15.4
FIGURE 4	590–690	Continuum plus lines	29.7	—			
—	700	Continuum only	7	$R=16.0$	$R=16.5$	15.1	13.8
FIGURE 5	700–760	Continuum plus lines	13.2	$R=15.3$			

NOTE: The F_λ's are of a 1 arcsec² area of a bright area in the Orion Nebula near the Trapezium (FIGURE 3). They are also expressed as equivalent stellar magnitudes by use of Johnson's BVR calibration.[18] The units of F_λ are erg s^{-1} cm^{-2} arcsec^{-2} nm^{-1}. For the wide passbands, the F_λ's tabulated are for continuum plus lines, averaged over the passband. The BVR magnitudes for AO, KO, and MO dwarf stars that have $E(B-V)=0.4$ and appear at $B=17.2$ (to match the "nebular equivalent magnitude" in that passband) are given in the last three columns. Comparison with the "nebular equivalent magnitudes" shows both how cool stars become more detectable above background at the longer wavelengths and the effect (in the B and R passbands) of excluding the emission lines.

FIGURE 5. The Orion Nebula photographed in the deep red passband 700–760 nm by R. J. Trumpler. The original negative was taken with the 36-in. Crossley Reflector, Lick Observatory, 1931 February 22. The emulsion was kryptocyanin-sensitized and ammoniated; the exposure time was 60 min behind

FIGURE 6. The Orion Nebula in the near-infrared passband 690–880 nm with the 120-in. reflector. This short exposure shows the brighter Trapezium Cluster stars.

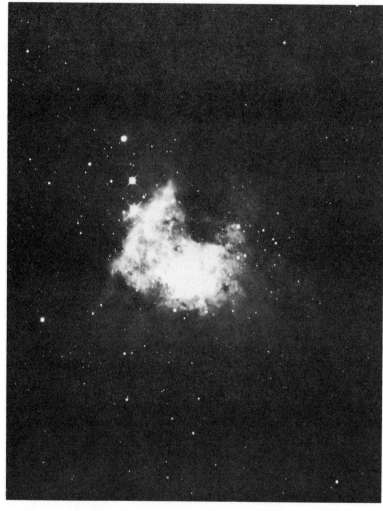

FIGURE 7. The Orion Nebula in the same passband as in FIGURE 6, but with a longer exposure. The inner region of the Trapezium Cluster is here lost in the bright nebulosity.

significant property of the cluster whose discovery far predated Trumpler's, namely the fact that a substantial fraction (about 20%) of its stars are irregularly variable in light.

Curiously, visual observation of faint stars against the brilliant nebular background has an advantage over conventional photography in that the response curve of the eye peaks in the yellow-green, between the stronger emission lines. Thus, it is not surprising that most of the faint variables in the immediate vicinity of the Trapezium were discovered before the introduction of photography. The first variable star, AF Ori, was discovered by W. C. Bond in about 1848. He noted several more, and a total of about nine were known following the work of (the first) Otto Struve.[3] Not long after the introduction of photography, an extensive search at Harvard resulted in the discovery of many more variables, but, because the Trapezium region is heavily overexposed on those conventional blue-violet exposures, those discoveries were predominantly in areas where the nebula is faint. In more recent years, Rosino has found a few more variables in the Trapezium area by exploiting the near-infrared advantage.[5] Aside from the

TABLE 2

STAR DENSITIES IN CLUSTERS

Cluster Name	Region to Which Counts Refer	Number of Stars pc^{-3}	Limiting M_v	Reference
Trapezium Cluster	Whole volume	560	+6.6	This paper
Tau-Aur clouds	Concentrations	4	+8.0:	15
Pleiades	Central 8 pc	1	+8.2	16
M37 (NGC 2099)	Central 7 pc	3	+6.5	17
M67 (NGC 2682)	Central 4 pc	8	+6.5	17

NOTE: The mass density at the centers of a number of globular clusters has been inferred by Bahcall and Hausman[13] and by Williams and Bahcall[14] from fits of star counts to cluster models. The range they obtained for $\rho(0)$ for eleven clusters is from 10^1 to 5×10^4 $M_\odot$ pc^{-3}, so these are lower limits on the total number of stars pc^{-3}.

bright eclipsing binaries θ^1 Ori A and D, there are now 17 named variables within the boundaries of the Trapezium Cluster, which are defined here by the edges of the bright nebulosity as shown on an infrared photograph like FIGURE 7.

Baade and Minkowski recognized that the space density of stars in the Trapezium Cluster was unusually high if conventionally estimated.[2] They preferred to avoid such a conclusion and, to do so, suggested that it was only a part of a much larger cluster seen through a window in the Orion obscuration. They speculated that this window was the result of a reduction in the opacity of the dust in front of the cluster, which was caused by the radiation field of the early-type Trapezium stars. The existence of a "window" is unquestioned, but, since the work of Wilson et al.,[6] an alternate interpretation has become generally accepted, namely that those hot stars have ionized a bubble in the front face of the molecular cloud out of which gas is streaming in our direction, while the back surface of the H II region is eating its way deeper into the neutral gas. A recent exposition of this view is given by Pankonin et al.[7]

My interpretation of the Trapezium Cluster is that it represents a sample of the stellar content of the Orion Molecular Cloud that has been rather suddenly revealed to us as a result of the gas and dust being swept out of the cavity around θ^1 Ori. The space density of stars in that cavity can be estimated in the following way. Star counts on direct plates taken in the 520–580 nm region show 67 stars down to about $V = 16.0$, not including the seven members of the θ^1 Ori multiple system, after subtraction of an allowance for field stars estimated from counts in a similar area just outside the cluster. The cluster area as defined here is a somewhat irregular region that can be approximated by a 185″ × 290″ rectangle, corresponding in the plane of the sky to 0.39 × 0.62 pc at 440 pc. If the depth of the cavity is 0.5 pc, the volume is 0.12 pc³. The color excesses of six early-type stars in the cavity (the components of θ^1 Ori and two fainter stars measured by Penston[8]) indicate no large departures from a mean $E(B - V)$ of 0.4 mag, which we therefore assume holds for the whole volume. Then the limiting absolute magnitude of the counts is about $M_v = +6.6$, to which limit the space density of stars is $67/0.12 = 560$ pc⁻³, with the components of θ^1 Ori not included. This is indeed an extraordinary density for a galactic cluster;‡ TABLE 2 gives densities for several ordinary clusters and for the Tau-Aur dark clouds to comparable limits.

There are two ways in which one might attempt to explain this space density as spuriously large. Both assume, in effect, that the Orion Nebula represents a hole that has been punctured in a thin obscuring screen by the Trapezium stars. If there is no obscuration behind, then the Trapezium Cluster would represent merely a sample of background stars. This seems an unlikely picture because (1) there is a substantial number of variables and Li-rich stars (see below) in the cluster, which would not be expected in the background, (2) there is a large column density of CO at the Trapezium that must be accounted for by material behind the Trapezium, and (3) the line of sight to the Becklin-Neugebauer object and its associated infrared sources passes through the Trapezium Cluster, and those objects are very heavily obscured.

The other explanation is through the Baade-Minkowski hypothesis that the cluster is real, but that we see only a part of it. If that were so, and the cluster was really x times larger than the aperture through which we view it, and if through that hole we are able to see across the cluster's full diameter, then the star density inferred above would be too large by a factor on the order of x. To reduce that density by two orders of magnitude, as required to match the densities in conventional older clusters (TABLE 2), would mean that we see only the central 4′ of a cluster that is really over 5° in diameter, and that our line of sight really extends for 50 pc behind the Trapezium, where the main mass of the molecular cloud lies. Such a geometry faces serious objections: (1) Why, then, should the ionized gas be issuing from the Trapezium region toward us, rather than in both directions? (2) Why does this huge cluster, containing many young stars, fill the low-density region between the front and rear clouds? (3) It is most improbable that the region of the BN object, supposedly in the far background, should be so closely aligned with the window at the Trapezium in the foreground cloud.

I conclude that there are major difficulties with these alternate geometries and that, although the density that is a consequence of the cavity model will certainly be revised when better photometry becomes available, there seems to be no escape from the conclusion that the star density in the Trapezium Cluster is very high.

---

‡Similarly high values have in the past been estimated for the Trapezium region by Strand,[22] by H. M. Johnson,[23] and by Blaauw.[24]

With respect to photometry of the Trapezium Cluster stars, no color-magnitude diagram exists because the background problem has frustrated all photometric observers to date.§ A new attack on this problem is now underway at the Lick Observatory, using narrow-band (10 nm) filters centered in line-free regions of the continuum at 530 and 800 nm and taking advantage of the proper background subtraction that is now possible with two-dimensional CCD arrays. We hope that some preliminary results of this program will be available at the end of the current observing season.

The brilliant background makes it equally difficult to obtain uncontaminated spectra of the fainter Trapezium Cluster stars, even between the stronger emission lines. Progress is possible at all by working only in good seeing and (with slit spectrographs) by using a background-suppression device to take maximum advantage of the fact that the star is a point source and the nebula a surface. There is a further advantage to working in the red, where, for later-type stars, the ratio of star:nebular continuum is relatively large (TABLE 1). The difficulties are such that, at the moment, adequate spectrograms have been obtained for only seven variables in the Trapezium Cluster proper and for five more around its periphery. Not very many more are within the grasp of the equipment now available, but one expects that, with the next generation of coudé spectrographs, it will be possible to go deeper, through the ability to subtract background more accurately.

The expectation was, of course, on the basis of their irregular variability and the fact that many T Tauri stars have been found farther out in the Orion Nebula, that these Trapezium variables would be T Tauri stars. T Tauri stars are readily recognizable as such on these 3.4 nm mm^{-1} coudé spectrograms, both from their metallic emission lines and the characteristically wide Hα emission that ought to protrude from under the narrow nebular Hα emission. Surprisingly, only one of the twelve stars observed appears to be a conventional T Tauri star: LL Ori, which alone shows the familiar P Cyg–like structure at Hα. There is no reason to doubt that these stars are relatively young: ten of the eleven with spectral types from G0 to K5 show a prominent Li I λ6707 line (the twelfth star is of type A, for which Li I is undetectable under any circumstances).

The membership of all these stars in the Trapezium Cluster will be placed on a firmer basis when proper motions and radial velocities become available in the next few years. But since irregular variables having such spectra are essentially unknown in the general field, it is much more plausible that most of this sample is, indeed, associated with the Orion Molecular Cloud. But they are clearly not like the conventional T Tauri stars found in the Tau-Aur dark clouds, for example. What can they be?

The conventional belief has been that stars that have passed through the T Tauri stage and are then in slow contraction toward the main sequence will look much like these Trapezium Cluster stars: a reduced level of variability in light, weak line emission, yet with substantial surface Li abundances.[10] Six such stars have been found as x-ray sources in the Tau-Aur clouds.[11] On the basis of that correspondence, I suggest that the Trapezium Cluster may be a relatively old, yet still pre-main-sequence,

§Trumpler[1] wrote, "The numerous fainter members of the Trapezium Cluster . . . would be expected to [be] F-, G- and K-type dwarf stars. Observations with color filters indeed show at a glance that they are yellow or reddish in color. Accurate color indices for these stars are now being determined, but the results are not yet completed." No further results were ever published.

population of low- or intermediate-mass stars that have been formed and trapped gravitationally for a long time in the Orion Molecular Cloud and that they are now revealed to us only because the newly formed high-luminosity stars of θ^1 Ori have opened up the cavity in the front face of the cloud.

If the average mass of a star in the Trapezium Cluster is taken to be 1 $M_\odot$, then the space density of detectable stars in the cluster (the components of θ^1 Ori still excluded; they would contribute about 80 $M_\odot$) is 560 $M_\odot$ pc^{-3}, equivalent to about 2 × 10^4 H atoms cm^{-3}. This is about an order of magnitude higher than the average density of the HII region around the Trapezium and suggests that, if those stars were formed in that volume, they must have been born when the gas density was very much higher before θ^1 Ori ionized the region and the present outflow began. There is no information on whether the optically impenetrable remainder of the cloud contains many more such stars. A proper infrared search is recommended. If the Trapezium Cluster star density is maintained over the volume of the molecular cloud and a radius of about 4 pc is assumed, then the total mass tied up in stars is on the order of 10^5 $M_\odot$, which is about the total gaseous mass of the southern Orion cloud complex as inferred from CO observations.[12]

Of course, there is an immediate objection to such a result: the conventional ratio of M(stars)$/M$(gas) in star-forming regions is 10% or less, so one suspects that an extra factor of perhaps ten has crept into this chain of assumptions. It will be interesting to re-examine the evidence and conjectures.

One assumes that stars formed in the Orion Molecular Cloud will remain trapped there until the cloud is disrupted. Blitz and Shu have estimated that the lifetime of such a complex is only $\lesssim 3 \times 10^7$ y.[19] Whatever the lifetime, it is clear that disruption will release a large number of low-mass stars into the general field. Unfortunately, one knows nothing about the mass spectrum of the small sample of these stars seen in the Trapezium Cluster, and so it remains only an interesting question whether the dissipation of such associations over the lifetime of the galactic disk has been the source of the large number of low-mass stars in the solar neighborhood, as has often been speculated.[20,21]

Summary

Conventional photographs of the Orion Nebula are dominated by the high-surface-brightness H II region surrounding the multiple star θ^1 Orionis. Specially filtered photographs in the red or near-infrared that avoid the stronger emission lines, first obtained by Trumpler in 1931, reveal a compact cluster of about 67 stars (brighter than $V = 16.0$) centered on θ^1 Ori: the Trapezium Cluster. These stars are seen only over the area corresponding to the shallow pocket, with a volume of about 0.12 pc^3, that has been excavated by θ^1 Ori in the front face of the molecular cloud. The star density in that volume is extraordinarily high, about 560 stars pc^{-3} brighter than $M_v = +6.6$. The spectra of twelve known variables in or very near the Trapezium Cluster show that they are not conventional T Tauri stars, although most have strong Li I λ6707, a reliable indicator of youth in types G–K. It is suggested that the Trapezium Cluster is a sample of post–T Tauri stars (*i.e.*, young, still-contracting stars of low to intermediate mass) that have been trapped gravitationally for a relatively long time in the massive

Orion Molecular Cloud and are revealed to us now only because θ^1 Ori has removed the obscuring dust from that volume. Since the masses of the cluster members are unknown, it is not possible to say whether or not it is through the past dissipation of such molecular clouds and the release of their stellar contents that the large number of low-mass stars in the solar neighborhood has been built up.

REFERENCES

1. TRUMPLER, R. J. 1931. Publ. Astron. Soc. Pac. **43**: 255–60.
2. BAADE, W. & R. MINKOWSKI. 1937. Astrophys. J. **86**: 119–22.
3. HUGGINS, W. 1866. Philos. Trans. R. Soc. London **156**: 381–97.
4. STRUVE, O. 1857. Mon. Not. R. Astron. Soc. **17**: 225–30.
5. ROSINO, L. 1956. Asiago Contr. No. 69; ROSINO, L. & A. CIAN. 1962. Asiago Contr. No. 125.
6. WILSON, O. C., G. MÜNCH, E. FLATHER & M. F. COFFEEN. 1959. Astrophys. J. Suppl. **4**(40) 199–256.
7. PANKONIN, V., C. M. WALMSLEY & M. HARWIT. 1979. Astron. Astrophys. **75**: 34–43.
8. PENSTON, M. V. 1973. Astrophys. J. **183**: 505–34.
9. HERBIG, G. H. 1950. Astrophys. J. **111**: 15–19; HARO, G. 1953. Astrophys. J. **117**: 73–82.
10. HERBIG, G. H. 1978. Problems of Physics and Evolution of the Universe. L. V. Mirzoyan, Ed.: 171–79. Armenian Academy of Sciences. Yerevan, USSR.
11. FEIGELSON, E. D. & G. A. KRISS. 1981. Astrophys. J. **248**: L35–38; WALTER, F. M. & L. V. KUHI. 1981. Astrophys. J. **250**: 254–61.
12. KUTNER, M. L., K. D. TUCKER, G. CHIN & P. THADDEUS. 1977. Astrophys. J. **215**: 521–28.
13. BAHCALL, N. A. & M. A. HAUSMAN. 1977. Astrophys. J. **213**: 93–99.
14. WILLIAMS, T. B. & N. A. BAHCALL. 1979. Astrophys. J. **232**: 754–60.
15. JONES, B. F. & G. H. HERBIG. 1979. Astron. J. **84**: 1872–89.
16. JONES, B. F. 1970. Astron. J. **75**: 563–74.
17. VAN DEN BERGH, S. & D. SHER. 1960. Publ. David Dunlap Obs. **2**: 203–51.
18. JOHNSON, H. L. 1966. Adv. Astron. Astrophys. **4**: 193–206.
19. BLITZ, L. & F. H. SHU. 1970. Astrophys. J. **238**: 148–57.
20. HERBIG, G. H. 1962. Symposium on Stellar Evolution. J. Sahade, Ed.: 45–59. Observatorio Astronómico. La Plata, Argentina.
21. MILLER, G. E. & J. M. SCALO. 1978. Publ. Astron. Soc. Pac. **90**: 506–13.
22. STRAND, K. A. 1958. Astrophys. J. **128**: 14–30.
23. JOHNSON, H. M. 1961. Publ. Astron. Soc. Pac. **73**: 148–52.
24. BLAAUW, A. 1964. Adv. Astron. Astrophys. **2**: 213–46.

DISCUSSION OF THE PAPER

P. THADDEUS (*Goddard Institute for Space Studies, New York, N.Y.*): There is no evidence for a large number of low-mass stars in the less-obscured region near the oldest subassociation Ori OB Ia.

HERBIG: There could be a large number of stars that have been missed in the Hα surveys.

P. SOLOMON (*State University of New York, Stony Brook, N.Y.*): Can you get a lower limit to the age of the Trapezium cluster from your observation that the low-mass stars do not have T Tauri characteristics?

HERBIG: This will be done when the photometry of the stars is completed, on the assumption that the stars are approaching the main sequence on radiative tracks.

T. MOUSCHOVIAS (*University of Illinois, Urbana, Ill.*): From the data you presented, isn't it true that (1) the Trapezium cluster is significantly younger than the other clusters, (2) the Trapezium cluster is smaller in diameter, and (3) the observed dispersion velocity is ~ 10 km s^{-1}, and that all three effects together imply that, if we wait for another $\sim 10^6$ y, the stellar density will be reduced by an order of magnitude?

HERBIG: It is true that the size of this cluster is smaller, about $\frac{1}{3}$ pc. However, we do not know the velocity dispersion of the low-mass stars. There is some older work by Strand, which includes some of these stars, which suggests 5 rather than 10 km s^{-1}. We have some new measurements in progress that should shed light on this problem.

M. HARWIT (*Cornell University, Ithaca, N.Y.*): You mention that the ionizing stars are moving at 6 rather than 11 km s^{-1}. This would suggest that they are coming out of the neutral background gas instead of moving into it, which would be an important difference.

HERBIG: The revised number is an improvement; however, it is still not sufficiently accurate to allow a definite conclusion to be drawn.

J. SILK (*University of California, Berkeley, Calif.*): You compared the density of T Tauri stars in concentrations in Taurus-Auriga with the total density of young stars in the Trapezium cluster. How much of a discrepancy do you think could remain if you allowed for post–T Tauri stars?

HERBIG: I do not know. The Jones-Herbig proper motion survey of the Tau-Aur clouds found very few nonemission, nonvariable stars that move with the T Tauri star population. But now Feigelson and Kriss, Walter and Kuhi, and Walter have found, as x-ray sources, six rather bright stars that seem to be post–T Tauri stars. They were missed completely in the Hα and variable-star surveys, so it is possible that a substantial number of such stars exist and should be added to the Tau-Aur star density that I have given.

INFRARED STUDIES OF STAR FORMATION IN ORION

Michael W. Werner

Space Sciences Division
Ames Research Center
Moffett Field, California 94035

INTRODUCTION
THE IMPORTANCE OF INFRARED OBSERVATIONS FOR
THE STUDY OF STAR FORMATION

The usefulness of infrared (1–1000 μm) observations for the study of star formation can be seen from the following considerations.

(1) The early stages of star formation are characterized by low temperatures, making them especially accessible to infrared observations.

(2) The formation of stars is observed in dusty regions. The dust absorbs optical or ultraviolet radiation from the young and forming stars and reradiates the absorbed energy in the infrared. Observations at infrared wavelengths can probe within such dust clouds to the immediate vicinity of the forming stars. Massive stars, which evolve rapidly, may spend most of their early careers within these dusty regions, remaining hidden from direct view at optical wavelengths.

(3) Atomic, ionic, molecular, and solid-state emission and absorption features that lie in the infrared can be used to probe the density and temperature regimes that occur during the star formation process. Spectroscopic observations of these features are providing valuable new information about the composition, excitation, and motion of these regions.

(4) High spatial resolution (≤1″) is readily accessible at infrared wavelengths between 1 μm and 10 μm and makes it possible to study the spatial structure of complex regions such as the infrared cluster in Orion.

Over the past decade, infrared observations exploiting these advantages, together with complementary observations at radio wavelengths, have greatly advanced our understanding of the earliest phases in the lives of massive stars. Much of this information has resulted from studies of the Orion Nebula and its associated molecular cloud, the nearest site of current massive star formation. This paper reviews these infrared studies of the Orion region, with particular emphasis on broad band continuum and low resolution spectrophotometric measurements and on observations made with high spatial resolution. Taken together, these observations present a vivid picture of the gross structure and energetics of the star-forming region in Orion. More general reviews of infrared astronomy and its application to the study of star formation have been presented recently[1,2] and discuss many interesting issues not dealt with here.

LARGE-SCALE FEATURES OF THE INFRARED EMISSION
FROM THE ORION NEBULA REGION

The Orion Nebula region, like other H II region–molecular cloud complexes in the Galaxy, emits copious amounts of power in the far-infrared (250 μm ≥ λ ≥ 25 μm)

79

0077–8923/82/0395–0079 $1.75/1 © 1982, NYAS

wavelength band. The power received in this wavelength band from the central 10' of the Orion complex is $3 \times 10^5 L_\odot$,[36,38,39] much greater than the power received at all other wavelengths combined. This far-infrared radiation is thermal emission from dust particles,[20] which are heated both by the Trapezium cluster—which excites the optical nebula—and by the Becklin-Neugebauer–Kleinmann-Low (BN-KL) cluster of infrared sources—which lie at the core of the adjacent molecular cloud. In FIGURE 1, we show contours of the continuum emission at 100 μm[36] superposed on an optical photograph of the nebula. The emission at 100 μm, as at all far-infrared wavelengths, peaks sharply at the position of BN-KL some 1' northwest of the Trapezium cluster. The ridge of far-infrared emission running through this peak coincides with the dense central ridge of Orion Molecular Cloud 1 (OMC1), as outlined by radioastronomical studies. The other principal feature of the 100 μm map, the bar of emission just outside the ionization front southeast of the Trapezium, has been discussed elsewhere,[3] but will not figure prominently in the present discussion. One important conclusion that is very apparent from FIGURE 1 is that the radiation at 100 μm arises from dust that lies primarily outside of the ionized gas. This appears to be the case for most H II region–molecular cloud complexes that have been studied with high spatial resolution in the far infrared.[4,5] At 30 μm, the emission from Orion also peaks at BN-KL, but, at this wavelength, the lower level contours follow the distribution of ionized gas more closely than at 100 μm (FIGURE 2), indicating the presence within the H II region of hot dust that is mixed with the ionized gas and heated by resonantly trapped Ly-α radiation.[6,7]

Measurements in two wavelength bands that lie close to the peak of the energy distribution may be compared to determine the color temperature and optical depth of the far-infrared emission. Although they are averages over often complicated lines of sight, these quantities are important because peaks and gradients in the temperature can be used to infer the location of the heat sources and the direction of energy flow, while the optical depth traces the geometrical distribution of heated dust. The color temperature is defined by a grey-body Planck curve fitted through the surface brightnesses measured at the two wavelengths. The optical depth is derived by comparing the surface brightness with that expected from a blackbody at the grain temperature, which may differ from the color temperature if a wavelength-dependent emissivity is adopted for the grains.

FIGURE 3 shows the distribution of 50 μm/100 μm color temperature and 100 μm optical depth superposed on the 100 μm flux map of the Orion Nebula. The optical depth peaks at the position of BN-KL and declines smoothly away from the peak; the ridge of dust apparent in the optical depth map coincides with the dense central ridge of OMC1.[8,9] The color temperature also peaks at BN-KL and falls smoothly in all directions; a subsidiary peak seen at a position of low optical depth in the direction of the Trapezium cluster is probably produced by the small amount of hot dust seen at this position at 30 μm. The peaks in column density and color temperature seen at the position of BN-KL are exactly what would be expected if the cluster were embedded in the dense portion of the molecular cloud where it presumably recently formed. The distribution of temperature in the vicinity of BN-KL shows that the cluster is responsible for heating the dust in the central regions of OMC1. The far-infrared observations can therefore be used to estimate the luminosity of the cluster, as is done below.

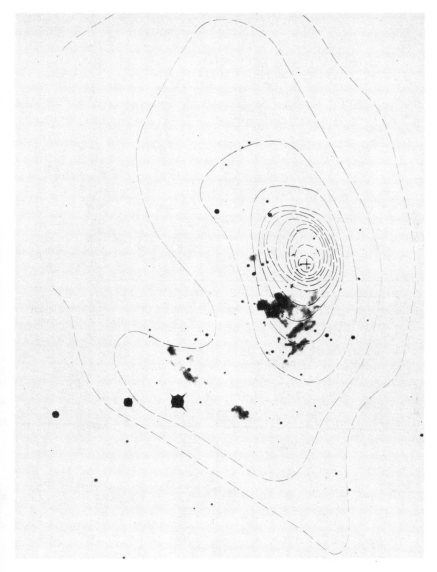

FIGURE 1. Contours of 100 μm continuum emission from the Orion Nebula and the central region of the Orion Molecular Cloud are shown superposed on an optical photograph. North is to the top and east to the left, and the north-south extent of the outer contour (FIGURE 2) is ∼10'. The cross marks the position of the BN-KL cluster. The Trapezium cluster is the group of bright stars just below and to the left of BN-KL.

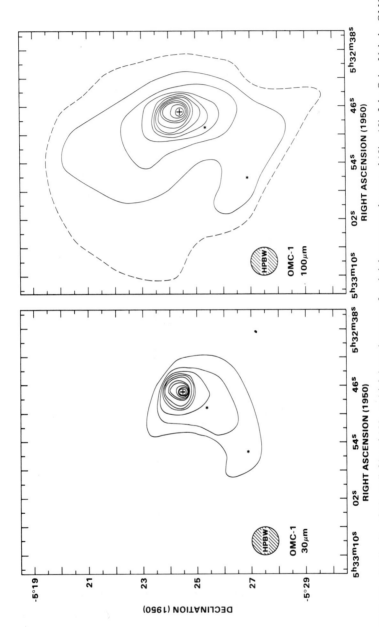

FIGURE 2. The distribution of 30 μm (left) and 100 μm (right) continuum surface brightness over the central $10' \times 10'$ of the Orion Nebula–OMC1 complex. On both figures, the asterisks mark the positions of θ^1 Ori C (closer to the emission peak) and θ^2 Ori A and the cross marks the position of the infrared cluster (BN-KL). The observations were made with $1'$ resolution from the Kuiper Airborne Observatory.[36] On the 30 μm map, the peak flux is 1.2×10^5 Jy into the $1'$ beam, the contour levels are $(0.05, 0.1, 0.2, \ldots, 1.0) \times (1.6 \times 10^{-14} \text{ Wm}^{-2}\text{sr}^{-1}\text{Hz}^{-1})$, and the total flux within the lowest contour is 3×10^5 Jy. On the 100 μm map, the peak flux is 1.0×10^5 Jy, the contour levels are $(0.025 \text{ (dashed)}, 0.05, 0.1, 0.2, \ldots 1.0) \times (1.3 \times 10^{-14} \text{ Wm}^{-2} \text{sr}^{-1}) \text{...} \text{total flux within the lowest contour is } 6 \times 10^5$ Jy.

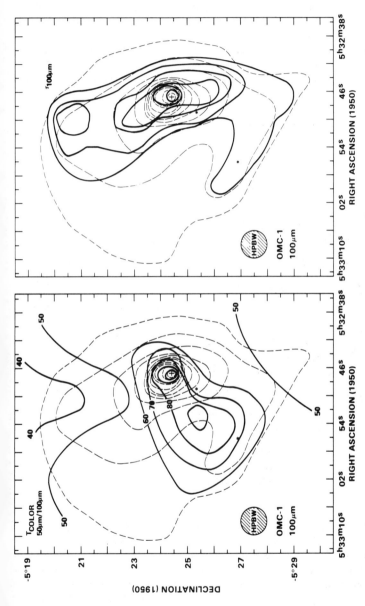

FIGURE 3. The 50/100 μm color temperature (left) and the 100 μm optical depth (right) are shown superposed on the distribution of 100 μm flux from the central regions of the Orion Nebula–OMC1 complex. The color temperature is derived from 1′ resolution 50 μm and 100 μm flux maps,[36] assuming grey emission. The outer contours are labeled in units of K. The innermost contours within the 80 K contour are at 90 and 100 K. The optical depth is derived assuming a λ^{-1} emissivity law for the grains. The contours are at levels corresponding to (0.1, 0.2, 0.4, . . . 1.0) times the peak optical depth, which is 0.3 at 100 μm. The secondary peak in the optical depth map north of BN-KL may be related to a feature seen on NH₃ emission maps of this region.[46]

The data shown above are consistent with a structure for the Orion Nebula complex such as that shown in FIGURE 4.[10,11] The optical nebula is located on the front surface of the molecular cloud as we view it; the opacity of the molecular cloud is so great that we could not see the nebula at optical wavelengths if it were behind the cloud. The BN-KL cluster is located in the dense molecular core and appears in projection immediately adjacent to the optical nebula.

It is interesting to note that, in spatial scale, luminosity, and number of objects, the BN-KL cluster is very similar to the nearby Trapezium cluster, which excites the optical nebula. The ionized gas flows observed optically and in the radio and discussed

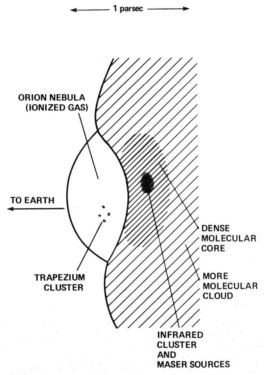

FIGURE 4. A schematic representation of the structures and orientation of the Orion Nebula–OMC1 region (after Zuckerman[10]).

elsewhere in this volume suggest that the gas seen at these wavelengths is boiling off the front surface of the molecular cloud in response to the energy input from the Trapezium stars, which are eating their way inward toward the infrared cluster. At the same time, the violent gas motions within the molecular cloud itself—which are revealed by the molecular spectroscopy results presented in this volume—show that the BN-KL cluster has begun to gnaw its way outward through the core of the molecular cloud. For these reasons, it seems likely that the objects in the BN-KL cluster will soon join those in the Trapezium cluster in the excitation of the nebula.

FIGURE 5 shows the infrared energy distributions for a 1' field of view centered on the BN-KL cluster and for several of the compact sources within the cluster. In the 1' field of view, the radiated power peaks at around 70 μm, suggesting a dust temperature of ~70 K, and the total luminosity is $10^5 L_\odot$. Note that observations at far-infrared wavelengths are possible only from above the atmosphere and are currently limited in angular resolution to $\gtrsim 30''$; these measurements therefore take in the cluster as a whole. At $\lambda \leq 35$ μm, observations from large ground-based telescopes using small fields of view isolate individual sources within the cluster. As shown in FIGURES 5 and 8, these compact sources have energy distributions showing emission from warm dust ($T \simeq 400$ K) as well as the deep ice and silicate absorption features at 3.1 μm and 9.7 μm characteristic of compact infrared sources in regions of star formation.[12] The directly observed ($\lambda \leq 25$ μm) luminosity of each of the point sources is only ~$10^3 L_\odot$, so attempts to attribute a definite fraction of the entire far-infrared luminosity ($L \gtrsim 10^5 L_\odot$ for $\lambda \geq 25$ μm) of the region to heating by a given point source will always be uncertain. Presumably, our view of these sources is being almost totally obscured by the very dust they are heating to produce the far-infrared emission.

The properties of the Orion Nebula region and of the central 1' region centered on BN-KL, as inferred from far-infrared observations, are summarized in TABLE 1. The mass, density, and temperature derived from the infrared observations are in agreement with those derived from molecular observations of this region,[8,9] emphasizing that the dust observed in the far infrared is well mixed with the molecular gas. The total infrared flux received from the central 1' corresponds to a luminosity of $10^5 L_\odot$ at an assumed distance of 500 pc. FIGURE 3 suggests that the BN-KL cluster is responsible for heating not only the central 1' but also material well outside of the central 1' of the cloud. Thus, we conclude from the far-infrared observations that the luminosity of the BN-KL cluster is at least $10^5 L_\odot$.

This estimate assumes isotropic emission in the far infrared; it has been pointed out that such an assumption may lead to an incorrect estimate of the cluster luminosity if the heating source is not at the center of the cloud,[13] but this effect is probably not

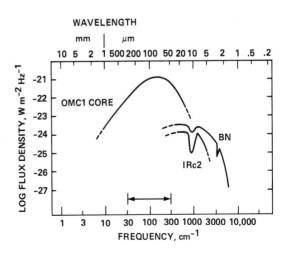

FIGURE 5. Infrared energy distributions for the 1' core of OMC1, centered on the BN-KL cluster, and for the compact sources BN and IRc2, adapted from Wynn-Williams.[37]

TABLE 1

THE FAR-INFRARED PROPERTIES OF THE ORION NEBULA REGION

Total 10′ × 10′ Central Region	
Luminosity (25–130 μm)	$3 \times 10^5 \, L_\odot$
Central Arcmin, Peaking on BN-KL Cluster	
Luminosity (20 μm–1 mm)	$10^5 \, L_\odot$
T_{color} (50/100 μm)	100 K
T_{grain} (50/100 μm)	65 K
Optical depth (100 μm)	0.3
Mass of gas (from 1 mm flux and dust: gas ratio)	200 $M_\odot$
Average density	2×10^6 cm^{-3}
Density distribution	$N(R) \propto R^{-2}$

qualitatively important in the present case. Regardless of the exact luminosity of the cluster, it is clear that the dust in its vicinity is heated by infrared radiation. This must be the case for such a thick dust cloud because optical and ultraviolet photons (whether from embedded or external stars) will travel only 10^{-3} pc within the cloud before being absorbed and reradiated in the infrared. It follows that it is very inefficient energetically to heat such a dense cloud with external optical and ultraviolet sources. This supports the identification of the embedded infrared cluster, rather than the Trapezium cluster, as the main energy source for the central region of OMC1. A consequence of the large column density through the center of the molecular cloud is that the far-infrared optical depth is on the order of unity; the far-infrared energy density within the cloud is therefore high enough that radiative excitation of molecules can become important.[14]

The near infrared ($\lambda \leq 25$ μm) is well suited for high spatial resolution studies and the far infrared (250 μm $\geq \lambda \geq 25$ μm) is the spectral region where the luminosity lies. The submillimeter wavelengths ($\lambda \geq 250$ μm) also have special significance. At these long wavelengths, even the densest clouds are optically thin and, for regions as warm as OMC1, the grain emission is only linearly dependent on temperature. Thus, this is the wavelength region best suited to the determination of the distribution of dust, which can be done with 30″ resolution at $\lambda \geq 350$ μm from large ground-based telescopes at dry mountaintop sites. The results of a recent study of the Orion Nebula region at $\lambda = 400$ μm with 35″ resolution are shown in FIGURE 6. At this resolution, two main condensations of dust are seen, each containing about 50 $M_\odot$ of dust and gas. The northern condensation, surrounding BN-KL, appears to be very compact even at this resolution, suggesting that the dust density $N(R)$ varies as R^{-2} with increasing distance R from the cluster.[15,44] This steep density distribution may be a residue of the collapse that led to the formation of the cluster. The southern condensation seen at 400 μm, which is also very compact, has been found to contain no 20 μm point sources down to a level of 10% of the brightness of the Becklin-Neugebauer source.[15] This condensation therefore appears to be a separate cloud fragment that has not yet begun to form stars, or within which the star formation process has been inhibited or terminated.

The Nature of the Infrared Cluster

FIGURE 7 shows a map of the BN-KL cluster at 20 μm with 2″ resolution and also the positions of the H_2O masers and the regions of 2 μm H_2 emission.[16] Recall that all the detail shown in this 20 μm map lies within the central resolution element of the far-infrared maps (FIGURE 2) and that radiation diffusing outward from this BN-KL cluster is believed to be responsible for heating the dust that radiates in the far infrared. The northern 20 μm peak is the Becklin-Neugebauer (BN) source.[17] The southern complex has, in the past, been referred to collectively as the Kleinmann-Low (KL) nebula;[18] in high-resolution maps,[16,19] this region is found to break up into a number of discrete sources superposed on a fairly uniform background. These sources will be referred to below by the labels (IRc2, etc.) used in FIGURE 7. At other infrared wavelengths, the cluster sources appear to be very different in their relative brightnesses; at wavelengths ≤ 5 μm, for example, BN is by far the brightest object in the cluster.[19] The nature and evolutionary status of the objects within this cluster is a crucial question in current studies of star formation. A variety of arguments reviewed elsewhere[1,16,20] and strengthened by the detection of ionized gas associated with BN[23,24]

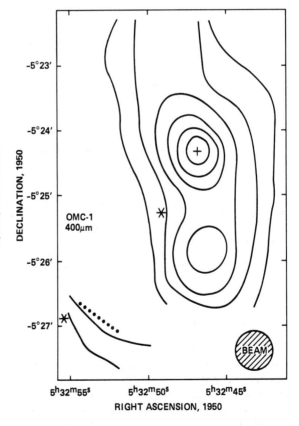

FIGURE 6. A submillimeter (400 μm) continuum map of the center of the Orion Nebula–OMC1 complex, adapted from Keene *et al.*[15] The contours are normalized to a peak value of 100, which corresponds to a 400 μm surface brightness of 6.6×10^{-16} $Wm^{-2}sr^{-1}Hz^{-1}$.

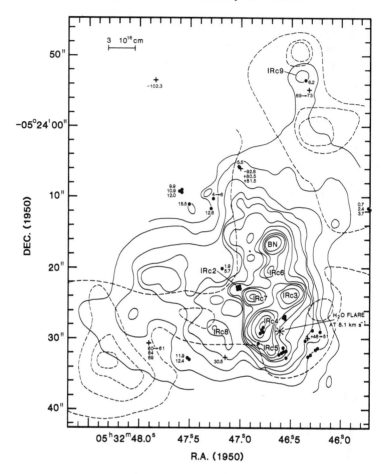

FIGURE 7. The solid contours show the 20 μm emission from the BN-KL region, measured with 2″ resolution. The contours are at levels of (1.2, 2.5, 5, 7.5, 10, 15, 20, 25, 30, 35, 40, 50, 60) × $(1.3 \times 10^{-15} \text{ Wm}^{-2}\text{sr}^{-1}\text{Hz}^{-1})$. The black dots show the VLBI positions of low-velocity H_2O masers, while the crosses mark the positions of weak high-velocity H_2O masers. The dashed contours are those of the $v = 1 \to 0$ S(1) emission from H_2 at 2 μm. The linear scale is for an assumed distance to Orion of 500 pc. (From D. Downes et al.,[16] by permission of the author and publisher).

show that this is, in fact, a group of young and forming stars; this interpretation is pursued below. The far-infrared observations previously described show that the luminosity of the cluster is $\gtrsim 10^5 \, L_\odot$. This luminosity suggests that several of the stars forming in the BN-KL cluster have $M > 10 \, M_\odot$, because the maximum luminosity reached by a 10 $M_\odot$ star as it forms is estimated to be only $\sim 10^4 \, L_\odot$.[21] Thus, the BN-KL cluster is truly a site of massive star formation.

The sources seen on the 20 μm map need not all be self-luminous; some may stand out only because they are density peaks. We discuss below evidence that shows that BN and IRc2 are clearly self-luminous, but that is less conclusive concerning IRc3 and IRc4. A second question to be asked is, Do any of the individual sources contribute a very large fraction of the total luminosity of the region? The supposition that all of the sources are comparably luminous cannot be excluded at present and, indeed, analysis of the 10 μm silicate features (FIGURE 8) suggests that this may be the case.[22] On the other hand, most of the luminosity of the Trapezium cluster is contributed by one star, namely θ^1 Ori C, and it is reasonable to search for a similarly dominant source within the BN-KL cluster.

The Becklin-Neugebauer Source

The best-studied object in the cluster is BN, and there is no doubt that it is self-luminous. This is indicated primarily by the large amount of short-wavelength infrared radiation, indicative of hot dust and, thus, a localized heating source, emitted from BN. Additional evidence comes from infrared spectroscopy showing small, very dense clouds of both ionized gas and hot ($T \cong 3000$ K) molecular gas associated with BN[23] and from the detection of radio emission from the ionized gas.[24] Measurements of the size of BN in the near infrared, summarized in TABLE 2, have also been important in establishing its nature. These measurements have been made at short wavelengths ($\lambda \leq 5$ μm) by speckle-interferometric techniques and at 10 μm using Michelson interferometry. The scatter in the results may be due, in part, to the fact that the observations were made at different position angles, because the source may not be circularly symmetric.[23] The observations do show, however, that infrared techniques

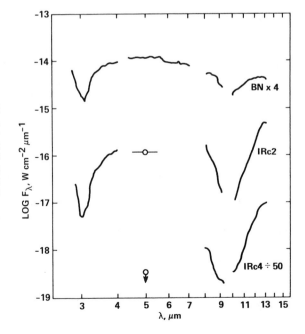

FIGURE 8. Near infrared (3–13 μm) spectra of BN, IRc2, and IRc4 in the BN-KL cluster, adapted from Aitken *et al.*[22] The beam size was 3".4 and the resolving power was 50 in the 3 μm region and 200 in the 10μm region.

TABLE 2

SUMMARY OF BN SIZE MEASUREMENTS

$\lambda(\mu m)$	Position Angle	Angular Diameter (Arcsec)	Linear Size (AU)	Reference
2.2	90°	≤0.15	≤75	40
3.5	90°	0.22	100	45
3.5	40°	0.08	40	41
4.8	40°	0.10	50	41
4.8	0°	≤0.15	≤75	42
10		0.80	400	43

NOTE: Blackbody diameter (T = 530 K) = 0."11 = 55 AU.[25]

are capable of measuring solar system–sized objects at the distance of Orion. They also show that the measured size of BN is quite comparable to the minimum angular size predicted from its flux and color temperature[25] and is in agreement with more detailed predictions based on models of the star formation process.[21,26] Thus, the observations present a picture of BN as a compact, optically thick dust cocoon heated from within by a recently formed star that has just started to ionize the surrounding gas. A much more complete model of the source, emphasizing important departures from circular symmetry, is presented in this volume[23] and suggests that the total luminosity of BN is 1–2 × 10⁴ $L_\odot$. We must therefore look elsewhere for the bulk of the luminosity of the cluster.

The Other Sources

Much less is known about the other main sources in the 20 μm map, IRc2, 3, 4, and 7. IRc2 is of particular interest, because it appears to be the source of the gas outflows seen in this region,[16,31] but IRc4 has also been suggested as an important luminosity source on the basis of its deep silicate feature[22] (FIGURE 8) and its association with highly excited NH₃ emission.[27] Three recent infrared observations described below provide new information about the nature of these objects.

Color Temperature between 8 and 12.6 μm

Infrared color temperature maps provide two ways of identifying heat sources: The color temperature tends to peak at the position of a heat source and color temperature gradients, indicative of the direction of energy flow, can be used to identify which sources are important for heating portions of a surrounding cloud. Preliminary analysis of a recently completed (Wynn-Williams and Becklin, in preparation) 8–12.6 μm color temperature map of the BN-KL cluster with 2" resolution suggests the following features.

(1) BN and IRc2 appear as color temperature peaks and are the only points on the map where the temperature rises above ≤250 K. This is consistent with the previous identification of BN as self-luminous and also draws our attention to IRc2 as a probable self-luminous object.

(2) No prominent feature is seen at the position of either IRc3 and IRc4, which

appear to have color temperatures of ~150 K and ~175 K, respectively, and the temperature appears to decrease fairly uniformly from the maxima at BN and IRc2 toward lower values at IRc3 and IRc4.

Polarization Measurements at 3.8 μm

FIGURE 9 shows the polarization of the emission at 3.8 μm from the region of the BN-KL cluster (Werner *et al.*, in preparation). This emission is highly polarized at many locations within the source. The percentage of polarization approaches 50% in many cases, but the polarization at BN and IRc2 is substantially less than this. The position angle of the polarization clearly shows a systematic variation with position. At the positions of high polarization, the total (polarized plus unpolarized) 3.8 μm flux into the 6″ beam is a small percentage of the flux from BN at this wavelength. The conclusions emerging from a preliminary analysis of these data and the associated distribution of total flux at 3.8 μm include the following.

(1) The high polarization observed over much of the map is attributed to

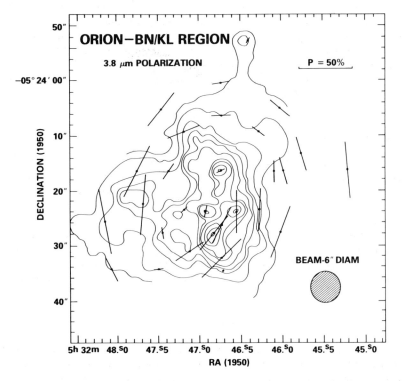

FIGURE 9. Measurements of the 3.8 μm polarization (Werner, *et al.*, 1982, in preparation) of the BN-KL region are shown superposed on the 20 μm contours.[16] The lengths of the lines show the percentage of polarization, while their orientation is that of the maximum electric vector at each position.

scattering by grains illuminated by 3.8 μm radiation originating from BN in the northern portion of the source and from the vicinity of IRc2 in the southern portion. Although several separate objects may, in fact, be involved, this southern source is identified as IRc2 in the following discussion.

(2) Strong 3.8 μm emission is not seen from IRc2 itself; in fact, this object is fainter at 3.8 μm than several nearby positions that appear to be scattering 3.8 μm radiation from IRc2. This suggests a geometry in which there is more extinction along our direct line of sight to IRc2 than from IRc2 to the scattering dust grains and then to us. The extra extinction required to IRc2 is ≥ 5 magnitudes at 3.8 μm and must be very close to the source. This dust may also be responsible for the deep 10 μm silicate feature seen in IRc2 (FIGURE 8).

(3) The polarization data suggest the presence of a substantial embedded luminosity source at IRc2; in the absence of the extra extinction described above, this source would be at least as bright at 3.8 μm as BN.

(4) The objects IRc3 and IRc4 are relatively bright and highly polarized at 3.8 μm. This suggests that, at this wavelength, they are seen as clumps of enhanced dust density. In fact, IRc3 and IRc4 are so highly polarized that all the 3.8 μm radiation coming from them may be scattered light, so the detection of 3.8 μm emission from these objects does not necessarily prove that they contain the hot dust that would indicate an embedded heat source. The much lower polarization of IRc2 is compatible with an embedded source polarized by dichroic absorption, as is the case for BN.[28]

(5) The distribution of the scattered light in the vicinity of IRc2 coincides with that of the OH masers[29] and the low-velocity H₂O masers (FIGURE 7); that is, the 3.8 μm radiation appears to be escaping from the source in the same directions (northeast-southwest) as this outflow.

Mapping at 34 μm

Several high spatial resolution 34 μm maps of the BN-KL cluster have been prepared recently (Gatley et al., 1982, in preparation, and Rieke, 1982, in preparation). These maps, one of which is shown in FIGURE 10, have the following features.

Peaks in the emission are seen at the positions of IRc4 and BN; the peak at IRc4 is the brighter of the two. No notable feature is seen at IRc2.

No previously unsuspected strong sources are seen, even though such objects could, in principle, lie so deeply embedded within the cloud as to be visible only at 34 μm or longer wavelengths. Therefore, we are encouraged to confine our search for the dominant energy sources in the cluster to those already known from the 20 μm data.

A Model for IRc2

A speculative and schematic model for IRc2 and its environs suggested by the data presented above is shown in FIGURE 11. The overall picture is similar to the structures of bipolar nebulae and also to that inferred for the protostellar source NGC 7538-IRS9 and its associated infrared reflection nebula.[5,30] Our direct view of IRc2 at 3.8 μm is blocked by a torus or disc of dust seen edge on; the 3.8 μm radiation that illuminates the scattering grains escapes freely through the poles of the disc. The molecular gas seen in

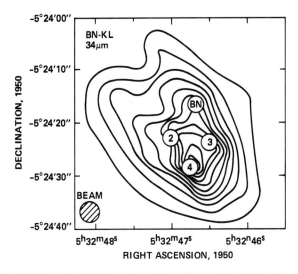

FIGURE 10. A map of the 34 μm emission from the BN-KL region, made with 4″ resolution (Gatley *et al.*, 1982, in preparation). The positions of the compact 20 μm sources BN, IRc2, IRc3, and IRc4 are shown by the circles.

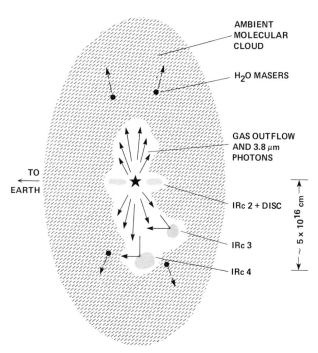

FIGURE 11. A schematic view of IRc2 and its environs.

the low-velocity OH and H_2O masers also escapes along the polar axes. In this picture, IRc2 and IRc4 are identified as dust condensations because of the efficiency with which they scatter 3.8 μm radiation from the central source.

The geometry described above is required to explain the polarization measurements, but it allows a quite luminous object to be hidden at the position of IRc2, because the bipolar geometry could hide it from direct view and cause it to emit its luminosity primarily perpendicular to the line of sight. If we wish to have the luminosity of IRc2 approach 10^5 $L_\odot$,[16] however, it will probably be necessary to argue that there is a decrease in the density of dust in its vicinity. Otherwise, as the most luminous object in the cluster, it would surely appear as a more prominent feature on the 34 μm map (FIGURE 10). It seems possible, as indicated in FIGURE 11, that the stellar wind associated with IRc2 has created a decrease in the dust distribution in its neighborhood. This simple picture does not, however, account in any natural way for the outflows that produce the shocked H_2 emission and appear to be perpendicular to those seen in the maser lines; perhaps the bipolar object shown in FIGURE 11 is embedded within a larger structure that channels the gas flows to produce the lobes of H_2 emission.[31] Further study of the polarized radiation and high spatial resolution measurements to determine the size and shape of IRc2 in the near infrared will provide useful tests of the picture shown in FIGURE 11.

The Nature of IRc4

The present infrared observations tell us less about IRc4 than about IRc2. The high dust density at IRc4 implied by the 3.8 μm and 34 μm observations supports the point of view that IRc4 is a dense condensation of dust and gas. It is not clear, however, whether it is heated primarily by an embedded heat source or by an external source such as IRc2. If the latter could be proven, it would raise the exciting possibility that IRc4 is a dense, compact fragment of roughly stellar mass that has separated out of its parent cloud but has not contracted sufficiently to be highly self-luminous. This would be a previously unobserved stage of star formation. The data presented here show no evidence for a substantial luminosity source at IRc4, but the absence of both intrinsic 3.8 μm emission and a peak in the 8–12.6 μm color temperature could be the result of extinction. Longer wavelength (20 to 34 μm) color temperature observations might be more successful in probing IRc4 and the data (Gatley et al., 1982, in preparation) appear to suggest a maximum in this temperature at its position. The high density, low temperature, and lack of associated ionized gas or neutral outflow suggest, in any case, that IRc4 is at an earlier evolutionary phase than IRc2 or BN. If the intrinsic luminosity of IRc4 is shown to be $\leq 10^4$ $L_\odot$, it will be interesting to consider the possibility that this luminosity is derived from collapse or accretion rather than from nuclear burning.[2,32]

INFRARED ASTRONOMY FROM SPACE

The discussion of the Orion infrared cluster shows that geometrical factors related to the structure and orientation of the individual sources have a strong influence on the appearance of the entire region. In addition, it appears that we are observing in BN-KL a late stage in the star formation process—one at which the young and forming stars have already begun to react strongly upon their environment. Observations of many other sources to the level of detail with which Orion has been studied will be required to average out the geometrical effects and to develop a detailed understanding of all stages of star formation. These studies will be pursued from current ground-based and airborne infrared telescopes using the advanced instrumentation—spectrometers and spatial arrays—now coming into use in the infrared. Within the next few years, moreover, the first space telescopes optimized for infrared studies will go into operation, and these will vastly increase our capabilities for the study of star formation. Major facilities for infrared space astronomy now being constructed or planned include the following three programs.

Infrared Astronomical Satellite (IRAS)[33]

IRAS is a joint United States–Netherlands–United Kingdom program that will carry out a sensitive all-sky survey in the infrared. It will be launched in 1982. The survey will be carried out with $\sim 1'$ angular resolution in broad spectral bands centered at 10, 20, 60, and 100 μm. The 60-cm diameter IRAS telescope will be cryogenically cooled. The increased sensitivity that results from using a cooled telescope in space will permit a (10σ) sensitivity for the sky survey of about 1 Jansky (1 Jy $= 10^{-26}$ Wm^{-2} Hz^{-1}) in each spectral band. (The effective sensitivity will be lower in the galactic plane because of confusion.) At 10 μm and 20 μm, this sensitivity level, to be achieved over the entire sky, is within an order of magnitude of what can be achieved with an hour-long integration on a 5″ diameter region of the sky from the ground with a background-limited detector on a large telescope. At 60 μm and 100 μm, the survey will reach lower flux levels than are attainable from current airborne and balloon-borne infrared telescopes.

For objects with characteristic temperatures, T, between 500 and 50 K, the 1 Jy sensitivity of the IRAS survey corresponds to the following relationship between readily detectable luminosity, L_{min}, and distance, R (in kpc):

$$L_{min}(R) \simeq 2R^2 \left(\frac{T}{500 \text{ K}}\right) L_{\odot}.$$

Thus, the IRAS survey could detect a $\sim 1 L_{\odot}$, 200 K object at a distance of 1 kpc, the BN object at ~ 25 kpc, or across the galaxy, and the central 10′ core of OMC1 at 1 Mpc, beyond the nearest external galaxies. IRAS will be a powerful tool for the study of star formation for several reasons.

(1) The all-sky survey will provide an enormous source list for detailed study, free of most selection effects, and the survey may also capture objects in short-lived and, therefore, rarely observed phases that are critical to our understanding of star formation.

(2) The good sensitivity to faint diffuse emission, particularly at long wavelengths, is well suited to studying both the early stages of star formation and the formation of low-mass, low-luminosity stars.

(3) Systematic infrared studies of star formation in external galaxies—particularly the Magellanic Clouds—will be possible for the first time with IRAS.

(4) The survey will identify isolated sources that can be used to refine theoretical models of the star-formation process. Complex regions such as Orion are not good testing grounds for calculations of the properties of a forming star because many sources may contribute to the luminosity of the region. A few apparently isolated protostars have been found that have energy distributions fairly well fit by current calculations throughout the infrared spectrum (FIGURE 12), but further tests of the models are clearly needed.

Shuttle Infrared Telescope Facility (SIRTF)[34]

SIRTF is a 1-m class, cryogenically cooled infrared observatory to be operated from the Space Shuttle, with the first flight in the late 1980's. SIRTF will carry multiple focal plane instruments that can be changed or upgraded between flights. It will be diffraction-limited at 5 μm or below, with ~3" angular resolution at 10 μm, and will be pointed and thus capable of long integrations. These factors will combine to make SIRTF $\geq$ 100 times as sensitive as IRAS.

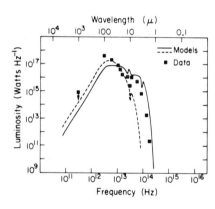

FIGURE 12. The 1–1000 μm infrared energy distribution of the source NGC 7538-IRS9,[5] given by the filled squares, is compared with two models for the spectral appearance of a protostellar envelope adapted from the work of Yorke.[26,32]

The low background and small beam size of SIRTF will make it well suited for the use of array detectors for forming infrared images of regions of star formation and the high sensitivity of SIRTF gives it impressive capabilities in this area. For example, with the type of infrared arrays foreseen for use on SIRTF, it will be possible to map the central 10' of Orion, or any other star-formation region, in several spectral bands in the 10 μm region down to a limiting flux density of ~ 20 mJy per 3" field of view in about 10 min.

Large Deployable Reflector (LDR)[35]

LDR is envisioned as a $\geq$10-m diameter ambient temperature telescope for far infrared and submillimeter observations to be launched in the 1990's. Its large aperture will provide, for perhaps the first time, high angular resolution (~3" at 100 μm) at wavelengths between 35 μm and 300 μm. The combination of high spatial resolution and high spectral resolution at these wavelengths, made possible by LDR, will be very important for the study of star formation.

SUMMARY

This paper reviews the results of infrared studies of the cluster of compact infrared sources in the Orion Molecular Cloud. The discussion is confined to broad band and low spectral resolution measurements (higher spectral resolution results are presented elsewhere in this volume). We discuss new measurements of the 8–12.5 μm color temperature and the 3.8 μm polarization of the infrared emission from the vicinity of the cluster, as well as new 34 μm maps of the region with high spatial resolution. These measurements are useful in identifying the principal luminosity sources within the cluster. Finally, we review the exciting prospects for future studies of star formation that will result from the use of space telescopes optimized for infrared observations.

ACKNOWLEDGMENTS

I thank Gareth Wynn-Williams for many helpful suggestions and discussions and for permitting me to report his data in advance of publication, Ian Gatley and George Rieke for sending me advance copies of the 34 μm maps, Don McCarthy for permitting me to quote his size measurement for BN, and Jocelyn Keene for allowing me to reproduce the 400 μm map in advance of publication. I thank Dave Black and Harriet Dinerstein for comments on the manuscript and Dora Willoughby and Shirley Shaifer for assistance with its preparation.

REFERENCES

1. WYNN-WILLIAMS, C. G. 1982. Annu. Rev. Astron. Astrophys. **20**. In press.
2. WYNN-WILLIAMS, C. G. & D. P. CRUIKSHANK, Eds. 1981. Int. Astron. Union Symp. 96, Infrared Astronomy. D. Reidel. Dordrecht.
3. BECKLIN, E. E., S. BECKWITH, I. GATLEY, K. MATTHEWS, G. NEUGEBAUER, C. SARAZIN & M. W. WERNER. 1976. Astrophys. J. **207**: 770–79.
4. GATLEY, I., E. E. BECKLIN, K. SELLGREN & M. W. WERNER. 1979. Astrophys. J. **223**: 575–83.
5. WERNER, M. W., E. E. BECKLIN, I. GATLEY, K. MATTHEWS, G. NEUGEBAUER & C. G. WYNN-WILLIAMS. 1979. Mon. Not. R. Astron. Soc. **188**: 463–79.
6. KRISHNA SWAMY, K. S. & C. R. O'DELL. 1968. Astrophys. J. **151**: L61–66.
7. PANAGIA, N. 1974. Astrophys. J. **192**: 221–32.
8. GOLDSMITH, P. F., W. D. LANGER, F. P. SCHLOERB & N. Z. SCOVILLE. 1980. Astrophys. J. **240**: 524–31.
9. LOREN, R. B. 1979. Astrophys. J. **234**: L207–11.
10. ZUCKERMAN, B. 1973. Astrophys. J. **183**: 863–69.
11. PANKONIN, V., C. M. WALMSLEY & M. HARWIT. 1979. Astron. Astrophys. **75**: 34–43.
12. AITKEN, D. K. 1981. In Int. Astron. Union Symp. 96, Infrared Astronomy. C. G. Wynn-Williams and D. P. Cruikshank, Eds.: 207–21. D. Reidel. Dordrecht.
13. NATTA, A., F. PALLA, N. PANAGIA & A. PREITE-MARTINEZ. 1981. Astron. Astrophys. **99**: 289–97.

14. MORRIS, M., P. PALMER & B. ZUCKERMAN. 1980. Astrophys. J. **237:** 1-8.
15. KEENE, J., R. H. HILDEBRAND & S. E. WHITCOMB. 1982. Astrophys. J. **252:** L11-15.
16. DOWNES, D., R. GENZEL, E. E. BECKLIN & C. G. WYNN-WILLIAMS. 1981. Astrophys. J. **244:** 869-83.
17. BECKLIN, E. E. & G. NEUGEBAUER. 1967. Astrophys. J. **147:** 799-802.
18. KLEINMANN, D. E. & F. J. LOW. 1967. Astrophys. J. **149:** L1-4.
19. RIEKE, G. H., F. J. LOW & D. E. KLEINMANN. 1973. Astrophys. J. **186:** L7-11.
20. WERNER, M. W., E. E. BECKLIN & G. NEUGEBAUER. 1977. Science **197:** 723-32.
21. YORKE, H. W. & B. M. SHUSTOV. 1981. Astron. Astrophys. **98:** 125-32.
22. AITKEN, D. K., P. F. ROCHE, P. M. SPENSER & B. JONES. 1981. Mon. Not. R. Astron. Soc. **195:** 921-30.
23. SCOVILLE, N. Z. 1981. *In* Int. Astron. Union Symp. 96, Infrared Astronomy. C. G. Wynn-Williams and D. P. Cruikshank, Eds.: 187-205. D. Reidel. Dordrecht. Also, this volume, pp. 125-35.
24. MORAN, J. M., G. GARAY, M. J. REID & R. GENZEL. 1981. Bull. Am. Astron. Soc. **13:** 852. Also, this volume, pp. 204-9.
25. BECKLIN, E. E., G. NEUGEBAUER & C. G. WYNN-WILLIAMS. 1973. Astrophys. J. **182:** L7-9.
26. YORKE, H. W. 1980. Astron. Astrophys. **85:** 215-20.
27. ZUCKERMAN, B., M. MORRIS & P. PALMER. 1981. Astrophys. J. **250:** L39-42.
28. DYCK, H. M. & C. J. LONSDALE. 1981. *In* Int. Astron. Union Symp. 96, Infrared Astronomy. C. G. Wynn-Williams and D. P. Cruikshank, Eds.: 223-33. D. Reidel. Dordrecht.
29. HANSEN, S. S. & K. J. JOHNSTON. 1982. Astrophys. J. In press.
30. TOKUNAGA, A. T., M. J. LEBOFSKY & G. H. RIEKE. 1981. Astron. Astrophys. **99:** 108-10.
31. PLAMBECK, R. L., M. C. H. WRIGHT, B. BAUD, J. BIEGING, P. T. P. HO, S. N. VOGEL & W. J. WELCH. 1981. Bull. Am. Astron. Soc. **13:** 865. Also, WELCH, W. J. 1982. This volume, pp. 154-61.
32. YORKE, H. W. 1979. Astron. Astrophys. **80:** 308-16.
33. VAN DUINEN, R. J. 1977. *In* Infrared and Submillimeter Astronomy. G. G. Fazio, Ed.: 173-83. D. Reidel. Dordrecht.
34. RANK, D. M. 1980. *In* Optical and Infrared Telescopes for the 1990's, Vol. 1. A. Hewitt, Ed.: 245-48. Kitt Peak National Observatory. Tucson, Ariz.
35. MURPHY, J. P., M. K. KIYA, M. WERNER, P. N. SWANSON, T. B. H. KUIPER & P. D. BATELAAN. 1980. Proc. SPIE **228:** 117-27.
36. WERNER, M. W., E. E. BECKLIN, I. GATLEY, G. NEUGEBAUER & K. W. SELLGREN. 1980. Paper presented at Int. Astron. Union Symp. 96, Kona, Hawaii, June 23-27, 1980.
37. WYNN-WILLIAMS, C. G. 1981. Sci. Am. **245**(2): 46-55.
38. HARPER, D. A. 1974. Astrophys. J. **192:** 557-71.
39. WERNER, M. W., I. GATLEY, D. A. HARPER, E. E. BECKLIN, R. F. LOEWENSTEIN, C. M. TELESCO & H. A. THRONSON. 1976. Astrophys J. **204:** 420-26.
40. HOWELL, R. R., D. W. McCARTHY & F. J. LOW. 1981. Astrophys. J. **251:** L21-25.
41. FOY, R., A. CHELLI, F. SIBILLE & P. LENA. 1979. Astron. Astrophys. **79:** L5-8. CHELLI, A., P. LENA & F. SIBILLE. 1979. Nature (London) **178:** 143-146.
42. HOWELL, R. R. & H. M. DYCK. 1982. Astrophys. J. In press.
43. LOW, F. J. 1980 *In* Optical and Infrared Telescopes for the 1990's, Vol. 2. A Hewitt, Ed.: 825-39. Kitt Peak National Observatory. Tucson, Ariz.
44. WESTBROOK, W. E., M. W. WERNER, J. H. ELIAS, D. Y. GEZARI, M. G. HAUSER, K. Y. LO & G. NEUGEBAUER. 1976. Astrophys. J. **209:** 94-101.
45. McCARTHY, D. W., S. G. KLEINMANN & F. J. LOW. 1982. Astrophys. J. In press.
46. ZIURYS, L. M., R. N. MARTIN, T. A. PAULS & T. L. WILSON. 1982. Astron. Astrophys. **104:** 288-95.

DISCUSSION OF THE PAPER

D. MATSAKIS (*U.S. Naval Observatory, Washington, D.C.*): There has been some speculation that the CH₃OH and other variable masers might be powered by variable ir sources. Do you have any data on time variations?

WERNER: We have no evidence yet for the certain variability of any sources.

P. GOLDSMITH (*University of Massachusetts, Amherst, Mass.*): What is the spatial resolution of the 12.5-μm/8-μm color temperature map?

WERNER: Two seconds of arc.

M. ZEILIK (*University of New Mexico, Albuquerque, N.M.*): Have you analyzed your polarization data sufficiently to extract the grain size responsible for the 3.8-μm scattering?

WERNER: No. However, we do see more polarization at 3.8 μm than at 2.2 or 5 μm. This result at 2.2 μm may be related to the size effect, but I am not sure at this stage.

N. SCOVILLE (*University of Massachusetts, Amherst, Mass.*): What is the percentage of polarization?

WERNER: The polarization goes up to 50%.

M. HARWIT (*Cornell University, Ithaca, N.Y.*): The 3.8-μm polarization is quite different from the spatially very uniform 10-μm absorption polarization measured by Dyck and Beichman. Together, the measurements may place useful constraints on infrared radiation transfer within the cloud.

MILLIMETER-WAVELENGTH LINES FROM
THE ORION PLATEAU SOURCE*

G. R. Knapp

Princeton University Observatory
Princeton, New Jersey 08544

INTRODUCTION

Emission lines from molecules in interstellar clouds are typically quite narrow, with $\Delta V \simeq$ 1–5 km s^{-1}. Early observations of the Becklin-Neugebauer–Kleinmann-Low infrared cluster in the Orion Molecular Cloud soon found an exception to this; in 1975, Zuckerman and Palmer pointed out that the line profiles of several molecules have a two-component structure; a "spike" feature superposed upon a broader "plateau" feature, with both components being centered at about the same velocity.[1] The spike emission is widespread and presumably originates in the large molecular cloud in which the BN-KL cluster is embedded. The plateau feature, on the other hand, is confined to a region with a diameter of less than a minute of arc centered on the BN-KL cluster. Furthermore, the relative chemical abundances in the two components differ; the plateau feature is anomalously strong in lines of oxygen- and sulphur-containing molecules, such as SO, OH, and H$_2$S.[1]

The great velocity extent of the plateau emission first became apparent from observations of the CO molecule; observations by Zuckerman *et al.* of the CO(1–0) line[2] showed that the velocity extent is at least ± 75 km s^{-1} (indeed, the high-velocity wings on the CO line are strong enough to be apparent for the first interstellar CO line emission observed[3]). Gas flows with large velocities are often seen in regions of active star formation; these flows manifest themselves in fast stellar winds,[4] in high-velocity H$_2$O maser components,[5-7] and in the fast-moving knots and high-velocity emission lines seen from Herbig-Haro objects.[8] That cold molecular gas also participates in these high-velocity flows is demonstrated by the presence of emission in the H$_2$O and CO lines; the observation of thermal emission in the CO line offers the possibility of measuring the mass and energy of such flows.

Since the discovery of the Orion plateau source, many interesting developments have occurred. These include: (1) the detection of H$_2$ quadrupole emission at 2 μm[9] and high-J transitions of CO[10,11] indicating the presence of hot ($T \sim 2000$ K) shocked molecular gas, with the presence of shocks being related to high-velocity gas flows in the region,[12,13] (2) the observations by Genzel *et al.* demonstrating the outflow of gas from the BN-KL region by measurement of the proper motions of H$_2$O maser components,[14] (3) the discovery of maser emission from the region in vibrationally excited SiO,[15] (4) the discovery of many other localized star-forming regions exhibiting similar gas flows, though none as energetic as the BN-KL flow have yet been

*This research was supported by a grant from the National Science Foundation, no. AST 80-09252, to Princeton University. The observations at the Owens Valley Radio Observatory were also supported by grants from the National Science Foundation, nos. AST 79-16815 and AST 80-07645, to the California Institute of Technology.

0077-8923/82/0395-0100 $1.75/1 © 1982, NYAS

found,[16] (5) observations of high-frequency transitions of several molecules, such as HCN, SO_2, and CO, which allow temperature estimates to be made,[17-26] (6) high spatial resolution observations of several molecular lines that provide estimates of the source size; these include single-dish observations of SO_2, CO, CS, SiO, and HCO^+,[22,25,27-31] very-long-baseline interferometry (VLBI) observations of the SiO maser,[32] and interferometric observations of lines of NH_3,[33] SiO,[34] SO, HCN, and HCO^+[35]—these last observations also give accurate positions for the line sources—(7) detailed mapping of the infrared emission from the region at several wavelengths,[36] and (8) high-sensitivity observations of several spectral lines in the region, revealing the presence of at least two other cloud components besides the spike and the plateau; these are the "hot core" component centered at $V \simeq +4$ km s^{-1} and likewise confined to the BN-KL region[37,38] and the maser "shell source" centered on IRc2.[39]

The present review deals with the plateau component. The available evidence on the position, size, mass, energy, and composition is summarized and the properties of

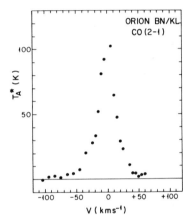

FIGURE 1. Profile of the CO(2–1) line towards the BN-KL cluster in Orion. The velocity is measured relative to an LSR velocity of +9 km s^{-1}.

the source are compared with those of other sources and with speculations about the origin and nature of the source.

THE PHYSICAL PROPERTIES OF THE HIGH-VELOCITY GAS

The Maximum Velocity

The spike emission feature from the Orion Molecular Cloud is centered at a velocity of 9 km s^{-1} with respect to the Local Standard of Rest (LSR), and the central velocity of the plateau emission is fairly close to this. Hereafter, all velocities will be measured relative to this velocity unless otherwise stated.

In the abundant and easily excited CO molecule, emission is detected to ±75 km s^{-1} [2,30,31,40] and to ~100 km s^{-1} in the CO(2–1) line.[22] This maximum velocity agrees well with that seen in the H_2 lines[41,42] and in the H_2O maser lines.[7,14] The CO(2–1) line profile observed over a velocity range from −105 to +60 km s^{-1} is shown in FIGURE 1 (these data were taken with the Owens Valley 10-m telescope; the range above $+60$ km s^{-1} was not covered because of a shortage of observing time[22]). This profile, together with other published CO(2–1) and (1–0) profiles,[2,22,31,40] shows that the CO emission is symmetric in these lines. For the H$_2$ lines, however, the lines are asymmetric; the blueshifted wings are much stronger than the redshifted wings,[41,42] which is attributed to internal absorption in the high-velocity gas source.[41]

Other molecular lines do not show such a wide velocity range; this is due, probably, to a combination of lack of excitation and low intensity in the line. It is also possible that there are compositional differences in the gas due to the superposition of more than one gas flow—this will be considered again later. The maximum observed velocity extents in various molecular lines are listed in TABLE 1. The velocity range given corresponds to half of the full width of the line at zero power.

The Radius

The first molecular line observations of the region showed that its size is ≤ 1 arcmin. More recently, higher spatial resolution observations have made possible an estimate of the source diameter. The single-dish diameters quoted in the following are all calculated on the assumption that the source is a disk of uniform brightness. The CO lines give the largest source diameter, with values of 40″–60″ found for the CO(1–0), (2–1), and (3–2) lines.[17,22,26,30] The CO(2–1) observations made at Owens Valley[22] show that the source size is roughly constant at all velocities (FIGURE 2); this is confirmed by high-resolution CO(1–0) observations.[30] The CO(2–1) source is also

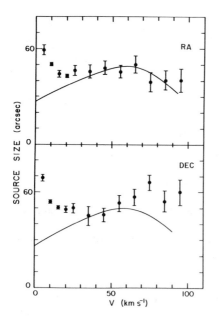

FIGURE 2. Observed source half-power size *versus* velocity for the CO (2–1) line. The solid line represents the expected source size for a model spherical cloud, expanding with $V \propto r$, with a maximum velocity of 100 km s^{-1} and a radius of 25″, smoothed by a beam of width 25″.

TABLE 1

MEASURED VELOCITY RANGE AND SOURCE SIZE
FOR MOLECULAR EMISSION FROM THE ORION PLATEAU SOURCE

Molecule	V_{max} (km s^{-1})	Diameter (arcsec)	References
CO (1–0)	75	40	30, 31, 50
CO (2–1)	100	50	22
CO (3–2)	40	40	17, 26
^{13}CO (2–1)	20	40	22, 31
CS	17	30–40	29
HCN	40	15	27, 31, 35
HCO$^+$	40	$\gtrsim$50	35, 59
H$_2$	100	80	41, 42
H$_2$O	40	—	58
H$_2$S	20	—	31
NH$_3$	45	<20	37
SO	45	15	35, 60
SO$_2$	25	20	22
SiO	50	15*	35, 43

*Some of the emission may be more extended, $\gtrsim 50''$.[43]

about 40″ in diameter.[22] Other molecules, which require higher densities for collisional excitation, give smaller source diameters. The CS source diameter is 30″–40″,[29] the SO$_2$ source about 20″,[22] the NH$_3$ source $\lesssim$ 20″,[37,38] and the thermal SiO $\lesssim$ 14″.[39,43]

Recent interferometer measurements[35] give more reliable results. The gaussian diameter of the SO emission is 15″, as is that of the thermal SiO and the HCN. The HCO source size appears to be $\gtrsim$ 50″, which is marginally consistent with single-dish results. The relatively large size of the HCO$^+$ plateau compared to that of the HCN is somewhat surprising in view of the similar densities required to excite the molecules. The source sizes are also listed in TABLE 1; the fact that they are smaller than the CO source could mean either that the density of the plateau source is larger towards its center or that the excitation conditions are different for different molecules in the plateau.

The observations show that the diameter of the CO source, 50″, is very similar to the internal diameter of the H$_2$ emission in the region;[44] the CO and H$_2$ distributions are compared in FIGURE 3. Since it is likely that the H$_2$ emission is produced by shock heating,[12,13] the distribution in FIGURE 3 suggests that the high-velocity gas is the cause of the shock and that the gas is flowing out of, and not into, the BN-KL region.[22,30]

At the distance of the Orion Nebula, 450 pc, the CO source radius is 1.7×10^{17} cm and the dynamical age of the region, with a maximum velocity of ~100 km s^{-1}, is 540 y.

The Shape

There is, at the moment, little detailed observational information about the projected shape of the plateau source. What there is suggests that the source is roughly circularly symmetric; this is particularly shown by the $CO(2-1)$[22] and SO[35] lines. The possibility that the source is bipolar is examined later in this paper; in the meantime, it is assumed that the outflow is spherically symmetric.

The Line Shapes and Nature of the Flow

The wide emission lines seen towards BN-KL are very likely due to expansion and not to rotation or collapse (the arguments for and against this point of view are summarized in References 2, 4, 14, 22, 30, 31, 35, 36, 39, and 45). As has been

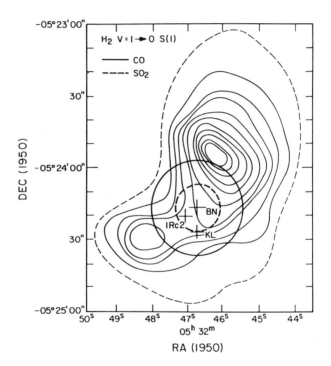

FIGURE 3. Comparison of the CO and SO_2 source sizes measured at Owens Valley with the 2 μm H_2 flux distribution of Beckwith et al.[44]

discussed many times, the line shapes for the plateau are very nongaussian; their shape is much more closely power-law or logarithmic.[31,46] Neither do they have the parabolic or flat-topped shapes produced by spherical expansion at a uniform velocity.[47] The line shapes, the constancy of source size with velocity,[22,30] and the strong evidence that the gas is flowing outwards all suggest that the gas motions are impulsive or ballistic; that is, the radial motions are the product of an initial impulse on an ensemble of clouds that then moves outwards, producing a velocity field with $V \propto r$.[14,22,40] (The effects on line shapes of different types of radial motion laws are discussed in References 46, 48, and 49.) If the expanding material is really in clouds, they are very numerous and small, since high-sensitivity CO profiles (see Reference 50) fail to reveal any of the velocity structure that could be attributed to individual clouds. A model with $V \propto r$ and density decreasing outwards will be adopted in the following to estimate the mass of the high-velocity gas.

The Mass, Energy, and Momentum in the High-Velocity Gas

With the above assumptions, and the further assumption that the gas temperature is uniform at 100 K throughout the cloud, the mass of the plateau gas can be estimated from the CO(2–1) observations. Comparison of the CO(1–0) and (2–1) lines and of the CO and ^{13}CO emission shows that the plateau emission in these lines is largely optically thin. If the element abundances are cosmic and one-tenth of the C is in CO, the total mass of the plateau is $\sim 5\ M_\odot$.[22,30] Of this gas, $\sim 2\ M_\odot$ is moving at velocities greater than 20 km s^{-1}, and $\sim 3\ M_\odot$ at lower velocities. This corresponds to a mass loss rate of $5 \times 10^{-3}\ M_\odot\ y^{-1}$, which is an order of magnitude greater than that observed for any evolved late-type star[51–53] and on the same order of magnitude as ejection from some early-type massive stars (see Reference 54).

The total kinetic energy in the flow under the above assumptions is 2.5×10^{46} erg s^{-1}, with a mean value of 1.5×10^{36} erg s^{-1}, or $\sim 350\ L_\odot$. The current kinetic energy of the highest-velocity gas (that at the greatest distance from the center) is 3×10^{36} erg s^{-1} or $10^3\ L_\odot$. This is close to the total H$_2$ luminosity.[44,55] Similarly, the total momentum in the flow is 2.3×10^{40} gm cm s^{-1}. If radiation pressure is driving the flow, so that

$$Mv/T = \langle L \rangle / c,$$

where T is the dynamical age of the region, then $\langle L \rangle \simeq 10^7\ L_\odot$, about 100 times greater than the total luminosity of the region.[56]

Temperature

The temperature of the plateau gas has been estimated several ways. The fast-moving shocked H$_2$ and the high-frequency CO emission give temperatures of ~ 2000 K. The SO$_2$ rotational temperature is 70 K,[21] while the CO(3–2) brightness temperature suggests that $T \gtrsim 100$ K. The temperature given by the CO lines increases with J,[24] showing the presence of gas having a wide range of temperatures. The inversion temperature inferred from NH$_3$ line ratios ranges from 45 K to well over 100 K.[38] Strong temperature variations are also suggested by the relative ratios of $T(3–2):T(1–0)$ for the HCO$^+$ and HCN lines.[18] As is the case for the CO(3–2) line, the HCN(3–2) line is dominated by plateau emission, which is, however, absent from the HCO$^+$ line.

Position

Finally, we come to the all-important question of the source position. Single-dish measurements are of limited accuracy, and can only place the source center somewhere in the Orion BN-KL cluster. The CO(3–2) source is centered between BN and KL, near IRc6, with the error bars including the positions of BN, IRc2, and IRc7.[17] The center of the HCO$^+$ source is 5″ north of BN.[25,28,31]

Interferometric observations show that the SiO maser is coincident with IRc2,[34] as is the H$_2$O shell source,[14] while the thermal SiO also seems to be centered on IRc2.[35,39] The OH masers are in two main clusters centered on IRc2 and IRc4.[57] The H$_2$O

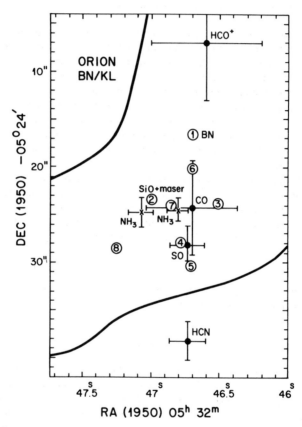

FIGURE 4. The distribution of the sources of high-velocity gas in BN-KL. The numbered circles are the positions of the IRc sources;[36,79] the solid contours show the edge of the 2 μm H$_2$ distribution.[44] The thermal and maser SiO sources are positionally coincident with IRc2. Also shown are the centers of the plateau in CO(3–2),[17] NH$_3$,[33] HCO$^+$,[31] SO,[35] and HCN.[35] The positional error bars are 1σ.

masers lie in a region about 40″ in diameter centered between BN and KL.[14] The high-velocity NH$_3$ emission has two centers whose positions straddle that of IRc7, with neither position being coincident with any of the infrared sources in the region.[33] The SO emission center is close to IRc4, while the HCN emission center is south of this position.[35] This displacement of the HCN center may be due to contamination by the low-velocity gas in the molecular ridge.[35]

These positions are plotted in Figure 4, along with the positions of the infrared sources in the region.[36] Examination of this figure suggests that (1) the apparent centers of the source may be affected by local excitation conditions, particularly by the infrared emission of the BN-KL cluster (see Reference 35), (2) the offsets of the HCN and HCO^+ plateau features may be due to contamination by the ridge feature, as suggested in Reference 35, (3) if these offsets are real, the difference between the ratio $T(3-2):T(1-0)$ for the HCO^+ and that for HCN lines[18] is due to the differences in beamsize and pointing used to make the observations, (4) the different centers found in NH_3, SO, and CO show that, if one source is responsible for all of the outflowing phenomena, the gas is not freely moving but may be expanding through density irregularities in the cloud (see Reference 31), and (5) there may be real local chemical differences, induced by the presence of shocks, uv radiation, etc.

CHEMISTRY

The CO:H₂ Ratio

In summarizing the chemical abundances in the high-velocity gas below, abundances relative to CO are given, since the available evidence on the CO/H_2 abundance does not at present provide a very firm number. In the above estimates of the plateau mass, it was assumed that 10% of the cosmic carbon is in the form of CO, so that $CO/H_2 \simeq 7 \times 10^{-5}$, in agreement with recent measurements for interstellar dust clouds.[61] It is alternatively suggested in the discussion by Phillips and Beckman,[62] that most of the C is in CO. This could be the case if dust in the region had been destroyed by shocks, since most of the carbon is depleted onto dust grains in the interstellar medium (see Reference 62).

The H_2 density in the central regions of the high-velocity gas is $\sim 2 \times 10^6$ cm^{-3}, as determined by measurements of line intensities of molecules such as CS[29] and SO_2;[21] also, the densities for the outer regions of the high-velocity gas flow are suggested to be several 10^3 cm^{-3}. These numbers support the lower values for the CO/H abundance used above and suggest that the relative CO abundance in the plateau is not greatly different from that elsewhere in the interstellar medium.

Since the high-velocity gas is optically thin in the CO(1–0) line and partly so in the CO(2–1) line,[17,22] the ^{13}CO emission is very weak in the plateau. The measured isotope ratios seem normal, with values of $C/^{13}C$ of 40[31] and 75 ± 20[22] being found.

Other Molecules

The abundances of several other species in the plateau relative to that of CO are listed in TABLE 2. Also listed is the range of observed relative abundances in cold dust clouds and in warm star-forming molecular clouds. Many more species have been observed in this region but are not listed in TABLE 2. These include HNC, which is not detected in the plateau and whose abundance varies widely in other molecular clouds, and CN, which has not been detected in the plateau. However, the upper limit on the CN/CO ratio of 7×10^{-5} is higher than the typical value of 10^{-5} found in molecular clouds.[64,65] The abundance of OCS relative to CO is 10^{-3}.[66]

TABLE 2

RELATIVE CHEMICAL ABUNDANCES, [X]/[CO], IN THE ORION PLATEAU

Species (X)	Plateau [X]/[CO]	Molecular Clouds [X]/[CO]	References
C I	<0.13	0.5	71, 72
CS	3×10^{-5}	1.5×10^{-5}	29
HCN	1.6×10^{-3}	10^{-5}–10^{-4}	31, 73
HCO$^+$	2.6×10^{-4}	10^{-5}–10^{-4}	31, 73
H$_2$O	~0.5	—	20, 58
H$_2$S	1.2×10^{-3}	3×10^{-6}	31, 74
SO	6×10^{-3}	2×10^{-6}	31, 75
SO$_2$	4×10^{-3}	$<5 \times 10^{-7}$	31, 75–77
SiO	~5×10^{-5}	2×10^{-7}	39, 66, 78

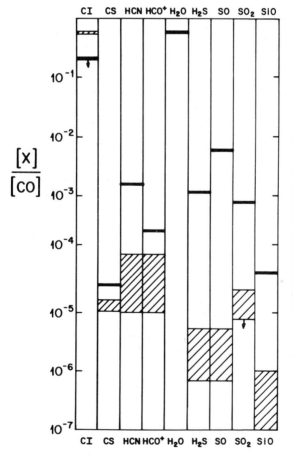

FIGURE 5. Comparisons of the relative abundances of several molecular species to the CO abundance in the plateau with those in interstellar clouds (after T.B.H. Kuiper, private communication). The plateau abundances are given as solid lines; the molecular cloud abundances (given as a range) are cross-hatched. Only upper limits are currently available for the C I plateau abundance and the SO$_2$ content of molecular clouds other than the plateau.

The abundances of TABLE 2 are presented graphically in FIGURE 5. Most striking are the great overabundances of H_2S, SO, SO_2, and SiO relative to the usual values found in molecular clouds. It is difficult to attribute these high values to excitation effects, since many other star-forming regions have similar or greater infrared luminosities but do not show strong emission in the sulphur-containing molecules.

The CS abundance is fairly close to normal values found in molecular clouds, *i.e.*, very little of the S is in CS. If the relative abundances of H_2S, SO, SO_2, and OCS are summed, the result is S/CO = 0.01; the cosmic S/C ratio is 0.05. It is known that carbon is depleted into dust grains by a factor between 5 and 10 in both high- and low-density interstellar clouds, while sulphur is essentially undepleted;[63,67] thus, in the Orion plateau source, essentially all the gas phase S is in these four molecules.

Available chemical models do not, as a rule, treat sulphur chemistry explicitly. Models that do (see References 68 and 69) find that the abundances of SO, H_2S, etc. are proportional to the total gas-phase S abundance and insensitive to the abundances of O, O_2, etc. The equilibrium models of Mitchell *et al.* predict large abundances of these sulphur-containing molecules.[68] The reactions producing these molecules,

$$H + S \rightleftharpoons HS + H,$$

$$HS + O \rightleftharpoons SO + H,$$

$$SO + O \rightleftharpoons SO + O,$$

are neutral molecule reactions and have very low rate coefficients.[69] The formation of SO and SO_2 under normal interstellar conditions, therefore, has a timescale much longer than the average lifetime of a molecular cloud. The presence of these molecules in Orion shows that they were formed under conditions of high temperature and density, such as those produced in a shock.[69] The shock velocities invoked are low, about 10 km s^{-1}, and it was shown by Kwan that shock velocities of 30 km s^{-1} dissociate the gas.[13] The impact of shocks moving faster than this may be mediated by magnetic fields.[70] However, the details of the mechanism that can simultaneously accelerate gas to 100 km s^{-1}, accelerate its chemical reactions, excite its H_2 emission, and manage not to destroy the molecules are not yet understood.

THE NATURE OF THE OUTFLOW

Is the Orion Flow Bipolar?

To date, some twenty other high-velocity gas flows have been found, and several of these have been mapped in some detail (see Reference 16 and J. Bally, this volume). While none of these flows is as energetic as that in the BN-KL region, they seem to be quite similar to it in kind. To date, all the other gas flows seen are bipolar, that is, the centers of the blue- and redshifted wings are not coincident, suggesting a double-jet type of outflow. Such bipolar flows have also been seen, at still lower energies, in some nearby molecular clouds not apparently associated with star formation[80] and in clouds containing T Tauri stars.[81] The Orion flow appears to be extreme in terms of its large energy, small physical size, small dynamical age, high ambient density, and chemical peculiarities. The Orion flow will be compared with others, and their effects on the cloud dynamics considered, by J. Bally later in this volume; in the present paper, the (currently rather conflicting) evidence relating to the bipolarity of the Orion BN-KL flow will be considered.

Bipolar structure is suggested by the observations of the SiO[43] and CO(3–2)[26] lines. These observations give conflicting evidence as to the relative displacements of the blue- and redshifted wings, with axes perpendicular to each other. The CO(3–2) observations suggest that the flow is channeled towards the two H_2 emission peaks.[44] Maps of the region in the 1 cm H_2CO lines[33] show two small dense "quiescent clouds" that may be channelling the flow (FIGURE 6).[44] However, the irregularities in the integrated H_2 emission could also be explained by variable extinction across the face of the distribution—the quiescent clouds have column densities corresponding to $A_v \cong$

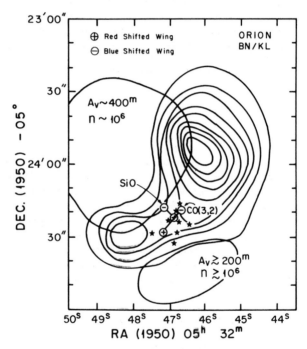

FIGURE 6. Directions of possible bipolar flows in the CO(3–2) and SiO(2–1) ($v = 0$) lines. The positions of the IRc stars, the H_2 contours, and the dense quiescent clouds[33] are also shown.

200^m–400^m, which is $\geq 10^m$ at 2 μ and $\geq 1^m$ at 20 μ.[33] This question can be settled by extinction measurements across the region.

The CO(2–1) and (1–0) lines do not show marked evidence of bipolar structure. In FIGURE 7 is shown the integral of T_A^* (2–1) over several velocity ranges;[22] there is no significant displacement of the plateau center as a function of velocity. This is also the case with the high-resolution CO(1–0) data.[30] However, the small angular size of the region makes it difficult to resolve the structure. In FIGURE 8 are shown Bally's CO(1–0) observations of the bipolar source NGC 2071 smoothed to the same effective

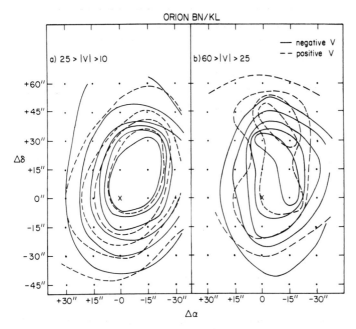

FIGURE 7. Maps of the integrated antenna temperature T_A^* of the CO(2–1) line in two positive and negative velocity ranges. The dots show the positions at which the observations were taken.

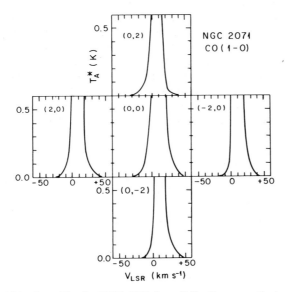

FIGURE 8. CO(1–0) profiles for NGC 2071 from Bally (in preparation) smoothed to an effective resolution equal to that used to map the BN-KL region in the CO(2–1) line (*cf.* FIGURE 7).

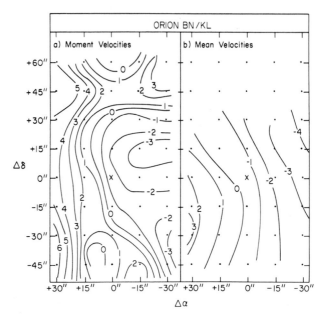

FIGURE 9. Plots of (a) moment velocities and (b) mean velocities for the CO(2–1) line in BN-KL (see text). The dots give the position at which the observations were taken.

resolution (beam size/source size) as was used for the CO(2–1) observations of BN-KL; it shows that the identification of a bipolar source using these observations would be quite uncertain. Perhaps a more sensitive way of examining the CO(2–1) data is to study the distribution of velocity across the region. This is shown in FIGURE 9, where the central velocity is defined two ways; first, by the velocity moment calculated between $|V| = 10$ and $|V| = 30$ km s^{-1}, and, second, by the mean velocity of the 20% power points (to avoid contamination by the spike in both cases). Both maps show a roughly east-west velocity gradient across the region. This may be due to bipolar structure or to the presence of more than one source of high-velocity emission in the region.

Thus, the available single-dish data on the bipolarity of the millimeter-wavelength line emission from the BN-KL region are ambiguous; this question can, however, be resolved by interferometer observations. The maps of the SO emission (Reference 35 and Welch, this volume) made interferometrically with high spatial resolution show little evidence of bipolarity. In FIGURE 10 is plotted the distribution of the H$_2$ emission over different velocity ranges;[42] this distribution shows, on a large scale, no sign of bipolar structure.

How Many Flow Centers?

The possibility that there are several different gas flows in the BN-KL region has been considered in many studies, and a recent summary was made by Genzel and

Downes.[45] The large number of such flows in other star-forming regions, plus the large number of infrared sources in the BN-KL cluster, makes this fairly likely.

The $+18$ km s^{-1} flow is indubitably associated with IRc2,[45] as shown by the positions of the SiO maser,[34] the $v = 0$, $J = 1$–0 SiO source,[39] and the H$_2$O shell source.[45] The flow appears to be similar to that around an evolved red giant star, in that its outflow velocity is roughly constant with radius; this produces parabolic or flat-topped lines; the former shape is seen in the CO(3–2) line[17] and the latter in the $v = 0$, $J = 1$–0 SiO line.[39] The central velocity of this source is about 4–5 km s^{-1} with respect to the LSR.

The higher-velocity gas, that having velocities of up to 100 km s^{-1}, appears to be a distinct flow and not an extension of the 18 km s^{-1} gas. This is shown by the change of shape of the molecular emission line profile at $V \gtrsim 25$ km s^{-1},[31,46] the different profiles for different molecular emission lines, and the apparent velocity shift across the source in the CO(2–1) map (FIGURE 9). FIGURE 9 suggests that the two flows have central velocities that differ by about 5 km s^{-1} and that they are not concentric. Several of the molecular species, notably SO and SO$_2$, may not be present in the "circumstellar" (18 km s^{-1}) flow. The apparent different position centers of the high-velocity gas may be real, *i.e.*, the gas may have a different source, or they may be due to motion of the gas through a clumpy surrounding medium.[31] The SO emission, however, appears to be centered on or near IRc2 (Welch, this volume), while the displacement of the HCO$^+$ center to the north of BN[31] suggests that the abundance of HCO$^+$ is shock enhanced here. The tentative conclusion to be drawn from all these conflicting pieces of evidence is that all the high-velocity gas has a common origin (IRc2), but that the chemistry, excitation, and flow pattern are modified by the structure of the surrounding region.

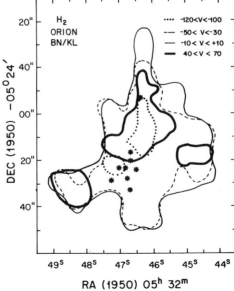

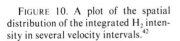

FIGURE 10. A plot of the spatial distribution of the integrated H$_2$ intensity in several velocity intervals.[42]

The Energy Source for the High-Velocity Gas

That high-velocity outflowing gas is associated with energetic events accompanying star formation seems to be well established. The total energy involved, even for the high-velocity source in the BN-KL region, is modest, and suggests that an energetic stellar wind (or winds) is the cause.[4,7,14,16,27,31,36,45] Winds are seen from many pre-main-sequence stars, with $\dot{M} \simeq 10^{-8}\ M_\odot\ y^{-1}$ at the 2–300 km s^{-1} typical for T Tauri stars[82] and about $10^{-5}\ M_\odot\ y^{-1}$ at the 2000 km s^{-1} typical for O stars.[83] The frequency of high-velocity molecular gas flows and the lifetimes of these flows have led to the suggestion that essentially all new-formed stars go through this stage and that the flows therefore provide an important source of internal energy, enough, in fact, to keep the clouds from collapsing, for the molecular clouds in which they are embedded[16,30] (see also Reference 84). However, the exact mechanism whereby the young star causes the high-velocity outflow is not yet understood at all.

Two types of models are considered; those in which the gas is accelerated by a shock produced by a stellar wind (see Reference 4 and Hollenbach, this volume) and those in which the molecular gas is accelerated directly by radiation pressure from an embedded star (see References 30 and 31). Both of these models suffer from several difficulties, including the probable dissociation of the molecules if they are directly accelerated by fast shocks to velocities greater than about 30 km s^{-1},[13] and the disparity between the luminosity available ($\sim 10^5\ L_\odot$) and that required ($\sim 10^7\ L_\odot$ for single scattering) (see Reference 31). A possible solution to this is that the optical depth of the dust is so high (several hundred) that each photon is scattered several hundred times (see Reference 30). This process is appealing because it results in a gradual acceleration in which the molecules are not destroyed. A similar process may be operating for evolved stars, for which, if radiation pressure is the driving mechanism, similar very large optical depths are required and in which molecular gas is accelerated to quite large velocities (up to about 80 km s^{-1}).[51-53]

The first mechanism, acceleration by a shock front caused by an ionized fast wind, has further difficulties. Such winds produce small H II regions, which are detectable in the radio continuum and whose detection provides a method of measuring the mass-loss rate in the wind (see the analysis by Panagia and Felli,[85]). Applying this analysis to the BN-KL region predicts a minimum radio flux of about 20 mJy at 2 cm for a wind velocity of 2000 km s^{-1}, which is incompatible with the limits on the radio flux from IRc2 and the observed intensity of BN (Moran, this volume).

A third possibility is the sporadic ejection of matter by a protostar still in the process of formation. While this could produce the apparent velocity field of the outflowing gas ($V \simeq r$, as opposed to $V =$ constant produced by radiation pressure), there is, as yet, little evidence that such a phenomenon actually takes place.

Thus, in summary, there is, as yet, no viable model for the mechanism causing the outflow. The present complex picture presented by the observations may change, and become simpler, once the different effects of excitation, chemistry, and velocity field are sorted out by high-resolution observations. The phenomenon of high-velocity molecular outflows from star-forming regions, of which the BN-KL region provides the most energetic example, is a newly discovered one, not predicted or understood by present star-formation theories, and holds the promise of aiding the advancement of these theories, as well as our understanding of the evolution of the interstellar medium.

ACKNOWLEDGMENTS

I am very grateful to Tom Kuiper and John Bally for providing data and ideas before publication.

REFERENCES

1. ZUCKERMAN, B. & P. PALMER. 1975. Astrophys. J. **199:** L35.
2. ZUCKERMAN, B., T. B. H. KUIPER & E. N. RODRÍGUEZ-KUIPER. 1976. Astrophys. J. **209:** L137.
3. WILSON, R. W., K. B. JEFFERTS & A. A. PENZIAS. 1970. Astrophys. J. **161:** L43.
4. SHULL, J. M. 1981. *In* Symposium on Neutral Clouds Near HII Regions. Penticton, B.C.
5. SULLIVAN, W. T., III. 1973. Astrophys. J. Suppl. Ser. **25:** 393.
6. MORRIS, M. 1976. Astrophys. J. **210:** 100.
7. GENZEL, R. & D. DOWNES. 1977. Astron. Astrophys. **61:** 117.
8. BÖHM, K.-H. 1978. *In* Protostars and Planets. T. Gehrels, Ed.: 632. University of Arizona Press. Tucson, Ariz.
9. GAUTIER, T. N., U. FINK, R. P. TREFFERS & H. P. LARSON. 1976. Astrophys. J. **207:** L129.
10. WATSON, D. M., J. W. V. STOREY, C. H. TOWNES, E. E. HALLER & W. L. HANSEN. 1980. Astrophys. J. **239:** L129.
11. STOREY, J. W. V., D. M. WATSON, C. H. TOWNES, E. E. HALLER & W. L. HANSEN. 1980. Astrophys. J. **247:** 136.
12. HOLLENBACH, D. J. & J. M. SHULL. 1977. Astrophys. J. **216:** 419.
13. KWAN, J. 1977. Astrophys. J. **216:** 713.
14. GENZEL, R., M. J. REID, J. M. MORAN & D. DOWNES. 1981. Astrophys. J. **244:** 884.
15. SNYDER, L. E. & D. BUHL. 1974. Astrophys. J. **189:** L31.
16. BALLY, J. & C. J. LADA. 1982. In preparation.
17. PHILLIPS, T. G., P. J. HUGGINS, G. NEUGEBAUER & M. W. WERNER. 1977. Astrophys. J. **217:** L161.
18. HUGGINS, P. J., T. G. PHILLIPS, G. NEUGEBAUER, M. W. WERNER & P. G. WANNIER. 1979. Astrophys. J. **227:** 441.
19. SCHLOERB, F. P., P. FRIBERG, Å. HJALMARSON, B. HÖGLUND & W. M. IRVINE. 1982. Preprint.
20. PHILLIPS, T. G., J. KWAN & P. J. HUGGINS. 1980. *In* Int. Astron. Union Symp. 87, Interstellar Molecules. B. H. Andrew, Ed.: 21. D. Reidel. Dordrecht.
21. PICKETT, H. M. & J. H. DAVIS. 1979. Astrophys. J. **227:** 446.
22. KNAPP, G. R., T. G. PHILLIPS, P. J. HUGGINS & R. O. REDMAN. 1981. Astrophys. J. **250:** 175.
23. VAN VLIET, A. H. F., DE GRAAUW, T. J. LEE, S. LIDHOLM & H. VAN DER STADT. 1981. Astron. Astrophys. **101:** L1.
24. GOLDSMITH, P. F., N. R. ERICKSON, H. R. FETTERMAN, B. J. CLIFTON, D. D. PECK, P. E. TANNEWALD, G. A. KOEPF, D. BUHL & N. MCCAVOY. 1981. Astrophys. J. **243:** L79.
25. ERICKSON, N., J. H. DAVIS, N. J. EVANS II, R. B. LOREN, L. MUNDY, W. L. PETERS III, M. SCHOLTES & P. A. VANDENBOUT. 1980. *In* Int. Astron. Union Symp. 87, Interstellar Molecules. B. H. Andrew, Ed.: 25. D. Reidel. Dordrecht.
26. PHILLIPS, J. P., G. J. WHITE & G. D. WATT. 1982. Mon. Not. R. Astron. Soc. In press.
27. RYDBECK, O. E. H., Å. HJALMARSON, G. RYDBECK, J. ELLDER, H. OLOFSSON & A. SUME. 1981. Astrophys. J. **243:** L41.
28. RYDBECK, O. E. H., Å. HJALMARSON, G. RYDBECK, J. ELLDER, A. SUME & S. LIDHOLM. 1980. *In* Int. Astron. Union Symp. 87, Interstellar Molecules. B. H. Andrew, Ed.: 39. D. Reidel. Dordrecht.
29. GOLDSMITH, P. F. W. D. LANGER, F. P. SCHLOERB & N. Z. SCOVILLE. 1980. Astrophys. J. **240:** 524.
30. SOLOMON, P. M., G. R. HUGUENIN & N. Z. SCOVILLE. 1981. Astrophys. J. **245:** L19.
31. KUIPER, T. B. H., B. ZUCKERMAN & E. N. RODRÍGUEZ-KUIPER. 1981. Astrophys. J. **251:** 88.

32. GENZEL, R., J. M. MORAN, A. P. LANE, C. R. PREDMORE, P. T. P. HO, S. S. HANSEN & M. J. REID. 1979. Astrophys. J. **231:** L73.
33. BASTIEN, R., J. BIEGING, C. HENKEL, R. N. MARTIN, T. PAULS, C. M. WALMSLEY, T. L WILSON & L. ZIURYS. 1981. Astron. Astrophys. **98:** L4.
34. BAUD, B., J. BIEGING, R. PLAMBECK, D. D. THORNTON, W. J. WELCH & M. C. H. WRIGHT 1980. *In* Int. Astron. Union Symp. 87, Interstellar Molecules. B. H. Andrew, Ed.: 545. D Reidel. Dordrecht.
35. WELCH, W. J., M. C. H. WRIGHT, R. L. PLAMBECK, J. BIEGING & B. BAUD. 1981 Astrophys. J. **245:** L87.
36. DOWNES, D., R. GENZEL, E. E. BECKLIN & C. G. WYNN-WILLIAMS. 1981. Astrophys. J **244:** 869.
37. WILSON, T. L., D. DOWNES & J. BIEGING. 1979. Astron. Astrophys. **71:** 275.
38. MORRIS, M., P. PALMER & B. ZUCKERMAN. 1980. Astrophys. J. **237:** 1.
39. GENZEL, R., D. DOWNES, P. R. SCHWARTZ, J. H. SPENCER, V. PANKONIN & J. W. M BAARS. 1980. Astrophys. J. **239:** 519.
40. KWAN, J. & N. Z. SCOVILLE. 1976. Astrophys. J. **210:** L39.
41. NADEAU, D. & T. R. GEBALLE. 1979. Astrophys. J. **230:** L169.
42. NADEAU, D., T. R. GEBALLE & G. NEUGEBAUER. 1982. Astrophys. J. **253:** 154.
43. OLOFSSON, L., A. HJALMARSON & O. E. H. RYDBECK. 1981. Astron. Astrophys. **100:** L30
44. BECKWITH, S., S. E. PERSSON, G. NEUGEBAUER & E. E. BECKLIN. 1978. Astrophys. J **223:** 464.
45. GENZEL, R. & D. DOWNES. 1981. *In* Symposium on Neutral Clouds near HII Regions Penticton, B.C.
46. KUIPER, T. B. H., B. ZUCKERMAN & E. N. RODRÍGUEZ-KUIPER. 1978. Astrophys. J **219:** 129.
47. MORRIS, M. 1975. Astrophys. J. **197:** 603.
48. BLUMENTHAL, G. R. & W. G. MATHEWS. 1975. Astrophys. J. **198:** 517.
49. CAPROTTI, E., C. FOLTZ & P. BYARD. 1980. Astrophys. J. **241:** 903.
50. LINKE, R. A. 1981. Private communication.
51. KNAPP, G. R., T. G. PHILLIPS, R. B. LEIGHTON, K.-Y. LO, P. G. WANNIER, H. A. WOOTTEN & P. J. HUGGINS. 1982. Astrophys. J. **252:** 616.
52. MORRIS, M., P. F. BOWERS & B. E. TURNER. 1982. Astrophys. J. In press.
53. KNAPP, G. R. 1981. Unpublished.
54. HOYLE, F., P. M. SOLOMON & N. J. WOOLF. 1973. Astrophys. J. **185:** L89.
55. FISCHER, J. 1981. Private communication.
56. WERNER, M. W., I. GATLEY, D. A. HARPER, E. E. BECKLIN, R. F. LOEWENSTEIN, C. M TELESCO & H. A. THRONSON. 1976. Astrophys. J. **204:** 420.
57. BOOTH, R. S. & R. P. NORRIS. 1980. *In* Int. Astron. Union Symp. 87, Interstellar Molecules. B. H. Andrew, Ed.: 579. D. Reidel. Dordrecht.
58. WATERS, J. W., J. J. GUSTINCIC, R. K. KAKAR, T. B. H. KUIPER, H. K. ROSCOE, P. N SWANSON, E. N. RODRÍGUEZ-KUIPER, A. R. KERR & P. THADDEUS. 1980. Astrophys. J **235:** 57.
59. STARK, A. A. 1981. Astrophys. J. **245:** 99.
60. CLARK, F. O., J. A. BIRETTA & H. M. MARTIN. 1979. Astrophys. J. **234:** 922.
61. FRERKING, M. A., W. D. LANGER & R. W. WILSON. 1982. Astrophys. J. In press.
62. PHILLIPS, J. P., & J. E. BECKMAN. 1980. Mon. Not. R. Astron. Soc. **193:** 245.
63. KNAPP, G. R., T. B. H. KUIPER & R. L. BROWN. 1976. Astrophys. J. **206:** 109.
64. ALEN, M. A. & G. R. KNAPP. 1978. Astrophys. J. **225:** 843.
65. WOOTTEN, H. A., S. P. LICHTEN, R. SAHAI & P. G. WANNIER. 1982. Astrophys. J **257:** 151.
66. GOLDSMITH, P. F. & R. A. LINKE. 1981. Astrophys. J. **245:** 482.
67. FIELD, G. B. 1974. Astrophys. J. **187:** 453.
68. MITCHELL, G. F., J. L. GINZBURG & P. J. KUNTZ. 1978. Astrophys. J. Suppl. Ser. **38:** 39.
69. HARTQUIST, T. W., M. OPPENHEIMER & A. DALGARNO. 1980. Astrophys. J. **236:** 182.
70. DRAINE, B. T. 1980. Astrophys. J. **241:** 1021.
71. BEICHMANN, C., T. G. PHILLIPS, H. A. WOOTTEN & M. A. FRERKING. 1982. Preprint.
72. PHILLIPS, T. G. & P. J. HUGGINS. 1981. Astrophys. J. **251:** 533.
73. WOOTTEN, H. A., N. J. EVANS, R. SNELL & P. A. VANDENBOUT. 1978. Astrophys. J **225:** L143.

74. THADDEUS, P., M. L. KUTNER, A. A. PENZIAS, R. W. WILSON & K. B. JEFFERTS. 1972. Astrophys. J. **176:** L73.
75. GOTTLIEB, C. W., E. W. GOTTLIEB, M. M. LITVAK, J. A. BALL & H. PENFIELD. 1978. Astrophys. J. **219:** L73.
76. SNYDER, L. E., J. M. HOLLIS, B. L. ULICH, F. J. LOVAS, D. R. JOHNSON & D. BUHL. 1975. Astrophys. J. **198:** L81.
77. SNYDER, L. E., W. D. WATSON & J. M. HOLLIS. 1977. Astrophys. J. **212:** 79.
78. DICKINSON, D. F., C. A. GOTTLIEB, E. W. GOTTLIEB & M. M. LITVAK. 1976. Astrophys. J. **206:** 79.
79. RIECKE, G. H., F. J. LOW & D. E. KLEINMANN. 1973. Astrophys. J. **186:** L7.
80. LANGER, W. D. & M. A. FRERKING. 1982. Astrophys. J. In press.
81. KNAPP, G. R. 1982. In preparation.
82. KUHI, L. V. 1964. Astrophys. J. **140:** 1409.
83. CONTI, P. S. & R. MCCRAY. 1980. Science **208:** 9.
84. NORMAN, C. A. & J. SILK. 1979. Astrophys. J. **228:** 197.
85. PANAGIA, N. & M. FELLI. 1975. Astron. Astrophys. **39:** 1.

DISCUSSION OF THE PAPER

T. MOUSCHOVIAS (*University of Illinois, Urbana, Ill.*): Is there a systematic velocity difference between HCO^+ and the neutral molecules?

KNAPP: I do not think so.

RECENT PROGRESS
ON MOLECULAR HYDROGEN IN ORION*

Steven Beckwith

Department of Astronomy
Cornell University
Ithaca, New York 14853

INTRODUCTION

The molecular hydrogen emission from the Orion Molecular Cloud is still the brightest (by a factor of about ten) known example of H_2 emission, making it the most thoroughly studied, as well. The emission lines come from vibrational-rotational levels with energies E/k corresponding to several thousand degrees Kelvin, levels that are never excited at the temperatures of 100 K or less that typically characterize the cloud material. The mere presence of these lines indicates regions of high-temperature (or at least highly excited) gas within the otherwise cool cloud. The discovery of the lines[1] generated a great deal of interest as a result, since there were no other indications of hot material at the time.

I think it is by now generally agreed that the H_2 is excited by the mass outflow from the infrared cluster, probably by some kind of shock heating as the outflowing gas accelerates ambient cloud material. The shock waves create regions of gas much hotter than the typical cloud temperatures (the H_2 lines are formed in the shocked gas, which is at 2000 K or so, compared to 100 K throughout most of the cloud). Shock heating was considered a plausible way to explain the H_2 emission at the time of its discovery, a suggestion that has been largely confirmed by theoretical calculations[2-4] and further observations.[5,6]

In Orion, the molecular hydrogen emission gives us a means of estimating quantities connected with the mass loss. The size, temperature, velocity, and intensity of the lines can be used to infer the mass-loss rate, the rate at which momentum is fed to the cloud through mass loss, and the energy-dissipation rate as the cloud is disrupted internally.[7,8] Because the molecules are excited in shocks, the lines provide us with a very useful probe of the physics of strong shock waves in molecular clouds as well. Different lines are formed at different distances behind the shock as the gas cools, so, at least in principle, we can infer the temperature and density profile, as well as abundances and, possibly, magnetic field strengths, of the shock wave from these observations. D. Hollenbach will discuss the theoretical modeling of the shock waves in his review, and you should turn to his paper for the interesting physics of molecular shocks. The theoretical results have some bearing on the estimates for the mass-loss rate, etc., mentioned above.

The estimates concerning the mass-loss phenomenon derived from H_2 observations give roughly the same numbers as those inferred from CO and H_2O maser observations. While the most recent H_2 observations change these estimates to some extent, I

*This research was supported, in part, by a grant from the National Aeronautics and Space Administration, no. NSG 2412.

0077–8923/82/0395–0118 $1.75/1 © 1982, NYAS

do not think that the major conclusions regarding the mass-loss phenomenon have changed. They are discussed in earlier reviews. The new observations have mainly refined the characterization of the H_2 velocity field, foreground and internal extinction, and excitation temperature.

The theories of H_2 excitation in shock waves have recently undergone major revisions, so there is a continuing interest in the most accurate line intensities and in quantities derived from the line intensities to compare with the latest calculations. Since this is a short review, I will concentrate on the latest observational results; I apologize to my colleagues for the elimination of historical perspective.

MORPHOLOGY

Molecular hydrogen emission extends over a few tenths of a parsec in the Orion Molecular Cloud (OMC1), almost symmetrically around the infrared cluster. The extent and location of the emission indicate an association with the mass outflow from the cluster and the low column densities of the excited molecules ($\sim 10^{20}$ cm^{-2}), coupled with the relatively high volume densities needed to thermalize the levels ($\sim 10^6$ cm^{-3}), imply that the excitation takes place in relatively thin ($\sim 10^{14}$ cm) layers within the cloud.[6]

The early observations thus indicated that the H_2 was excited in a thin but extended region that is probably at the boundary between the outflowing gas from the cluster and the ambient molecular cloud.

The recent maps of the H_2 velocity field[9-11] have changed this simple scenario. While it is still true that the vibrationally excited H_2 occupies a small fraction of the cloud's volume, it is probably not confined to a thin shell, as might be expected at the edge of an expanding bubble. Typical lines are 60 km s^{-1} wide (FWHM), extending to more than 100 km s^{-1} at zero intensity.[12] The width of the lines, together with the lifetime of the source, suggests that the H_2 is distributed (in some geometry consistent with the obscured column density) throughout much of the volume of the expanding region.

Most of the observations are of the vibrational-rotational lines at 2 μm, so I will discuss those first. The 2-μm line velocities are consistent with predominantly radial outflow from the infrared cluster. The linewidths tend to be small at large angular displacements from the infrared cluster and larger toward the center of the maps. The velocities are systematically blueshifted with respect to the local standard of rest (LSR) velocity of the molecular cloud, with the lines from the central portions of the nebula (in the direction of the infrared cluster) having the most blueshifted average velocities and greatest full widths at half maximum. The predominance of blueshifted velocities probably results from extinction within the expanding region, whereby the intensities of the lines farthest from the observer in the line of sight (the redshifted lines) are diminished by extinction within the source.[9]

The widest lines (full width at zero intensity) are seen northwest of the cluster away from the center of the map. We don't yet know if this is the result of the velocity field itself or of observational uncertainty. If it comes from the velocity field, then it may indicate that the flow is channeled in such a way as to give very high radial velocities at certain points in the region where they would normally be expected to be tangential. It might also be caused by a contribution to the velocity field of different sources of mass outflow; for example, from the infrared source IRS2. It is my feeling that the latter is unlikely to be the case simply because the short lifetime for the H_2 phenomenon in Orion[7] makes it unlikely that two different objects would be undergoing high mass loss simultaneously. One should note that there is a strong observational bias in the results. The widest lines are seen in the same direction as the most intense lines. Since the full width at zero intensity is determined to a large extent by how far out the lines can be seen relative to the noise, wide lines might be masked in the central direction if the gas densities are low or if the foreground extinction is unusually high.

The interpretation of these data is, unfortunately, hampered by the extinction in front of and within the source. I will discuss the extinction measurements shortly. For now, it is sufficient to note that the foreground extinction is on the order of 1–3 mag at 2 μm over most of the emission region. Therefore, variations in this extinction might easily be misinterpreted as variations in the velocity field, so, while the overall character of the line profiles is probably an accurate representation of the gas velocities, detailed changes in the line shape from point to point may well be misleading.

It is possible, at least in principle, to overcome some of the effects of extinction by observing lines at wavelengths where the interstellar attenuation is much reduced. Purely rotational lines of H_2 occur at wavelengths much longer than 2 μm from levels with energies similar to those of the vibrational levels. Beck and her collaborators obtained extensive spatial coverage of the $v = 0$–0 S(2) line at 12.3 μm with relatively good (20 km s^{-1}) velocity resolution to compare with the 2-μm results.[11,13] The 12-μm line profiles differ in detail from those at 2 μm, but give the same overall picture: The cloud is undergoing radial expansion about the infrared cluster with velocities as high as 100 km s^{-1} or more. There is no strong red/blue asymmetry in the 12-μm lines, although it is present in some of the central spectra. There are rapid variations in line shape between nearby spatial positions, which may imply emission from clumps moving relative to one another within the cloud. This interpretation reconciles the relatively small filling factor implied by the low column densities with the large radial distribution of the gas implied by the large linewidths. If the H_2 emission is excited primarily in clumps, then the interpretation of the velocity field is further complicated by the unknown distribution of clumps.

Knacke and Young mapped the $v = 0$–0 S(9) line at 4.68 μm.[14] This map looks quite different from the maps of the 2-μm lines. Part of the discrepancy is probably explained by the different spatial resolutions used for the long- and short-wavelength maps. The S(9) line may also be affected by telluric interference, which makes the observations sensitive to variations of the line velocity and to atmospheric changes at the time of the observations. The kind of observations made by Knacke and Young should be a very useful tool for understanding the effects of extinction, but further observations at different spatial resolution and, especially, higher spectral resolution should be made to properly interpret their data.

EXTINCTION

The foreground extinction to the H_2 emission is substantial; its absolute value and wavelength dependence must be determined to derive accurate line intensities and intensity ratios from the observed lines. There are several discrepant values for the foreground extinction ranging from one to four magnitudes at 2 μm, corresponding to visual extinctions between 10 and 40 mag.[10,14-18] The measurements are based on the reddening between lines at different wavelengths extrapolated to give a total extinction based on a ratio of total-to-selective extinction derived from a standard reddening curve.

Most of the extinction estimates are based on measurements of the $v = 1-0$ Q(3) to $v = 1-0$ S(1) line ratio, the lines occurring at 2.42 and 2.12 μm, respectively. The most accurate measure of the reddening is made using velocity-resolved profiles of the lines to correct for telluric interference;[10] the extinction is based on an extrapolation from these values. This method indicates a total extinction of one to two magnitudes in the direction of the brightest emission lines. There is evidence for extinction variations near the position of the most intense lines.[10] There is surprisingly little variation of reddening with velocity, as would be expected if the redshifted portions of the lines are more heavily attenuated than the blueshifted sides. The signal-to-noise ratios are low in the directions where the most-pronounced variations in line profiles should occur, however.

It is difficult to estimate the uncertainty in the extinction inferred from the 2-μm measurements, since it lies mainly in the uncertainty of the extrapolation. This problem might be overcome, at least in principle, by measuring lines occurring at widely separated wavelengths. Measurements of the extinction based on the reddening between 2.1 and 5.5 μm[17] indicate 2 to 2.5 mag of extinction at 2 μm, whereas the measurement of several lines between 3 and 4 μm indicates a 2-μm extinction on the order of four magnitudes.[14] These measurements are plagued by uncertainties of their own, however. In the former case, the measurements were made with different instruments on different telescopes, so beamsize corrections dominate the uncertainties. The 3- to 4-μm reddening measurement is very sensitive to an assumed excitation temperature that is not very well determined for those lines.

Furthermore, all reddening measurements are affected by the internal extinction in the line-emitting region itself. The red/blue asymmetry of the 2-μm lines indicates an internal extinction on the order of 2–3 mag. The measurement of two lines where the lines are attenuated to different degrees by extinction is a measurement of two different regions of gas. At the longer wavelengths, we see further in to the emission region, so our estimate of the foreground reddening depends on some kind of model for the internal extinction. This problem is not overcome by measuring lines close together in wavelength because the smaller effect of internal extinction is amplified by the large extrapolation. It may, therefore, be very difficult to measure an accurate value for the foreground extinction.

Although the different extinction measurements appear to disagree, I think it is actually surprising that the agreement is as good as it is, considering the uncertainties in the methods and the probable extinction variations across the object. The measurements mentioned here were made with different spatial and spectral resolutions, often of different parts of the nebula. It seems likely that there will be extinction variations across the face of the emission region, giving rise to discrepant measurements at different positions, just as the line profiles vary greatly from one position to another. Equally uncertain are the assumptions used to extrapolate from selective to total extinction. We know of other clouds where the ratio of total-to-selective extinction at near-infrared wavelengths varies within the cloud,[19,20] so it seems dangerous to place too much confidence in the standard reddening curve when we derive total extinction. There are even measurements of the H_2 lines in Orion that indicate that the extinction curve has unexpected fluctuations: The O branch lines at 3 μm are much weaker than would be expected, based on the normal reddening curve.[16,17] In view of the uncertainties, it is probably prudent to say the extinction is about two magnitudes, where "about" means within a factor of two or so. The reader is cautioned against taking any value too seriously.

The effects of internal extinction are most clearly seen in the 12-μm profiles.[11] Here the red-blue asymmetry is seen only in profiles toward the central region. There are positions in the nebula where the emission at 12 μm is predominantly redshifted with respect to the velocity of most of the cloud material. It is possible that an accurate characterization of the internal extinction could be made by comparing of 12- and 2-μm lines for various spatial positions in the cloud. It would be a difficult experiment; the pointing would have to be carefully controlled, since the profiles are known to change rapidly across the face of the nebula.

Because of the importance of accurate line intensities and intensity ratios to compare with theoretical calculations, it would be helpful to know the extinction more accurately than the number quoted above. It should be possible to improve the observations by measuring a large number of lines at different wavelengths with high spectral resolution at a series of positions. In principle, we could measure enough lines to determine the extinction curve and excitation temperature (which may depend on the level energy) reasonably accurately. It does not seem very likely that the average extinction toward H_2 will be greatly different from two magnitudes, however, since several different methods give similar results.

LEVEL POPULATIONS AND EXCITATION TEMPERATURE

It was apparent from the first measurements of the $v = 1-0$ band line ratios that the excitation temperatures are on the order of 1000 K or more. The excitation temperature for the hydrogen emission is usually quoted as 2000 K, based on the ratio of the $v = 2-1$ S(1) to $v = 1-0$ S(1) line intensities.[6] If the hydrogen molecules are excited in shock waves, then a range of temperatures should occur behind the shock as the gas cools through line emission. Different levels, or different molecules, should be sensitive to different parts of the shock, so we might trace the temperature structure through observations of many lines.

Measurements of the 12-μm $v = 0-0$ S(2) line gave the first indication of temperature structure.[13] Even with an assumed foreground extinction of four magnitudes for the 2-μm lines, which value is probably too high and which will enhance the corrected 2-μm intensities, the derived column densities in the $v = 0$, $J = 4$ level are much higher than they should be, based on a 2000 K excitation temperature. The 12-μm line comes primarily from gas at lower temperatures (1000 K or less) but higher column densities than the vibrationally excited 2-μm lines. The result is exactly what is expected for the cooling of shocked gas.

If the level populations among many rotational-vibrational lines are compared, the populations show an increasing excitation temperature with increasing energy levels.[17] The range of temperatures is approximately 1000 to 3000 K for these levels. Once again, this result supports the notion of a shock-heated gas giving rise to the molecular hydrogen emission. The cooling time of the gas from 3000 to 1000 K, for example, is simply too short (on the order of a few years) to be explained in a simple way by any other hypothesis.

The most recent shock models[21,22] invoke magnetic field strengths of a few milligauss to explain the H_2 line intensities and intensity ratios. The models predict that the gas temperatures will never rise above 2500 K or so because of the effects of the magnetic field. Field strengths of milligauss are approximately one thousand times the strongest fields anybody has observed in Orion. Support for large magnetic fields might come from an observed upper limit to the gas temperatures in the shock, which is in harmony with the theoretical predictions. The observational uncertainty for the highest observed excitation temperatures is large enough that 2500 K is not precluded as an upper limit to the excitation temperature. Other lines are much more sensitive to high-temperature gas. The 3- to 4-μm lines observed by Knacke and Young[14] should be very sensitive to high-temperature gas; unfortunately, they are spread out in wavelength and are also sensitive to the absolute value and shape of the extinction curve. All the 3- to 4-μm line intensities can be explained with a single temperature and assumed extinction. Although use of a single temperature is an oversimplification, it is unlikely that such a good fit would be obtained if there were substantial quantities of very high temperature gas (greater than 3000 K, for example), so the observations probably support the theoretical models. Other shocked molecules, such as CO, might be even better tracers of hot material.

Molecular hydrogen has many transitions at different wavelengths from levels of widely differing energy. It should be possible to separate the effects of excitation temperature and extinction when comparing different line intensities if enough lines are observed with the same spatial and spectral resolution. It is tempting to think that here is one example of a molecular cloud where careful observations and refined theory may eventually work together to unravel the remaining questions about the H_2 excitation.

ACKNOWLEDGMENTS

It is a pleasure to thank S. Beck, B. Draine, N. Evans, J. Guiliani, E. Salpeter, and N. Scoville for stimulating discussions that helped in the preparation of this article.

REFERENCES

1. GAUTIER, T. N., III, U. FINK, R. R. TREFFERS & H. P. LARSON. 1976. Astrophys. J. **207:** L129.
2. HOLLENBACH, D. J. & J. M. SHULL. 1977. Astrophys. J. **216:** 419.
3. KWAN, J. 1977. Astrophys. J. **216:** 713.
4. LONDON, R., R. McCRAY & S. I. CHU. 1977. Astrophys. J. **217:** 442.
5. GRASDALEN, G. & R. R. JOYCE. 1976. Bull. Am. Astron. Soc. **8:** 349.
6. BECKWITH, S., S. E. PERSSON, G. NEUGEBAUER & E. E. BECKLIN. 1978. Astrophys. J. **223:** 464.

7. BECKWITH, S. 1981. *In* Infrared Astronomy. D. P. Cruikshank and C. G. Wynn-Williams, Eds.: 167. D. Reidel. Dordrecht.
8. SHULL, J. M. & S. BECKWITH. 1982. Annu. Rev. Astron. Astrophys. In press.
9. NADEAU, D., T. R. GEBALLE & G. NEUGEBAUER. 1982. Astrophys. J. In press.
10. SCOVILLE, N. Z., D. N. B. HALL, S. G. KLEINMAN & S. T. RIDGWAY. 1982. Astrophys. J. **253**: 136.
11. BECK, S. C. 1981. Ph.D. Thesis, University of California at Berkeley.
12. NADEAU, D. & T. R. GEBALLE. 1979. Astrophys. J. **230**: L169.
13. BECK, S. C., J. H. LACY & T. R. GEBALLE. 1979. Astrophys. J. **234**: L213.
14. KNACKE, R. F. & E. T YOUNG. 1982. Astrophys. J. **249**: L65.
15. BECKWITH, S., S. E. PERSSON & G. NEUGEBAUER. 1979. Astrophys. J. **227**: 436.
16. SIMON, M., G. RIGHINI-COHEN, R. R. JOYCE & T. SIMON. 1979. Astrophys. J. **230**: L175.
17. BECKWITH, S., N. J. EVANS II, I. GATLEY, G. GULL & R. W. RUSSELL. 1982. Astrophys. J. In press.
18. DAVIS, D. S., H. P. LARSON & H. A. SMITH. 1982. Preprint.
19. CARRASCO, L., S. E. STROM & K. M. STROM. 1973. Astrophys. J. **182**: 95.
20. MCMILLAN, R. S. 1978. Astrophys. J. **225**: 880.
21. DRAINE, B. T. 1980. Astrophys. J. **241**: 1021.
22. DRAINE, B. T., W. G. ROBERGE & A. DALGARNO. 1981. Preprint.

DISCUSSION OF THE PAPER

B. DRAINE (*Institute for Advanced Study, Princeton, N.J.*): Could you state how certain you are regarding the H_2 $v = 2$ rotational temperature estimate of 3000 K?

BECKWITH: The one-sigma uncertainties are about ± 500–600 K, which is caused mainly by uncertainties in the extinction. A better experiment, in principle, is to measure the 3–2 lines, and this is in progress.

GAS DYNAMICS IN THE CIRCUMSTELLAR NEBULA
OF THE BECKLIN-NEUGEBAUER SOURCE*

N. Z. Scoville,† D. N. B. Hall,‡ S. G. Kleinmann,§
and S. T. Ridgway‡

†*Department of Physics and Astronomy*
University of Massachusetts
Amherst, Massachusetts 01003

‡*Kitt Peak National Observatory*
Tucson, Arizona 85726

§*Space Sciences Laboratory*
Massachusetts Institute of Technology
Cambridge, Massachusetts 02139

INTRODUCTION

The Becklin-Neugebauer (BN) object in Orion is located near the center of the dynamic activity of the Kleinmann-Low (KL) infrared nebula. Evidence for enormous energy release ($>10^{47}$ erg) over the last 10^3 y in the cloud core is provided by radio observations of CO thermal emission and H_2O masers, which indicate that the molecular gas is expanding at ~ 50 km s^{-1}, and by near-infrared observations of H_2 emission, which indicate the existence of shock fronts of 2000 K gas. Most recently, near-infrared spectroscopy has been shown to be the most promising method of defining the nature of the star inside BN and of probing the physical conditions and gas dynamics in its immediate vicinity. Emission lines of H II and highly excited CO have been detected at $\lambda \simeq 2.3$ and 4.1 μm; absorption lines from low-excitation CO have also been measured at $\lambda = 4.6$ μm. The fluxes in the Brackett lines from the H II region suggest an overlying extinction equivalent to $A_V \simeq 30$ mag and require the exciting star to have a uv luminosity equal to a B0.5 main sequence star.[1-3] Evidence that the H II region is probably not a simple Strömgren sphere is provided by the CO emission lines, which probably arise from extremely dense molecular gas ($n_{H+H_2} > 10^{10}$ cm^{-3}) in a region of size less than a few AU.[4] Perhaps most surprising are the dynamics indicated by the H II line profiles: Emission is detected in the wings of the Br α line extending to ± 100 km s^{-1} and the peak velocity of 21 km s^{-1} implies a 12 km s^{-1} redshift relative to the Orion Molecular Cloud.[5] The early data has settled neither the origin of this velocity shift nor the issue of whether or not the H II lines properly indicate the motion of BN. In the CO absorption lines, two velocity systems are seen—one at $+9$ km s^{-1}, attributable to OMC1, and a second at -16 km s^{-1}, which had not been previously anticipated.[3] Hall *et al.* surmised that the blue absorption component could arise in the high-velocity flows of the cloud core if BN was at or near the center of expansion![3]

*This research was supported, in part, by grants from the National Science Foundation, nos. AST 80-26702 and AST 78-27639, for N. Z. Scoville and S. G. Kleinmann, respectively.

125

TABLE 1

SPECTRAL FEATURES IN BN

Transition	Rest Wavenumber (cm^{-1})	Emisson/Absorption
$\lambda = 2.1$–2.4 μm Spectrum		
H$_2$ $v = 1 \rightarrow 0$ S(2)	4917.007	E
Unidentified	4785.5	E
H$_2$ $v = 1 \rightarrow 0$ S(1)	4712.905	E
H I Br γ $n = 7 \rightarrow 4$	4616.560	E
Na I ^{2}S$-^2$P^0	4532.65	E
	4527.02	E
H$_2$ $v = 1 \rightarrow 0$ S(0)	4497.839	E
H$_2$ $v = 2 \rightarrow 1$ S(1)	4448.96	E
CO $v = 2 \rightarrow 0$ Bandhead	4360.09	E
$v = 3 \rightarrow 1$ Bandhead	4305.42	E
$v = 0 \rightarrow 2$ R-branch	4364–4300	A
$v = 0 \rightarrow 2$ P-branch	4220–4256	A
$v = 4 \rightarrow 2$ Bandhead	4250.87	E
$v = 5 \rightarrow 3$ Bandhead	4196.48	E
$\lambda = 3.8$–4.3 μm Spectrum		
H I Pf γ $n = 8 \rightarrow 5$	2673.404	E
H I Br α n $= 5 \rightarrow 4$	2467.765	E
$\lambda = 4.5$–5.0 μm Spectrum		
CO $v = 1 \rightarrow 0$ R + P-branches	2030–2180	A/E
^{13}CO $v = 1 \rightarrow 0$ R + P-branches	2070–2125	A/E
H I Pf β $n = 7 \rightarrow 5$	2148.795	E

Here we would like to summarize partial results of new observations of BN we obtained[5] during observing sessions between December 1977 and February 1981 using a rapid scanning 1.4-m FTS[12] at the Coudé focus for the Mayall 4-m telescope on Kitt Peak. Data were taken at resolutions between 0.08 cm^{-1} and 0.30 cm^{-1}, chosen either to maximize the signal-to-noise ratio on the broad features (*e.g.*, the 2.3-μm CO bandheads and the H II lines) or to resolve the separate Doppler components in the narrow CO absorption and emission lines. The velocity resolutions at band center in the spectra range between 7 and 20 km s^{-1}. Aperture sizes of 2–4″ were used. Complete observations and analysis are discussed elsewhere.[5]

SPECTROSCOPY OF BN

TABLE 1 lists the lines identified in the $\lambda = 2$–5 μm spectra of BN. The lines believed to be closely associated with BN include those of H I, Na I, and the CO emission features at $\lambda = 2.3$ μm and 4.6 μm. The CO absorption and H$_2$ emission probably arise in gas further out along the line of sight to BN.

CO

The portion of the K-band spectrum covering the CO ($\Delta v = 2$) overtone bands is shown in FIGURE 1. At low J in the $v = 0 \rightarrow 2$ band, a series of well-resolved absorption lines are seen; at high J on the R-branch of each of the first three bands ($v = 2 \rightarrow 0$, $3 \rightarrow 1$, and $4 \rightarrow 2$), broad emission feature are seen at the bandheads. In the fundamental band ($v = 0 \rightarrow 1$), the same absorbing gas is seen but, due to the factor of 135 increase in the absorption coefficients, the ^{12}CO lines are very heavily saturated except at $J > 15$ (FIGURE 2). On the other hand, the entire ^{13}CO band is still unsaturated at 4.6 μm and can be measured with high precision since there is no significant telluric ^{13}CO absorption.

FIGURE 3 shows the kinematic profiles of the CO obtained by averaging all the detectable rotation-vibration lines in the CO $v = 0 \rightarrow 1$, $v = 0 \rightarrow 2$ bands and the ^{13}CO $v = 0 \rightarrow 1$ band. At the high resolution of this data ($\Delta V = 7$ km s^{-1}), the absorbing gas

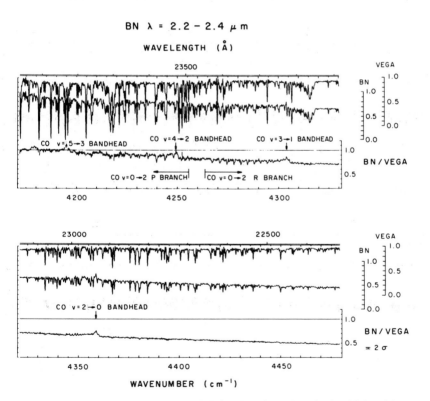

FIGURE 1. Spectra of BN and the comparison star (Vega) are shown in the vicinity of the $\lambda = 2.3$ μm CO overtone bands. Absorption in the Earth's atmosphere is removed by ratioing BN to Vega. The ratio spectrum (bottom trace in each panel) shows CO absorption at low J in the $v = 0$–2 band and emission features at the bandheads of $v = 2 \rightarrow 0$, $3 \rightarrow 1$, and $4 \rightarrow 2$. The frequency resolution was 0.3 cm^{-1}, equivalent to 20 km s^{-1}.

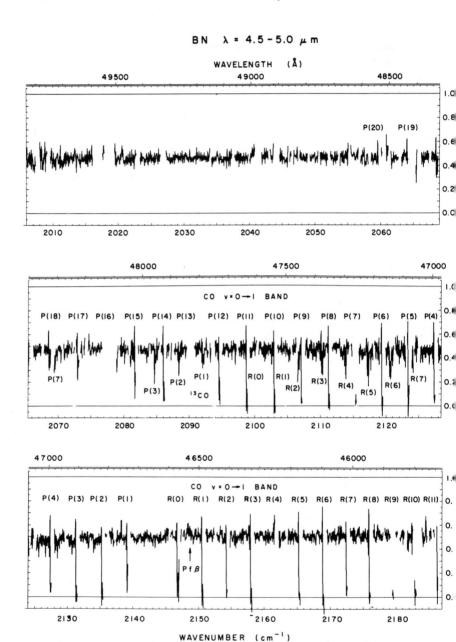

FIGURE 2. Ratioed spectrum of BN at λ = 4.5–5.0 μm covering the CO and ^{13}CO fundamental bands and the Pf β recombination line. (Gaps in the ratio spectrum correspond to telluric absorption >55%.) The resolution is 0.07 cm^{-1}, corresponding to 7 km s^{-1}.

is seen to consist of two discrete velocities, V_{LSR} = 8.6 and V_{LSR} = −17.5 km s^{-1}. Neither of these strong absorptions is likely to be associated closely with BN. The former is at the same velocity as OMC1 as determined from mm-line data; this absorption thus samples the foreground column in the quiescent cloud. Hall *et al.*[3] surmised that the blue component is formed in an expanding shell situated in front of BN but still within the cloud core—perhaps the blue component is related to the high-velocity "plateau" source seen in the mm CO line.

Emission at +20 km s^{-1} and Absorption at +30 km s^{-1}

The other Doppler components seen in FIGURES 2 and 3, the emission feature at V_{LSR} = 20 km s^{-1}, and weak absorption at V_{LSR} +30 km s^{-1} are probably intimately

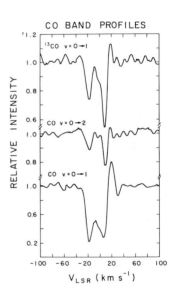

CO BAND PROFILES

FIGURE 3. Kinematic profiles for the v = 0–1 CO and ^{13}CO bands and the v = 0–2 CO ban obtained by averaging all lines detected in each band. The velocity resolution is 7 km s^{-1}. The fundamental bands of CO and ^{13}CO show two absorption velocities corresponding to OMC1 (+9 km s^{-1}) and the "expanding shell" (−17 km s^{-1}), plus emission line at +20 km s^{-1} produced close to BN.

associated with BN. Both the distribution of line intensities as a function of J and the possible detection of ^{13}CO emission (FIGURE 3) suggest that the ^{12}CO emission is optically thick. The rate at which the emission falls off with increasing J (between P17 and P20 in FIGURE 2), suggests a rotational temperature of ~600 K for this gas. The absorption appears to have a similarly high temperature. The fact that this temperature is much greater than that of the absorption lines suggests that these features are formed much closer to BN.

Since the 4.6-μm CO emission is optically thick and has a known temperature of T_R ≃ 600 K, the flux observed in the thick lines can yield an estimate of the size of the emitting region. For a nominal extinction $A_{4.6\,\mu m}$ ≃ 0.5 mag, we find

$$R \simeq 3 \times 10^{14} \text{ cm,}$$

which is about 20 times the upper limit to the size of the very-high-excitation region producing the CO headbands (discussed below). Though the position of the $+30$ km s^{-1} absorption cannot be so tightly constrained, it is, of course, necessary that this gas be further out in the circumstellar nebula than the dust producing the 4.6-μm continuum. The color temperature of this continuum is $T_c = 500$–600 K;[6] the radius of this "photosphere" must therefore be greater than ~25 AU (3.8×10^{14} cm), corresponding to the size of a blackbody with the observed temperature and luminosity. Recent speckle interferometry at 3.5 and 4.8 μm indicates that the actual size is within a factor of two of this lower limit.[7] The fact that the absorption component at $+30$ km s^{-1} has a rotational temperature of approximately 600 K suggests that it arises from about the same radius as does the 20 km s^{-1} emission.

Bandhead Emission at 2.3 μm

The last CO feature, perhaps the most unusual, is the bandhead emission seen at $\lambda = 2.3$ μm. Rotational temperatures greater than 2000 K are required to significantly populate $J = 50$ and give the observed shapes of the bands; the gas producing this emission is, therefore, quite distinct from that producing the narrow 4.6-μm emission lines. If the bandheads are optically thin, then the relative intensities of the separate bands imply a vibrational temperature of 3500 ± 500 K. Since the radiative decay rate out of the highest observed level ($v = 4$, $J = 50$) at $E/k \simeq 19\,000$ K is 100 s^{-1}, the density required to excite these levels is $n_{H+H_2} \gtrsim 10^{10}$–$10^{11}$ cm^{-3}. Based upon our upper limits to the fundamental band emission from the same rotationally hot CO (FIGURE 2), we set a lower limit to the optical depth such that the size of the emission region must be less than 1.5 AU.[4] Theoretical analyses of the possible collisional and ir pumping mechanisms for CO also imply a size $R \lesssim 5$ AU.[8] The width of the individual lines within the bandheads are in the range 50–100 km s^{-1}; the mean velocity is estimated to be $V_{LSR} = +20 \pm 10$ km s^{-1}.

Ionized Gas

FIGURE 4 shows the profiles of the Br α and Br γ lines at $\lambda = 4.05$ and 2.16 μm. Each line exhibits two components: a narrow core at $V_{LSR} = +21$ km s^{-1} with $\Delta V_{FWHM} = 38$ km s^{-1} and broad wings detectable over a full extent of ~350 km s^{-1}. Both componen' hould be identified with BN, since the observed Br α flux in the 2" aperture is too high by a factor of 30 to be from M42 and their center velocity is redshifted 24 km s^{-1} relative to M42.[3] The most recent VLA measurements indicate that the size of at least one of the H II components is $R \simeq 15$ AU.[13]

Comparing the observed Br γ/Br α flux ratio of 0.096 with the ratio of 0.42 expected for recombination in a dense plasma, we deduce a reddening of E (2.16–4.05 μm) = 1.6 mag, implying an extinction at Br α of $A_{4.05\,\mu m} = 0.7$ mag for the standard van de Hulst No. 15 dust extinction law. The photoionization of this H II region with Lyman continuum photons requires the equivalent of a main sequence B0–B0.5 star. The total luminosity of such a star is 1–2×10^4 $L_\odot$.[9] This luminosity is an order of magnitude higher than that obtained by directly integrating the infrared photometry at $\lambda \lesssim 20$ μm for BN[6] and about a factor of three higher than the observed luminosity corrected for extinction inferred from the 10-μm silicate absorption.[10] The discrepancy between the photometric luminosity and the implied ionization rate can be removed if substantial ionization occurs *via* absorption of the Balmer continuum by H I in $n = 2$, as suggested by Simon *et al.* for other embedded stars.[14]

It is interesting to note that the Br γ/Br α ratio shows no significant change between the line core and the high-velocity wing on the red side. The ratio of areas for the two components is 2.26 $\pm$ 0.08 and 2.54 $\pm$ 0.2, from which we infer that the reddening does not differ by more than 10% to 20% between the two. This fact suggests rather strongly that the high- and low-velocity H II are physically associated at the same depth in the cloud.

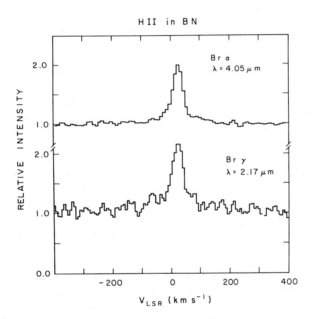

FIGURE 4. Profiles of the hydrogen Br α ($\lambda = 4.05$ μm) and Br γ ($\lambda = 2.17$ μm) lines in BN. The velocity resolution is 20 km s^{-1}.

Krolik and Smith have recently pointed out that the near-infrared hydrogen lines detected in stars like BN may often originate as thermal line emission in massive stellar winds rather than in circumstellar H II regions.[11] In the dense stellar wind model, one expects the emission rate to be greatest for the line that is optically thick to the greatest radius in the flow; the observed flux ratios will therefore not necessarily resemble a normal recombination spectrum. For several reasons, we believe that these phenomena do not dominate the BN spectrum: (1) Optical depth effects will distort the Pf β/Br γ and Br α/Br γ ratios in opposite directions, yet both observed ratios are consistent (albeit not stringently) with a single reddening, and (2) no difference is seen in the width or shape of the line wings of the Br α and Br γ lines, implying that they sample similar depths in the flow. Lastly, we note that, for the narrow-line core component, the linewidth is consistent with a "static" circumstellar H II region, *i.e.,* very much less than that expected for a stellar wind.

THE NATURE OF BN

The immediate circumstellar nebula is most fascinating. We find compelling evidence for both an ultracompact H II region and high-density molecular gas within a few AU of the star. A natural situation in which both may arise would be a dense neutral disk, presumably a prestellar nebula with the young star located at the center. The high-velocity H II could be a wind moving out perpendicularly to the disk along the polar axis, where outflow is relatively unobstructed. If the dust around BN is primarily confined to a disk, one may then understand the fact that the luminosity derived from near-infrared photometry is three to ten times less than that expected for the star required to ionize the H II region. The similarity of the density estimates required for the CO bandhead emission and the emission measure of the high-velocity H II region (if it is only a few AU in size) are consistent with the notion that the H II region borders the CO emission region. The neutral gas associated with BN probably extends out to at least 25 AU, since the 4.6-μm CO emission is best modeled as arising just outside the dust "photosphere." The low-velocity H II gas may possibly be situated outside the high-velocity H II gas, where the ionized wind impacts the inside of this outer dust and gas envelope.

The conditions in the molecular gas are constrained on the inside by the CO bandhead emission data ($R \lesssim 0.3$ AU) and on the outside by the 4.6-μm emission and high-temperature absorption. Thus, we deduce a temperature gradient of 5000 K to 3000 K within the bandhead region and a further drop to 600 K at 20–25 AU. The gas density probably drops from $n_{H_2} > 10^{11}$ cm^{-3} to 10^7 cm^{-3} over this region. Whether there is a smooth falloff or a fairly constant density over most of the disk is not presently known. But since the density is so much higher at smaller radii, a reasonable upper limit to the mass of the disk is the estimate $M_{H_2} \simeq 1.6 \times 10^{-3} M_\odot$ obtained for the bandhead region. Since the dust will sublimate at temperatures above $T_D \simeq 1500$ K, we expect that essentially no grains will exist inside $R \simeq 2$ AU, *i.e.,* in the bandhead emission region. The velocity dispersion of the disk gas varies from 50 to 100 km s^{-1} at the inner edge to $\lesssim 4$ km s^{-1} at the outer boundary. Given the fact that the disk is not seen edge on, the observed velocity widths are consistent with both orbital motion and gravitational free-fall of gas at 0.5 to 25 AU around a star of a few solar masses. (We are also exploring the possibility of other models not involving a disk.)

Despite the fact that the CO fundamental band emission must come from a much larger radius than either the H II region or the bandhead emission region, the velocities of all these tracers are $V_{LSR} = +21$ km s^{-1}. We feel that this is compelling evidence that the systemic velocity of BN and its immediate circumstellar nebula is, indeed, $+21$ km s^{-1}, redshifted 12 km s^{-1} with respect to the center of mass of the OMC1 cloud. The possible absorption at $+30$ km s^{-1} would then be gas on the near side of the dust photosphere, which is infalling at ~ 9 km s^{-1}.

We may thus interpret the $+20$ km s^{-1} emission together with the $+30$ km s^{-1}

absorption as an inverse P Cygni profile formed in an accreting gas envelope just outside the dust photosphere. The accretion rate ($\dot{M} \simeq - 10^{-6} M_\odot \, y^{-1}$) can be crudely estimated from the observed velocity, the size, and the column density deduced for the emission line.

The 12 km s^{-1} redshift of BN relative to OMC1, the cloud out of which this star recently condensed, remains an enigma. There can be little doubt that BN did recently form here (as opposed to being simply a field star passing through the cloud core), since the circumstellar H II region implies an early spectral type. It is also not clear how an interloping star could accumulate the dense circumstellar gas ($n_{H_2} > 10^{11}$ cm^{-3}) we detect within 1 AU of BN. This circumstellar matter could quite plausibly be the remnant of a prestellar nebula that has not had sufficient time for dispersal. The 12 km s^{-1} must then be explained as either gravitationally-bound orbital motion of BN or unbounded motion that will quickly result in the departure of BN from the cloud core. In the latter situation, escape from the core region will occur on the timescale $R/V \lesssim$ 3000 y, assuming $R \lesssim 10^{17}$ cm. Binding this motion would require $10^3 \, M_\odot$ in the core, assuming once again that BN is at $R = 10^{17}$ cm (corresponding to 15″ at a 500 pc distance). The required mass is approximately a factor of 10 to 100 higher than the mass of gas estimated to reside within this radius by millimeter line measurements. However, a significant mass of stars in a young cluster should not be ruled out, given the high observed luminosity ($L_R = 10^5 \, L_\odot$) for the region. (This requires the formation of many stars with $M < 3$ M$_\odot$ to get a sufficiently high overall mass:luminosity ratio.) For the binary system hypothesis, we may apply an additional constraint: the absence of change $\gtrsim 1$ km s^{-1} in the velocity (of the BN H II region) over the three-year interval between the Br α spectra we took. Thus, BN cannot have moved through more than ~1/10 of its orbit within three years and the orbital period must exceed 30 y. Applying Kepler's law to these parameters, we find that the orbit must have a radius $\gtrsim$10 AU and that the mass of the system must be $\gtrsim$2 $M_\odot$.

SUMMARY

High-resolution spectroscopy in the near infrared can provide a very specific probe of the compact high-excitation circumstellar nebulae immediately surrounding young stars and protostars. In this review, we summarized the results from our recent observations of the Becklin-Neugebauer source in Orion. The windows observed between $\lambda = 2 \, \mu$m and $\lambda = 5 \, \mu$m show molecular bands of CO, ^{13}CO, and H$_2$, a doublet of Na I, and four recombination lines of hydrogen originating from two distinct ionized and molecular zones near BN: (1) an ultrahigh-density H II region ($n_e \simeq 10^8$ cm^{-3}), with a radius ≤ 15 AU and a total mass of ~$10^{-7} \, M_\odot$, and (2) a high-excitation molecular region extending from 0.3 to 25 AU with physical conditions $n_{H_2} \gtrsim 10^{11}$ cm^{-3} and $T \simeq 3500$ K down to $n_{H_2} \simeq 10^7$ cm^{-3} and $T \simeq 600$ K. (Continuity between these two molecular regimes is not required by the data.)

To reconcile the existence of a large circumstellar H II region ($R \simeq 15$ AU)[13] with dense neutral gas much closer to BN ($R \leq 1$ AU), we proposed a possible model in which the neutral gas close to BN is confined to a circumstellar disk and the H II gas is situated in hemispheres above and below the disk. Outside the ionized region is a surrounding envelope of lower-temperature dust and molecular gas. The recombination lines suggest two kinematic components in the H II gas: a supersonic wind ($V >$ 100 km s^{-1}) possibly originating from the star and a low-velocity region perhaps occurring at $R \simeq 15$ AU, where the wind is halted by exterior neutral gas. The CO lines from this envelope show possible inverse P Cygni profiles, providing evidence of an accretion flow at $V \simeq 10$ km s^{-1} in the molecular material. The estimated flow rates outwards in the H II wind and inwards in the neutral gas are in the range $10^{-6} - 10^{-8}$ $M_\odot$ y^{-1}.

The central star in BN must have a Lyman continuum emission rate equivalent to a main sequence B0.5 star if the H II gas is photoionized. The luminosity of such a star is $1-2 \times 10^4$ $L_\odot$. From four separate spectroscopic tracers, the systematic velocity of BN and the immediate circumstellar nebulae (H II and molecular) between 1 and 25 AU is $V_{LSR} = 21$ km s^{-1}, implying that BN is moving into OMC1 at $+12$ km s^{-1} along the line of sight. If BN is not gravitationally bound, it will leave the cloud core ($R \lesssim 10^{17}$ cm) in only a few thousand years!

ACKNOWLEDGMENTS

N. Z. Scoville is grateful to the University of Hawaii's Institute for Astronomy for a sabbatical leave appointment and support from the Guggenheim Foundation during the period in which much of the analysis was done. This is contribution No. 506 of the Five College Astronomy Department.

REFERENCES

1. GRASDALEN, G. L. 1976. Astrophys. J. **205**: L83.
2. JOYCE R. R., M. SIMON & T. SIMON. 1978. Astrophys. J. **220**: 156.
3. HALL, D. N. B., S. G. KLEINMANN, S. T. RIDGWAY & F. C. GILLETT. 1978. Astrophys. J. **223**: L47.
4. SCOVILLE, N. Z., D. N. B. HALL, S. G. KLEINMANN & S. T. RIDGWAY. 1979. Astrophys. J. **232**: L121.
5. SCOVILLE, N. Z., D. N. B. HALL, S. G. KLEINMANN & S. T. RIDGWAY. 1982. In preparation.
6. BECKLIN, E. E., G. NEUGEBAUER & C. G. WYNN-WILLIAMS. 1973. Astrophys. J. **182**: L7.
7. FOY, R., A. CHELLI, F. SIBILLE & P. LENA. 1979. Astron. Astrophys. **79**: L5.
8. SCOVILLE, N.Z., R. KROTKOV & D. WANG. 1980. Astrophys. J. **240**: 929.
9. PANAGIA, N. 1973. Astron. J. **78**: 929.
10. DOWNES, D., R. GENZEL, E. E. BECKLIN & C. G. WYNN-WILLIAMS. 1981. Astrophys. J. **244**: 869.
11. KROLIK, J. H. & H. A. SMITH. 1981. Preprint.
12. HALL, D. N. B., S. T. RIDGWAY, E. A. BELL & J. M. YARBOROUGH. 1979. Proc. Soc. Photo Opt. Instrum. Eng. **172**: 121.
13. MORAN, J. M. 1982. This volume.
14. SIMON, M., M. FELLI, L. CASSAR, J. FISCHER & M. MASSI. 1982. In preparation.

DISCUSSION OF THE PAPER

B. ZUCKERMAN (*University of Maryland, College Park, Md.*): Do you have enough measurements to rule out the binary hypothesis?

SCOVILLE: No. We have taken measurements of the Brackett lines over four years; they put a limit to velocity changes of ≤ 0.5 km s^{-1}. This cannot rule out wide binaries.

R. GENZEL (*University of California, Berkeley, Calif.*): Doesn't the detection of radio continuum emission from BN (Moran *et al.*, pp. 204–9, this volume) rule out a circumstellar H II region as small as 3 AU and suggest that its size is more like 30 AU?

SCOVILLE: The H II region is probably not spherically symmetric. It could be confined to as little as 2 AU in one direction and be much more extensive out of the disc. Of course, the suggestion that there is a disc is cute, but the geometry may well be different. I should also point out that the observed ν^2 spectrum for the radio free-free emission suggests that the H II detected in the radio is not a simple stellar wind emission. It must be either a wind terminated at some outer radius by recombination (as modeled by Simon and coworkers) or a quiescent, optically thick H II region on the outside of the wind H II region.

C. G. WYNN-WILLIAMS (*University of Hawaii, Honolulu, Hi.*): You suggest that the band head emission is affected by extinction, but isn't the dust destroyed at 3500 K?

SCOVILLE: I think that that is correct. If the band heads are emitted from a region at 3500 K, then it seems likely that the dust will have sublimated here and the dust-to-gas abundance ratio would be very low. The extinction of the band head emission could come from more exterior dust.

FAR-INFRARED CO LINE EMISSION FROM ORION-KL*

Dan M. Watson

Department of Physics
University of California, Berkeley
Berkeley, California 94720

THE FAR-INFRARED SPECTRUM OF CO

The hot, dense shocked region surrounding the Kleinmann-Low (KL) nebula, which is associated with strong near-infrared emission from molecular hydrogen[1] and high-velocity emission from millimeter-wavelength molecular lines,[2] is also a strong source of far-infrared and submillimeter molecular line emission. Submillimeter observation may be said to have started with the ground-based detection of CO $J = 3 \rightarrow 2$ at 870 μm;[3] seven additional transitions of CO have now been observed, ranging to rotational quantum numbers as high as $J = 30$ and to wavelengths as short as 87 μm. Emission from H_2O[4] and OH[5] in Orion-KL has also been reported. These rotational lines of CO, OH, and H_2O are very important in cooling the shocked gas[6] and these and other far-infrared and submillimeter lines are potentially useful as probes of the unusual physical and chemical conditions there. Being largely free from the problems of extinction and contamination by emission or absorption from overlying ambient gas that affect the interpretation of near-infrared and millimeter-wave observations, respectively, the far-infrared lines seem uniquely well suited for the study of regions of active star formation such as that in Orion.

The far-infrared lines of CO, in particular, have proven very useful. The cross-sections for collisional excitation of the CO rotational levels can be computed to a fairly high accuracy ($\sim$30%) even for J as high as 60.[7,8] It turns out that the collisional transition rates and radiation rates are approximately equal when $J \simeq 25$ for the density and temperature conditions likely in the shocked gas, so the present observations include lines produced by molecular states that should be fairly far from thermal equilibrium as well as some lines from states that have populations very close to equilibrium. Thus, a comparison between CO observations and multilevel collisional excitation calculations allows one to determine the H_2 density and temperature, whence the CO and near-infrared H_2 intensities can be combined to yield the CO/H_2 relative abundance and the total mass of the hot shocked gas. In this paper, we review the CO observations made to date and use the most recent H_2 observations to determine the above parameters.

OBSERVATIONS AND CALCULATIONS

The present submillimeter and far-infrared observations of CO in Orion-KL are summarized in TABLE 1. All of the lines listed, except for the $J = 6 \rightarrow 5$ line, lie in

*This research was supported, in part, by a contract from the National Aeronautics and Space Administration, no. NGR 05-003-511.

0077–8923/82/0395–0136 $1.75/1 © 1982, NYAS

TABLE 1

SUBMILLIMETER AND FAR-INFRARED CO LINES FROM THE ORION-KL SHOCKED REGION*

Transition $(J \to J - 1)$	Wavelength (μm)	Intensity† $(W\ cm^{-2}\ sr^{-1})$	Reference
$6 \to 5$	433.6	1.1×10^{-10}	15
$17 \to 16$	153.3	8.2×10^{-10}	14
$21 \to 20$	124.2	1.0×10^{-9}	9, 10
$22 \to 21$	118.6	1.7×10^{-9}	10
$27 \to 26$	96.8	6.5×10^{-10}	9
$30 \to 29$	87.2	2.4×10^{-10}	9
$34 \to 33$	77.1	$<3.3 \times 10^{-11}\ (3\sigma)$	—

*All these measurements were made in the direction of KL ($\alpha_{1950} = 5^h32^m46^s7$, $\delta_{1950} = 5°24'28''$).

†Averaged over the beams, which were about 1' in diameter in each case.

regions of the spectrum in which the atmosphere is opaque and were observed with the NASA Kuiper Airborne Observatory. Except for the longer-wavelength $J = 6 \to 5$ line, the instrumental resolution has been insufficient to obtain useful velocity information. The $J = 21 \to 20$ line (FIGURE 1) has been observed with a 40'' beam at several points along a strip passing through the two main H_2 emission peaks and has been shown to extend about 1.5 along this direction. The emission is uniform over this range; the line from KL is just as strong as that from the brightest H_2 peak IRS2.[9]

The calculation of density, temperature and CO abundance from the CO $J = 21 \to 20$, $27 \to 26$, and $30 \to 29$ intensities and observations of H_2 lines is discussed in References 8 and 9; we will describe it briefly here. Since the CO lines are unextinguished and optically thin,[10] their intensities may be expressed simply as

$$I_J = \frac{hc}{4\pi\lambda} A_J N_J, \tag{1}$$

where A_J is the A-coefficient for the transition $J \to J - 1$ and N_J is the column density of CO in the Jth state. The relative populations of the rotational states may be determined for a given H_2 density, $n(H_2)$, and kinetic temperature, T, by solving the

FIGURE 1. The CO $J = 21 \to 20$ line[9] observed in the direction of the Kleinmann-Low nebula.

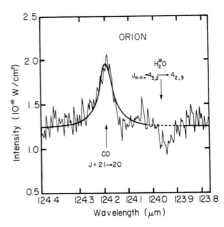

following system of equations:

$$N_J\left[A_J + n(H_2)\sum_{J'} \gamma_{JJ'}\right] = N_{J+1} A_{J+1} + n(H_2)\sum_J N_J\gamma_{JJ},$$

$$\sum_J N_J = N_{CO},$$
(2)

where $\gamma_{JJ'}$ is the collisional rate coefficient[8] for the transition $J \rightarrow J'$ and N_{CO} is the total CO column density; stimulated emission and absorption has been ignored (it can also be shown that radiative excitation is unimportant). Values of $n(H_2)$ and T are obtained by finding the solution to (2) that has the same shape as the observed intensity distribution. Once $n(H_2)$ and T are known, N_J/N_{CO} is determined for all J and N_{CO} may be obtained from any of the observed intensities by using (1).

In the case of Orion, much is already known about the shocked gas from H_2 observations. The latest observations can be summarized as follows: Rotation-vibration transitions at 2 μm arise from matter at about 2000 K; this gas lies behind an extinction of two magnitudes at 2 μm[11] and its column density, averaged over a 1' diameter area centered on IRc2, is 2.4×10^{19} cm^{-2}. The pure rotational $J = 4 \rightarrow 2$ line at 12.3 μm has also been observed in this region.[12] A value for the extinction at 12 μm of 0.75 mag may be inferred from the 2-μm extinction using standard interstellar extinction curves.[13] From this and the observed intensities of the 12.3 μm line, we infer a column density in the $v = 0$, $J = 4$ state of 1.5×10^{20} cm^{-2}. This obviously cannot be due to the 2000 K gas and indicates an additional amount of somewhat cooler gas.

Storey et al.[9] modeled the region by two components, one at 2000 K (designated "hot") and the other one cooler (designated "warm"). By assuming that each component had the same CO fractional abundance and that they were in pressure equilibrium (so that $n(H_2) \propto 1/T$), the H_2 column densities derived from the 2-μm and 12-μm observations were used along with solutions of (2) to fit the CO intensity distribution and thus derive the temperature of the cooler component, the CO fractional abundance, and the molecular hydrogen density. In their calculations, Storey et al. used an older value of the 2-μm extinction of 4 mag; in light of the recent redetermination, resulting in $A_{2\,\mu m} = 2$ mag, we have repeated the calculation using the more recent H_2 results mentioned above. Acceptable fits to the data are obtained for the following ranges of parameters:

$$T_{warm} = 500-1000 \text{ K},$$

$$N_{H_2,warm} = 5.9-2.1 \times 10^{21} \text{ cm}^{-2},$$

$$n_{H_2,hot} = 3.5 \times 10^6-5.5 \times 10^5 \text{ cm}^{-3},$$
(3)

$$CO/H_2 = 8.1 \times 10^{-5}-1.6 \times 10^{-4}.$$

The parameter combination resulting in the best fit† is displayed in FIGURE 2. It is worth noting that the 2000 K model component no longer has much influence on the observed CO intensities (cf. Reference 9), rendering unnecessary the assumption of

†Defined as the curve that comes closest to fitting the $J = 27 \rightarrow 26$ line and $J = 34 \rightarrow 33$ upper limit while still passing through the $J = 21 \rightarrow 20$ and $J = 30 \rightarrow 29$ points. The $J = 22 \rightarrow 21$ line was not used in the fit because it was not measured very precisely.

pressure equilibrium. Owing to slight modifications of the collision rates between the earlier and present calculations, which almost made up for the change in 12-μm extinction, the present results are fairly close to those of Storey *et al.*

Included in FIGURE 2 and in TABLE 1 is a 3σ upper limit for the intensity of the $J = 34 \rightarrow 33$ line, obtained in February 1981 with the U.C. Berkeley tandem Fabry-Perot spectrometer and the 91.4-cm telescope of the NASA Kuiper Airborne Observatory. This upper limit is substantially lower than what one would extrapolate from the longer-wavelength lines. It is not clear whether this indicates a fault in the observations or in the collisional excitation rates used in the extrapolation, so a remeasurement of this line is desirable.

MASS AND DISTRIBUTION OF THE SHOCKED GAS

The CO $J = 17 \rightarrow 16$ line at 153.3 μm, recently detected in the direction of KL by the Cornell far-infrared group,[14] also appears in FIGURE 2. The Cornell group points

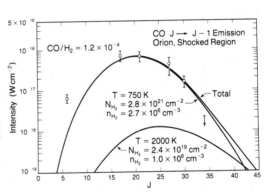

FIGURE 2. The distribution of CO line flux in a 1' beam resulting from the best fit of a two-component model to the data (solid lines) and the data themselves (marked points and error bars). The physical parameters of the shocked region derived from the fit are also shown in the figure.

out that the intensity of this line is very insensitive to density and temperature in the ranges appropriate to the shocked gas and is therefore a direct measure of the total column density of hot CO. Their observations indicate a total CO column density of 4.0 $\times 10^{17}$ cm^{-2}, in good agreement with the calculation based on the shorter wavelength lines. Assuming a 1.5 source diameter and a CO/H$_2$ relative abundance of 1.7×10^{-4},[9] they infer a mass of 1.5 $M_\odot$ for the hot shocked gas. As indicated by FIGURE 2, this mass would apply to the "warm" gas; the mass of the "hot" 2000 K component would be a factor of about 100 smaller.

Because both H$_2$ densities and H$_2$ column densities are derived from these measurements, we gain some insight into the distribution of the shocked gas, as mentioned by Storey *et al.*[9] By simply dividing the column density by the volume density, one obtains a length scale for the emitting region along the line of sight; for the "hot" and "warm" components, this produces scale lengths of 2.4×10^{13} cm and 1.0×10^{15} cm, respectively. The position of the $J = 6 \rightarrow 5$ measurement on FIGURE 2 indicates that it is mostly produced by gas still cooler than the "warm" component; a

CO column density of 1.2×10^{18} cm^{-2} can be inferred for this cooler component from the line intensity.[15] If we assume the same CO abundance and H_2 density for the "cool" component as for the "warm" one, a scale length of $l \simeq 3.7 \times 10^{15}$ cm is obtained. Thus, the scale length of gas of a given temperature increases as the temperature decreases, just as expected for postshock relaxation. However, each of these scale lengths is much smaller than the projected length of the emitting region, $L \simeq 6.5 \times 10^{17}$ cm ($1\rlap{.}{'}5$ at a distance of 480 pc). Hence, the shocked gas—even its cooler components (possibly even the "plateau" source)—must exist in the form of thin shells or small clumps, a configuration that is accepted as appropriate for the hot hydrogen.

CONCLUSIONS

The results of present far-infrared and submillimeter observations can be summarized as follows.

In addition to a 2000 K component with an average column density of 2.4×10^{19} cm^{-2} known from 2-μm observations, the hot shocked gas includes a larger component with $T \simeq 750$ K and an H_2 column density of 2.8×10^{21} cm^{-2} that dominates the 12-μm H_2 line and far-infrared CO line emission. The mass inferred for this gas is 1.5 $M_\odot$.

The shocked gas has an H_2 density of about 10^6 cm^{-3}.

The CO/H_2 abundance ratio inferred for the shocked gas is 1.2×10^{-4}, corresponding to 20% of the cosmic abundance of carbon existing in the form of CO. This result of a comparison of optically thin CO and H_2 lines arising in shielded matter constitutes the first *direct* measurement of the CO/H_2 relative abundance.

All the postshock gas, possibly even the plateau source, exists in the form of thin sheets perpendicular to the line of sight or in the form of clumps with a typical size of $10^{13} - 10^{15}$ cm.

It should be possible to improve on the precision of the derived parameters with measurements of CO lines involving J-values in the range from 30 to 40 and with more extensive observations of the longer-wavelength H_2 lines. However, even with the present precision, the far-infrared and submillimeter CO lines provide important constraints on the nature of the Orion-KL shock.[16]

ACKNOWLEDGMENTS

The observations made with the U.C. Berkeley tandem Fabry-Perot spectrometer were done in collaboration with J. W. V. Storey of the Anglo-Australian Observatory and R. L. Genzel and C. H. Townes of U.C. Berkeley. I am indebted to Drs. Storey, Genzel, and Townes for numerous discussions and to S. C. Beck, S. Beckwith, and M. O. Harwit for access to their unpublished data.

[NOTE ADDED IN PROOF: In February 1982, the CO $J = 34 \rightarrow 33$ line was detected; its intensity is 8.6×10^{-11} W cm^{-2} sr^{-1} (Watson, Genzel, Townes, and Storey, 1982, in preparation). This result agrees quite well with the value extrapolated from the longer-wavelength lines (*cf.* FIGURE 2) and it is now clear that the previous upper limit was the result of a faulty measurement.]

REFERENCES

1. BECKWITH, S. 1982. This volume.
2. KNAPP, G. 1982. This volume.
3. PHILLIPS, T. G., P. J. HUGGINS, G. NEUGEBAUER & M. W. WERNER. 1977. Astrophys. J. 217: L161.
4. PHILLIPS, T. G., J. KWAN & P. J. HUGGINS. 1980. In Int. Astron. Union Symp. 87, Interstellar Molecules. B. H. Andrew, Ed.:21. D. Reidel. Dordrecht.
5. STOREY, J. W. V., D. M. WATSON & C. H. TOWNES. 1981. Astrophys. J. 244: L27.
6. HOLLENBACH, D. J. & C. F. MCKEE. 1979. Astrophys. J. Suppl. Ser. 41: 555.
7. GREEN, S. & L. D. THOMAS. 1980. J. Chem. Phys. 73: 5391.
8. MCKEE, C. F., J. W. V. STOREY, D. M. WATSON & S. GREEN. 1982. Astrophys. J. In press.
9. STOREY, J. W. V., D. M. WATSON, C. H. TOWNES, E. E. HALLER & W. L. HANSEN. 1981. Astrophys. J. 247: 136.
10. WATSON, D. M., J. W. V. STOREY, C. H. TOWNES, E. E. HALLER & W. L. HANSEN. 1980. Astrophys. J. 239: L129.
11. BECKWITH, S., N. J. EVANS II, I. GATLEY, G. E. GULL & R. W. RUSSELL. 1982. Astrophys. J. Submitted.
12. BECK, S. C., E. E. BLOEMHOF, E. SERABYN, C. H. TOWNES, A. T. TOKUNAGA, J. H. LACY & H. A. SMITH. 1982. Astrophys. J. 253: L83.
13. BECKLIN, E. E., K. MATTHEWS, G. NEUGEBAUER & S. P. WILLNER. 1978. Astrophys. J. 220: 831.
14. STACEY, G. J., N. T. KURTZ, S. D. SMYERS, M. HARWIT, R. W. RUSSELL & G. MELNICK. 1982. Astrophys. J. 257: L37.
15. GOLDSMITH, P. F., N. R. ERICKSON, H. R. FETTERMAN, B. J. CLIFTON, D. D. PECK, P. E. TANNENWALD, G. A. KOEPF, D. BUHL & N. MCAVOY. 1981. Astrophys. J. 243: L79.
16. HOLLENBACH, D. J. 1982. This volume.

DISCUSSION OF THE PAPER

N. SCOVILLE (*University of Massachusetts, Amherst, Mass.*): Your conclusion that a fair fraction of carbon is in CO agrees with what we found for BN.

MASER SOURCES IN THE ORION-KL REGION*

R. Genzel,† M. J. Reid,‡ J. M. Moran,‡
D. Downes,§ and P. T. P. Ho.‖

†*Department of Physics*
‖*Department of Astronomy*
University of California, Berkeley
Berkeley, California 94720

‡*Harvard-Smithsonian Center for Astrophysics*
Cambridge, Massachusetts 02138

§*Institut de Radio Astronomie Millimetrique*
Grenoble, France

INTRODUCTION

In the past few years, molecular gas with an unexpected high-velocity dispersion has been found at the center of the Orion-KL region, along with quiescent gas from the dense core of the molecular cloud. This high-velocity gas was first detected as broad wings in "thermal" molecular lines in the mm range and as high-velocity maser features in the 22 GHz line of H_2O. It is now quite certain that the high-velocity motions are caused by mass outflow from one or more newly formed stars at the center of the infrared nebula. The maser emisson from the water molecule and that from OH and SiO have proven particularly useful in the investigation of the kinematics and structure of the outflow. In the following, we will summarize the observations and discuss the relationships among the masers, the highly excited thermal gas, and the compact infrared objects in the Orion-KL region.

THE 18 KM S⁻¹ OUTFLOW FROM IRc2

FIGURE 1 shows the spectra of the three strong maser molecules toward BN-KL: the $v = 1, 2, J = 1 \rightarrow 0$ SiO transition at 43 GHz, the $6_{16} \rightarrow 5_{23}$ line of H_2O at 22 GHz, and the 1.7 GHz OH masers.[1-3] The SiO maser spectrum is double peaked with a separation of about 24 km s⁻¹ between the strongest maser spikes and has a total velocity spread of 35 ± 3 km s⁻¹ (full width at zero power, FWZP). The velocity centroid is 5.5 ± 1 km s⁻¹, which is shifted by 3.5 km s⁻¹ from the velocity of the molecular cloud. The H_2O and OH maser profiles also show these "shell features," which are reminiscent of the maser spectra from the expanding envelopes of evolved late-type giants and supergiants. Interferometric observations[4,5] have shown that these shell-type maser features are associated with the envelope of the warm compact infrared source IRc2, to within about ±0.″7. The SiO maser is a probe of the conditions

*R. Genzel is supported by the Miller Society for Basic Research in Science.

142

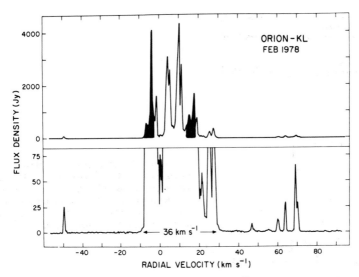

FIGURE 1a. The 22 GHz H_2O maser spectrum toward KL,[10] showing the strong low-velocity masers and the high-velocity flow. The H_2O "shell features" (shaded black) are coincident with IRc2 and are about ten times larger than the other H_2O features.

very close (R of about a few 10^{14} cm) to the infrared source. The double-shell profile can be formed only if velocity gradients are small; it can be best interpreted as radiative transport in a shell expanding (or contracting) at a constant velocity.[1,4] Expansion seems more probable (see the arguments in Reference 1). Pure rotation is excluded, since it would produce a triple-line profile. The data are, however, compatible with a rotating and expanding shell where $0 \lesssim v_{rot}/v_{exp} \lesssim 1$. The physical conditions in the envelope of IRc2 at $R \simeq 2 \times 10^{14}$ cm seem to be different from those in most evolved stars. In these stars—except for VY CMa—the SiO masers do not show the large

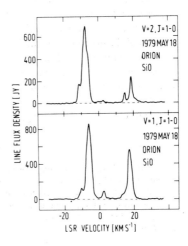

FIGURE 1b. The SiO $J = 1 \rightarrow 0$ (43 GHz) masers toward IRc2.[1] The strongest peaks are separated by ~24 km s^{-1}, but the total (FWZP) width of the SiO emission is 35 ± 3 km s^{-1}.

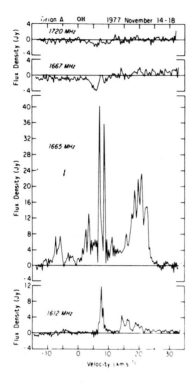

FIGURE 1c. The 1612 and 1665 MHz OH masers associated with IRc2.[3] The total width is 36 km s^{-1}, but the centroid is shifted by ~2 km s^{-1} to the red of the SiO profile (the H$_2$O spectrum is shifted by ~4 km s^{-1}). The 1667 and 1720 MHz spectra (top) show absorption by thermal OH gas.

velocity range (20–60 km s^{-1}) of the H$_2$O and OH masers further out in the envelope.[6,7]

The H$_2$O shell features (shaded black in FIGURE 1) probe radii of $\leq 10^{15}$ cm around IRc2 and have about the same velocities as the SiO masers. Finally, the OH masers at $R \simeq 1$–3×10^{16} cm have strong maser features between $v_{\mathrm{LSR}} = 3$ and $v_{\mathrm{LSR}} = 10$ km s^{-1} in addition to the blue- and redshifted shell features. The redshifted features are at larger velocities (17–24 km s^{-1}, rather than 13–21 km s^{-1}) and the LSR centroid of the spectrum is 7 ± 1 km s^{-1}, shifted from the IRc2 centroid of 5.5 km s^{-1} toward the velocity of the molecular cloud. The OH masers lie among position angle (p.a.) 68° east of north, within $\pm 4''$ of IRc2, and the blue- and redshifted shell features are on opposite sides of the source.[8] The emission of the thermal $v = 0$ rotational lines of SiO[1,5,9] are also centered on IRc2, to within ± 1–$2''$. The total velocity width and centroid of the $v = 0$ SiO spectra are the same as those of the $v = 1$ and $v = 2$ maser transitions. The source size is ~10″, with a brightness temperature of 150–200 K, estimated from the

FIGURE 2. Thermal molecular spectra toward the KL region of Orion. The CO (top[18]), the SiO (middle[2]), and the NH$_3$ (bottom[25]) lines show the different kinematic components: the quiescent gas (spike), the 18 km s^{-1} and hot core components, and the high-velocity plateau. Note the varying widths of the plateau component in different molecules and the similarity between hot core and 18 km s^{-1} components. The CO profile may also show this component as an asymmetry on the blue wing of the line.

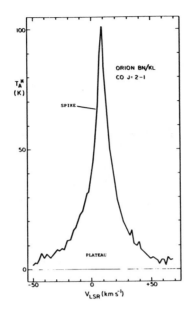

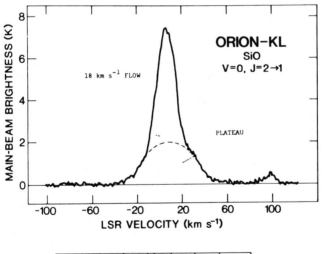

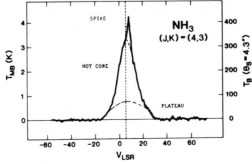

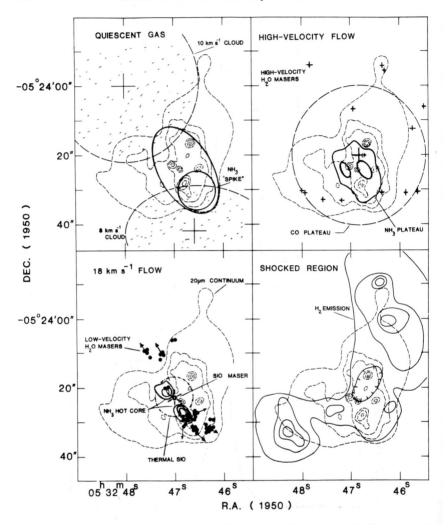

FIGURE 3. The angular distribution of the different kinematic components in KL superposed on the 20 μm map (dashed[26]). (a) The quiescent gas: H_2CO,[13] NH_3,[14,15] and HCN.[27] (b) The 18 km s^{-1} flow in H_2O masers (dark dots[10]), SiO masers, and thermal SiO (dashed dotted[1,5]); the hot core NH_3 emission (heavy lines[13,14]). The OH masers[8] are located ±4″ on either side of IRc2 (≙ SiO maser). (c) The high-velocity (plateau) gas in CO (dashed[18,28]), NH_3 (heavy lines[14]) and H_2O masers (little crosses[10]). Only the HPW and centroid (cross) for CO and the characteristic dimension of the two NH_3 knots and the extended NH_3 emission are indicated. (d) The 2-μm S(1) line brightness distribution from Reference 29.

optically thick $J = 2 \to 1$, $J = 3 \to 2$ lines (see the profile in FIGURE 2). The velocity ranges of the SiO maser, the OH maser, and the $v = 0$ SiO lines are the same, indicating a nearly constant velocity of expansion (12–20 km s^{-1}) over almost three orders of magnitude in radius from the star.

THE LOW-VELOCITY H₂O MASER OUTFLOW

The strong low-velocity H_2O features ($-8 \lesssim v_{LSR} \lesssim 28$ km s^{-1}, FIGURE 1)[10] are spread out over about $35'' \times 10''$ in p.a. 30° east of north around the center of the infrared cluster (FIGURE 3b). The total velocity spread is thus the same, within a few km s^{-1}, as that of the 18 km s^{-1} flow around IRc2. However, the velocity centroid is 10 ± 2 km s^{-1} and the profile as a whole appears to be shifted toward the red by ~4

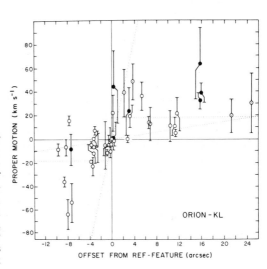

FIGURE 4. Proper motions of H_2O maser features in KL (from Reference 10). Top: Relative proper motions (km s^{-1}, for a distance of 450 pc) plotted against the angular offset from the reference feature. Open circles are RA motions, open circles with a dot are Dec motions, and filled circles are high-velocity masers. Almost all maser features occupy the first and third quadrant of the diagram only. The data indicate a large-scale outflow of the H_2O cloudlets from a central point. Bottom: Two-dimensional proper motion and the center of expansion in KL (only the features for which reliable Dec motions were available are shown). The length of the vectors is proportional to the transverse velocity and the error bars in direction are shown by error cones. The best estimate for the center of the low-velocity maser flow is shown by 1σ error bars.

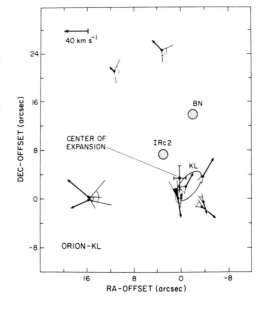

km s^{-1}. The H$_2$O features are concentrated in "centers of activity" $\sim$2–4" in size that have an average velocity spread of about half the total spread. There is, in particular, a large concentration of maser features near IRc4/IRc5. The H$_2$O masers are dense clumps of gas ($n_{H_2} \gtrsim 10^8$–10^{10} cm^{-3}) with a size of $\sim 10^{13}$ cm. The detection of proper motions[10] has shown that these maser cloudlets are moving at velocities $\lesssim$20 km s^{-1}. The measurements indicate a large-scale expansion for all masers from a common centroid somewhere between IRc2 and IRc4, and the data are fit best by a constant expansion velocity between radii of 2" and 25" (FIGURE 4). The three-dimensional model suggests that the H$_2$O centers of activity may be due to spatial clumping on scale sizes of $\sim$4".

The radial velocity spread in these 4" clumps is large and about 40–90% of the expansion velocity, but the velocity centroids vary only by a few km s^{-1} from center to center. This behavior can be explained by the geometry of a clumpy, expanding source. We think that the most likely interpretation of these measurements is that the low-velocity H$_2$O flow is also associated with IRc2. The general shift to the red by $\sim$4 km s^{-1} from the flow directly around IRc2, and possibly the offset of the centroid of the proper motions from IRc2, may be caused by interaction with the molecular cloud. This point of view is actually supported by the OH maser observations. As noted above, the OH maser features are also slightly redshifted from the IRc2 center velocity (by about half the amount by which the H$_2$O maser is redshifted), although they are still clearly associated with IRc2. The clustering of H$_2$O spots near IRc4, and also toward the northeast, may be correlated with the distribution of the dense molecular gas shown in FIGURE 3a. Hence, the excitation (and/or creation) of H$_2$O masers may be strongly coupled with the interaction of the flow with the molecular cloud.

THE 18 KMS^{-1} FLOW AND ITS INTERACTION WITH THE QUIESCENT GAS IN BN-KL: THE HOT CORE

The 1.2-cm inversion lines of NH$_3$ (the $J \gtrsim 3$ transitions, in particular, FIGURE 2) and other high-excitation molecular lines show a feature of $v_{LSR} \simeq 5$ km s^{-1}, with FWHM $\simeq$ 10–15 km s^{-1} (corresponding to FWZP $\sim$ 25 to $\gtrsim$30 km s^{-1}). An excitation analysis of several nonmetastable ($J \neq K$) NH$_3$ lines[11,12] indicates that the kinetic temperature of this component is high ($\sim$200 K; hence the designation "hot core"). Recent high-resolution very large array (VLA) observations of these NH$_3$ inversion lines give a detailed picture of the angular distribution of the hot core gas. The (3,3) map (FIGURE 3b)[13,14] shows that the brightest spot (T_b = 190 K) is near IRc4. In addition, there is a second peak $\sim$3" northeast of IRc2 and an arc or ridge bending around to the southeast of IRc2 and connecting these two knots. The (3,3) emission is, almost certainly, optically thick and marks the distribution of the highest kinetic temperature. Maps of the hot core component in the (4,4), (3,2), and (4,2) lines, however, show a different distribution (FIGURE 5), which could be due to variations in excitation conditions (*e.g.*, density) across the source. The maps in FIGURE 5 show that the emission of the nonmetastable NH$_3$ gas, as well as that of the (4,4) transition peak, is within $\lesssim$2" of IRc2. The line emission comes mainly from the region between IRc2 and IRc4, elongated northeast-southwest, and is located within the arc of emission delineated by the (3,3) hot core component. Also note that the (4,4) and (3,2) hot core emission seems to avoid the positions of the infrared sources IRc3 and IRc4 and sharply bends around the infrared sources. Maps of the NH$_3$ spike component indicate

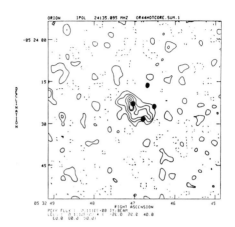

FIGURE 5. VLA maps of the NH$_3$ hot core emission (from Reference 15). Top: Map of the (4,4) emission, integrated over a v_{LSR} of $-10-+15$ km s^{-1}, with the spike and plateau subtracted. The contour levels are -20, $+20$, 40, 60, 80, and 90% of 110 mJy per beam area (2$''$3 × 2$''$6 FWHM). Middle: Map of the (4,2) line between v_{LSR} of 1 and 6.5 km s^{-1}. The resolution is 2$''$1 × 2$''$6 (FWHM) and the contour units are -50, -25, 25, 50, 75, and 95% of 160 mJy per beam area. Bottom : Map of the (3,2) line between v_{LSR} of 1 and 6.5 km s^{-1}. The resolution is 1$''$4 (FWHM) and the contour units are -50, -25, 25, 50, 75, and 95% of 140 Jy per beam area. On all the maps the dots indicate the compact infrared sources (right to left): IRc3, BN, IRc4, and IRc2.

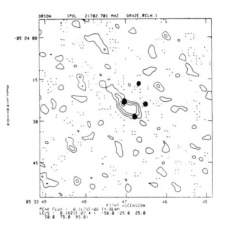

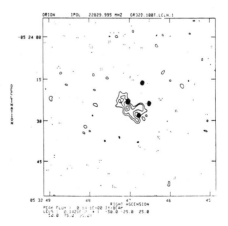

Annals New York Academy of Sciences

that the region around IRc3, IRc4, and IRc5 is the peak of warm quiescent gas in KL (FIGURE 3a).[15] A comparison of the (2,2) and (1,1) lines suggest that this peak is also a peak in kinetic temperature of the gas.[24] The angular distribution of the NH_3 hot core emission, the characteristic velocity centroid of ~ 5 km s^{-1}, and the linewidth are reminiscent of the 18 km s^{-1} flow associated with IRc2. Hence, we suggest that the hot core component in NH_3 and other high-excitation molecular lines may be gas within the 18 km s^{-1} flow or formerly quiescent gas that has now been pushed along with the flow (creating a shocked swept-up shell). The various lines of NH_3 and of other high-excitation molecules (see the discussions by G. Knapp and J. Welch in this volume) react in different ways to collisional or shock excitation in dense gas and to radiative excitation by the intense infrared radiation field at the center of the nebula. While the nonmetastable levels—in particular, the $J - K > 1$ lines (e.g, (4,2))—may be excited by the infrared radiation field within a few arcsec of the central source, the emission in the optically thick (3,3) line may be found at the interface between the streaming gas and the dense quiescent gas.

THE HIGH-VELOCITY PLATEAU

As can be seen from the spectra (FIGURES 1 and 2), the H_2O maser spectrum and thermal molecular lines show emission over a range between 50 and 190 km s^{-1} (the high-velocity plateau). The high-velocity H_2O maser features are spread over a large region (4×10^{17} cm or $\sim 60''$, FIGURE 3c). The proper motions detected for a few of these weak masers are again consistent with expansion, but the centroid could be anywhere between IRc4 and BN. The high-velocity plateau emission of most of the high-excitation ($\mu \gtrsim 1$ Debye) molecules is confined to about 10–15$''$ near the center of the infrared cluster. Unfortunately, the plateau emission in the extreme wings is weak and the position centroids determined from the emission between -10 and $+30$ km s^{-1} may be contaminated by the 18 km s^{-1} and hot core components. The positional centroid of the high-resolution NH_3 maps (FIGURE 3c) is defined best and is $\sim 1.5''$ west and 2$''$ south of IRc2.[13,15] The NH_3 plateau gas comes from two compact knots and a more extended ($\theta \simeq 10''$) structure. The high-velocity NH_3 gas shows a similar angular distribution for all velocities and an anticorrelation with the 20 μm infrared sources.¶ The high-velocity emissions of SiO and SO, on the other hand, are centered a few arcsec northwest of IRc2, toward BN.[9,16,17] This latter position is close to the centroid of the CO "plateau" source, which is extended over $\sim 40''$.[18] The SiO data also indicate that the centroids of the blue- and redshifted wings are separated by about 6$''$ in the northwest-southeast direction.[9] As with the low-velocity flow(s), the velocity centroid may be a "fingerprint" of the central source(s), and, in fact, the situation is very similar to that of the low-velocity gas: The spatially confined plateau gas (NH_3 and, possibly, HCN) has a velocity centroid of ~ 5 km s^{-1} and a spatial centroid within a few arcsec of IRc2, while the more extended plateau components have a velocity centroid of ~ 9 km s^{-1} and a spatial centroid slightly further north. Again, this may indicate either two or more sources or one dominant flow interacting with the inhomogeneous quiescent gas further out.

The proper motions of the high-velocity masers indicate that the high-velocity plateau—as well as the 18 km s^{-1} flow—represents expanding motion. This is also

¶See, however, the note added in proof.

evident from dynamical arguments. To keep the gas gravitationally bound at radii of 10–20″ and velocities $\gtrsim 50$ km s^{-1} would require a central mass $>10^4 M_\odot$, while estimates of the mass actually contained in the central 1′ are two orders of magnitude lower. Rotation is also inconsistent with the angular distributions of most of the molecular lines. The high-velocity expansion is clumpy and the flow cannot be an expansion with constant velocity. The velocities of high-velocity masers increase with distance from the center ($v_{exp} \propto R^{0.3}$). Shapes of the CO plateau and the angular size of the CO source as a function of radial velocity also suggest that velocities must increase with distance from the center.[18,19] An increase of velocities with radius could be the result of a large intrinsic velocity dispersion in the flow. The fastest particles would then automatically be furthest out. A large intrinsic velocity dispersion may be caused by the interaction of a stellar wind from the central star(s) with a surrounding cloud or by variability in the mass-loss rate or in the expulsion velocity of the wind. Alternatively, the high-velocity gas may still be accelerated at $R \simeq 10^{17}$ cm. This idea may be supported by the fact that the CO plateau and the H$_2$O masers (at radii of $\gtrsim 20″$) have a velocity range about a factor of two larger than the NH$_3$, SO, and SiO plateaus (radii 5–10″; but note that lack of sensitivity may prevent us from detecting gas at $v \gtrsim 110$ km s^{-1}). The acceleration mechanism may then be a hydrodynamic effect of the expansion of a channeled flow in a medium of steep density gradients. The acceleration at radii of 10^{17} cm cannot be due to radiation pressure or stellar wind pressure in a spherically symmetric situation (in the optically thin limit). In this case, the velocities of accelerated clumps would rapidly reach a terminal value if the radii were much larger than the initial radius, which may be about 10^{14}–10^{15} cm.

Discussion and Conclusions

There are at least two different outflows at the center of the KL cluster in Orion. The 18 km s^{-1} outflow is almost certainly associated with IRc2 (within the error bars of $\pm 1″$). The velocity centroid of the 18 km s^{-1} flow component is 5.5 ± 1 km s^{-1}, which may be the velocity of IRc2 itself. The low-velocity H$_2$O maser features and the hot core component identified in the spectra of the NH$_3$ inversion lines may basically represent the same phenomenon: outflow from IRc2 at low velocities. The observational differences (angular distribution, clumping of the maser features, velocity centroid, and total velocity range) may then be caused by the interaction of the flow with the material in the vicinity of the source of outflow.

The origin of the second, high-velocity, flow is, at present, uncertain. The brightness centroids of the maps of the different molecules give different answers, ranging from IRc4 to BN. The data do show, however, that the high-velocity gas can also be found close to IRc2/IRc4, and probably represents—as does the 18 km s^{-1} flow—a more or less continuous mass outflow. The low- and high-velocity flows may actually be close to each other at the center of the cluster.

There are some indications of anisotropy. The low-velocity outflow is elongated in the northeast-southwest direction, along the ridge of the quiescent molecular cloud. The high-velocity gas, on the other hand, is distributed mainly in the perpendicular direction (northwest-southeast), possibly along the directions of the steepest density gradients. If the low- and high-velocity flows come from the same object, then this possible anticorrelation may indicate anisotropic mass loss.

The masses contained within the flows, or the implied mass-loss rates, are high and about the same for the two flows (~ 1–20 $M_\odot$ or 10^{-3}–10^{-2} $M_\odot y^{-1}$; for a more detailed discussion, see Reference 20). These large mass-loss rates immediately exclude the Becklin-Neugebauer object (BN) as a source of any important fraction of the high-velocity flow, since the current mass loss at the surface of the star is two to three orders of magnitude too low.[21,22] The momentum supplied by the radiation of the infrared cluster is also probably not sufficient (by one to two orders of magnitude) to drive the flows by single scattering. A way out has been given in the model of Reference 23, in which the photons are scattered and absorbed several times, thereby enhancing the momentum available to $\tau L_*/c$. This may be sufficient to drive the flow for opacities $\tau_{IR} > 10$ (for other explanations, see Reference 20).

As mentioned before, the observed brightness distributions vary from molecule to molecule and transition to transition. This is probably caused by the varying sensitivity of the molecular lines to infrared and collisional excitation and by the differences in opacity.

ACKNOWLEDGMENTS

The authors thank J. Bieging, R. Martin, T. Pauls, and T. Wilson for permission to use data before publication. The National Radio Astronomy Observatory is operated by the Associated Universities, Inc., under a contract with the National Science Foundation.

[NOTE ADDED IN PROOF: A careful analysis of the NH_3 inversion lines suggests that the high-velocity wings visible in the metastable lines is not due to gas actually moving of velocities >20 km s^{-1}, but is probably the blended satellite hyperfine structure of the central hot core component. Hence, the NH_3 lines are highly optically thick ($\tau > 10$) and NH_3 column densities reach 5×10^{18} cm^{-2}.[15]]

REFERENCES

1. GENZEL, R., D. DOWNES, P. SCHWARTZ, J. SPENCER, V. PANKONIN & J. BAARS. 1980. Astrophys. J. **239**: 519.
2. SNYDER, L. & D. BUHL. 1974. Astrophys. J. **189**: L31.
3. HANSEN, S. 1980. Ph.D. Thesis. University of Massachusetts.
4. GENZEL, R., J. MORAN, A. LANE, C. PREDMORE, P. HO, S. HANSEN & M. REID. 1979. Astrophys. J. **231**: L73.
5. BAUD, B., J. BIEGING, R. PLAMBECK, D. THORNTON, W. WELCH & M. WRIGHT. 1980. In Int. Astron. Union Symp. 87, Interstellar Molecules. B. Andrew, Ed.: 545. D. Reidel. Dordrecht.
6. REID, M. & J. MORAN. 1981. Annu. Rev. Astron. Astrophys. In press.
7. SPENCER, J., A. WINNBERG, F. OLNON, P. SCHWARTZ, H. MATTHEWS & D. DOWNES. 1981. Astron. J. In press.
8. HANSEN, S. & K. JOHNSTON. 1980. Bull. Am. Astron. Soc. **12**: 824.
9. DOWNES, D., R. GENZEL, Å. HJALMARSON, L. Å. NYMAN, H. OLOFSSON & B. RÖNNÄNG. 1982. Astrophys. J. **252**: L29.
10. GENZEL, R., M. REID, J. MORAN, & D. DOWNES. 1981. Astrophys. J. **244**: 884.
11. MORRIS, M., P. PALMER & B. ZUCKERMAN. 1980. Astrophys. J **237**: 1.
12. ZIURYS, L., R. MARTIN, T. PAULS & T. WILSON. 1981. Astron. Astrophys. **104**: 288.

13. BASTIEN, P., J. BIEGING, C. HENKEL, R. MARTIN, T. PAULS, C. WALMSLEY, T. WILSON & L. ZIURYS. 1981. Astron. Astrophys. In press.
14. BIEGING, J., R. MARTIN, T. PAULS & T. WILSON. 1982. In preparation.
15. GENZEL, R., D. DOWNES, P. HO & J. BIEGING. 1982. Astrophys. J. In press.
16. WELCH, W., M. WRIGHT, R. PLAMBECK, J. BIEGING & B. BAUD. 1981. Astrophys. J. 245: L187.
17. PLAMBECK, R., M. WRIGHT, W. WELCH, J. BIEGING, B. BAUD, P. HO & S. VOGEL. 1982. Astrophys. J. Submitted.
18. KNAPP, G., T. PHILLIPS, P. HUGGINS & R. REDMAN. 1981. Astrophys. J. In press.
19. KWAN, J. & N. SCOVILLE. 1976. Astrophys. J. 210: L39.
20. GENZEL, R. & D. DOWNES. 1982. *In* Regions of Recent Star Formation. Roger and Dewdney, Eds. D. Reidel. Dordrecht.
21. SCOVILLE, N. 1981. *In* Int. Astron. Union Symp. 96, Infrared Astronomy. C. G. Wynn-Williams and D. Cruikshank, Eds. D. Reidel. Dordrecht.
22. MORAN, J., G. GARAY, M. REID & R. GENZEL. 1982. In preparation and this volume.
23. PHILLIPS, J. & J. BECKMAN. 1980. Mon. Not. R. Astron. Soc. 193: 245.
24. HO, P. & R. MARTIN. 1982. In preparation.
25. ZUCKERMAN, B., T. KUIPER & E. RODRIGUEZ-KUIPER. 1976. Astrophys. J. 209: L137.
26. DOWNES, D., R. GENZEL, E. BECKLIN & C. WYNN-WILLIAMS. 1981. Astrophys. J. 244: 869.
27. RYDBECK, O., Å. HJALMARSON, G. RYDBECK, J. ELLDÉR, H. OLOFSSON & A. SUME. 1981. Astrophys. J. In press.
28. SOLOMON, P., G. HUGUENIN & N. SCOVILLE. 1981. Astrophys. J. 245: L19.
29. BECKWITH, S., E. PERSSON, G. NEUGEBAUER & E. BECKLIN. 1978. Astrophys. J. 223: 464.

DISCUSSION OF THE PAPER

P. GOLDSMITH (*University of Massachusetts, Amherst, Mass.*): In a recent study at the University of Massachusetts, we determined that vibrationally excited HC_3N, which is centered approximately at IRc2, peaks at 5 km s^{-1} with a linewidth of ~15 km s^{-1}. How does this fit in with your model for this region?

GENZEL: The 5 km s^{-1} is very important because it is the central velocity of the maser peaks and of the peak of the hot core of NH_3. Thus, it is probably the central velocity of IRc2.

SYNTHESIS MAPS OF MILLIMETER MOLECULAR LINES

W. J. Welch, R. L. Plambeck, M. C. H. Wright, J. H. Bieging,
B. Baud, P. T. P. Ho, and S. N. Vogel

Radio Astronomy Laboratory
University of California, Berkeley
Berkeley, California 94720

INTRODUCTION

Molecular line profiles observed toward the Kleinmann-Low (KL) Nebula in Orion exhibit very broad wings, extending over velocity intervals as large as 190 km s^{-1}.[1] Because the region in which the broad pedestal feature originates is small (<1'), mapping with an angular resolution of a few arcsec is required to show the distribution of this fast-moving gas. Using the Hat Creek Interferometer, Plambeck *et al.* have recently obtained maps of the 2_2–1_1 rotational transition of SO at 86 GHz with a spatial resolution of 6", finding systematic motions in the gas.[2] In an earlier study with the same instrument, Welch *et al.* reported some information about the distribution of the pedestal feature in HCN, HCO$^+$, and SiO ($v = 0$), although with an angular resolution of 15–20".[3] Though SO (and presumably the other sulfur compounds) has a higher-than-usual abundance in the pedestal feature, the important ion HCO$^+$ has a very low abundance. A comparison between the SO maps and those of excited molecular hydrogen,[4] 20-μm infrared emission,[5] and the flow of H$_2$O masers[6] suggests a model for the pedestal feature.

THE DISTRIBUTION OF SO

Plambeck *et al.* obtained maps of the pedestal SO distribution in KL with velocity resolutions of 4.2 km s^{-1} and an angular resolution of 6" × 6".[2] FIGURE 1 is from their paper: FIGURE 1a shows the map of the integrated spectrum, centered on IRc2; BN and IRc4 are also shown. The peak brightness temperature is 50 K. FIGURES 1b, 1c, and 1d are the distributions of H$_2$O masers, 20-μm emission, and molecular hydrogen line emission, respectively. FIGURE 2 contains the individual channel maps; channel widths are 4.2 km s^{-1}. Considerable structure is apparent over all channels, and one does not see an isotropic flow from a single object. Two characteristics of the flow emerge, a broad component in nearly every part of the source having roughly the width of the whole source, superposed on a large-scale systematic motion. FIGURE 3 is the broad spectrum in a 6" × 6" pixel toward IRc3, along with the spectrum of the entire source. FIGURE 4 shows the spectrum toward other parts of the region. The systematic motion becomes evident from a comparison of the distributions of the redshifted and blueshifted gas. The former arches to the northwest and the latter to the southeast.

154

0077–8923/82/0395–0154 $1.75/1 © 1982, NYAS

These shapes, plus the double-peaked appearance of the integrated spectrum (FIGURE 1a), suggest that the SO is distributed in an expanding ring of gas seen nearly edge on, but tilted slightly with respect to the line of sight. A model of this expanding ring is shown in FIGURE 5.

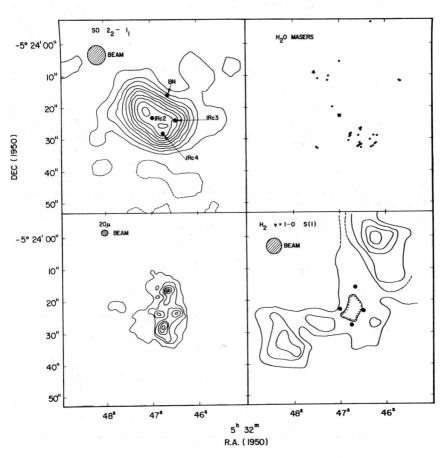

FIGURE 1. (a) Map of the integrated SO spectrum. (b) The distribution of H_2O masers. (c) The 20-μm emission. (d) Molecular hydrogen line emission.

DISTRIBUTION OF OTHER MOLECULES

TABLE 1 lists fringe visibilities for three other molecules in addition to SO, for a 12 m east-west interferometer spacing. The pedestal components of HCN and SiO ($v = 0$) are smaller than ~50″, like the SO, whereas the HCO^+ is less concentrated. Rydbeck *et al.* found that $n(HCO^+)/n(HCN)$ in the pedestal distribution is 0.1 of its

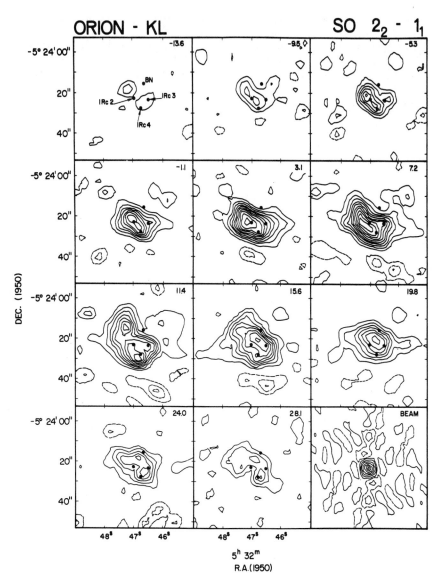

FIGURE 2. The individual SO channel maps. Channel widths are 4.2 km s^{-1}. The beam half width is 6".

value in the surrounding cloud, assuming that the pedestal source sizes are the same.[7] The HCO$^+$ visibility in TABLE 1, which shows the HCO$^+$ distribution to be broader, then gives an upper limit of about 3% for the ion HCO$^+$ density relative to what it is in the surrounding cloud.

Other Sulfur-Bearing Compounds

A number of other sulfur-bearing compounds are observed in the pedestal feature, also with somewhat higher abundances than usual. It is tempting to suppose, for the moment, that their distributions are also given by the maps of FIGURE 2. In a number of cases, more than one line of a molecule is observed. The line ratios invariably prove to correspond to local thermodynamic equilibrium and low optical depth. Thus, it is

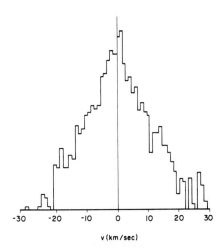

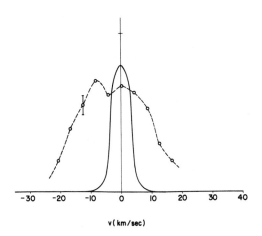

FIGURE 3. (a) The overall source spectrum. (b) The spectrum in a 6″ × 6″ pixel toward IRc3. The model spectrum is for a spherically collapsing protostar with a core mass of ~15 $M_\odot$. The SO is optically thick.

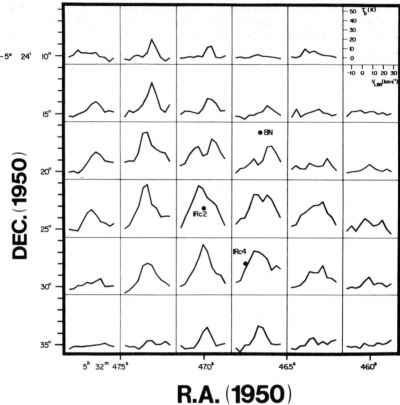

FIGURE 4. The SO spectrum toward other parts of the source.

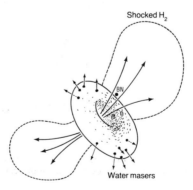

FIGURE 5. A model for the SO distribution: a ring of gas swept up by the wind from the core.

TABLE 1

SPECTRAL LINES OBSERVED

Molecule	Transition	Frequency (GHz)	Single Dish T_A^* (K)		Fringe Visibility*	
			Total	Pedestal	Total	Pedestal
SO	2,2–1,1	86.094	0.55	0.3	0.7	1.0
SiO	$v = 0, J = 2-1$	86.847	0.55	0.25	0.6	0.7
HCN	$J = 1-0$	88.632	9.5	1.0	0.2	0.7
HCO$^+$	$J = 1-0$	89.189	7.5	0.7	0.1	0.2

*Fringe visibilities measured using a two-hour average of the data taken at a 12-m east-west spacing; we smoothed the single-dish data to the same spectral resolution as the interferometer data before computing visibilities.

relatively easy to estimate the total abundance of each species from the original low spatial resolution observations and a guess at the excitation temperature. TABLE 2 contains the resulting total abundances for an excitation temperature of 100 K. The SO data is from Gottlieb et al.,[8] SO$_2$ from Pickett and Davis,[9] CS from Goldsmith et al.,[10] H$_2$S from Thaddeus et al.,[11] and OCS from Lada et al.[12]

The total number of sulfur molecules observed is 7.8×10^{51}. The cosmic abundance of sulfur is $n(S)/n(H) \simeq 1.6 \times 10^{-5}$. Hence, the minimum implied abundance of molecular hydrogen is 2.4×10^{56}, with a corresponding mass of 0.4 $M_\odot$. From excitation considerations, Plambeck et al. estimate a hydrogen mass of ~ 15 $M_\odot$.[2]

THE MODEL FLOW

In the model of FIGURE 5, Plambeck et al. propose that an energetic wind flowing from one or more sources near IRc2 has swept up and compressed the gas in the irregular ring-like structure. Where the ambient density is lower, above and below, the wind escapes to produce the shocked H$_2$ (and probably hot CO) seen as lobes of molecular hydrogen emission. The ring is clumpy with a mean density of $\sim 10^7$ cm^{-3}. The masers appear at the edge of the ring, where small dense bullets accelerated by the source of the wind strike the ambient cloud clumps near the outer surface of the ring.

The RMS velocity in the overall SO spectrum is about 25 km s^{-1}. For a hydrogen mass of 15 $M_\odot$, the implied kinetic energy in the gas of the ring is about 1.4×10^{47} erg. With a maximum BN-KL luminosity of $\sim 10^5$ $L_\odot$, the energy available over the ~ 600 y crossing time is 7×10^{48} erg, and the wind energy must be a substantial fraction of this.

Sulfur compounds appear to be enhanced, and there is a lack of the ion HCO$^+$. The implication is that ion-molecule chemistry is not important here. Rather, shock

TABLE 2

ABUNDANCES OF SULFUR COMPOUNDS

SO	CS	SO$_2$	OCS	H$_2$S
4.1×10^{51}	1.5×10^{50}	2.1×10^{51}	1.38×10^{51}	1.03×10^{50}

chemistry may dominate[13] as a result of the shocks that must form as the supersonic wind sweeps up the ambient gas.

SUMMARY

Aperture synthesis maps of the emission of SO at λ 3.4 mm in Orion, made with an angular resolution of 6″ and a velocity resolution of 0.8 km s^{-1}, reveal a combination of turbulent and systematic motions over a veolocity range of at least 50 km s^{-1}. A plausible model is an expanding, irregular ring of gas seen approximately edge on, the ring being dense material swept up by winds from one or more objects in the KL core. Lower resolution observations of HCN, SiO ($v = 0$), and HCO$^+$ show that HCN and SiO have a substantial component in the ring but that the HCO$^+$ is very underabundant there. The observed abundance of several sulfur molecules at low resolution yields a minimum hydrogen mass of 0.4 $M_\odot$ and a minimum kinetic energy in the ring of 4×10^{45} erg. For the more probable mass of 15 $M_\odot$ based on excitation considerations, the kinetic energy is 1.4×10^{47} erg.

REFERENCES

1. KNAPP, G. R., T. G. PHILLIPS, P. J. HUGGINS & R. O. REDMAN. 1981. Astrophys. J. **250:** 175.
2. PLAMBECK, R. L., M. C. H. WRIGHT, W. J. WELCH, J. H. BIEGING, B. BAUD, P. T. P. HO & S. N. VOGEL. 1981. Astrophys. J. In press.
3. WELCH, W. J., M. C. H. WRIGHT, R. L. PLAMBECK, J. H. BIEGING & B. BAUD. 1981. Astrophys. J. **245:** L87.
4. BECKWITH, S., S. E. PERSSON, G. NEUGEBAUER & E. E. BECKLIN. 1978. Astrophys. J. **223:** 464.
5. GENZEL, R., M. J. REID, J. M. MORAN & D. DOWNES. 1981. Astrophys. J. **244:** 884.
6. DOWNES, D., R. GENZEL, E. E. BECKLIN & C. G. WYNN-WILLIAMS. 1981. Astrophys. J. **244:** 869.
7. RYDBECK, O. E. H., W. M. IRVINE, A. HJALMARSON, G. RYDBECK, J. ELLDER & E. KOLLBERG. 1979. Astrophys. J. **235:** L171.
8. GOTTLIEB, C. A., E. W. GOTTLIEB, M. M. LITVAK, J. A. BALL & H. PENFIELD. 1978. Astrophys. J. **219:** 77.
9. PICKETT, H. M. & J. H. DAVIS. 1979. Astrophys. J. **227:** 446.
10. GOLDSMITH, P. F., W. D. LANGER, F. P. SCHLOERB & N. Z. SCOVILLE. 1980. Astrophys. J. **240:** 524.
11. THADDEUS, P., M. L. KUTNER, A. A. PENZIAS, R. W. WILSON & K. B. JEFFERTS. 1972. Astrophys. J. **176:** L73.
12. LADA, C. J., M. OPPENHEIMER & T. W. HARTQUIST. 1978. Astrophys. J. **226:** L153.
13. HARTQUIST, T. W., M. OPPENHEIMER & A. DALGARNO. 1980. Astrophys. J. **236:** 182.

DISCUSSION OF THE PAPER

M. W. WERNER (*Ames Research Center, Moffett Field, Calif.*): The size you measure seems a little smaller than that reported by others. If you integrate up the map, does it contain all the flux seen in single-dish measurements?

WELCH: It integrates up to about 75% of the single-dish measurements for SO.

W. M. IRVINE (*Onsala Space Observatory, Onsala, Sweden*): It is comforting that we (Schloerb, Friberg, Hjalmarson, Höglund, and Irvine, to be published) have recently used observations of 14 lines of SO_2 and $^{34}SO_2$ to derive an absolute abundance in the plateau source that is within 50% of your value. I believe that we must look carefully at the extent to which sulfur-bearing molecules are or are not chemical tracers of shocks, however. The relative abundance of SO in cold, dark clouds, for example, may not be much less than it is in Orion (Rydbeck *et al.* 1979. Astrophys. J. **235**: L171). Also, the chemical models of Prasad and Huntress (1981. Astrophys. J. Suppl. Ser. **43**:1) predict fractional abundances of SO_2 and SO between 10^{-5} and 10^{-6} for quiescent clouds with $T = 50$ K and $n \simeq 10^4$–10^5.

SUMMARY OF OBSERVATIONS
OF THE STAR-FORMING REGION

B. Zuckerman

Astronomy Program
University of Maryland
College Park, Maryland 20742

Orion is a fine example of the opportunities available to students considering a career in astronomy. Unlike well-tilled fields such as chemistry and physics, it is still possible to make basic advances relatively painlessly. In spite of the fact that Orion is one of the two or three best-studied astronomical sources outside of the solar system, I was able, some eight years ago, to advance from knowing nothing about Orion to being one of the world's experts by publishing a six-page paper in the *Astrophysical Journal*. The immediate payoff was an invitation to speak on Orion at a meeting at a ski resort in the Austrian Alps. So I jetted over to Europe. A few years later I jetted to Mexico City to give another review talk on Orion. And now I have taken Amtrak to this meeting. I think that it is time to change fields.

Is Orion really unusual or does it just appear so because it is so close to Earth? Zeroing in on the Becklin-Neugebauer–Kleinmann-Low infrared cluster, a picture of a truly unusual region of star formation is beginning to emerge. Although the total luminosity of BN-KL is not exceptionally high ($\sim 10^5 \, L_\odot$) compared with other H II region/infrared complexes, many other aspects do appear to be unique. In the microwave domain, some of the molecules detected toward BN-KL have been seen nowhere else. The peak temperature of the narrow ("spike") emission feature is unmatched, as is the velocity dispersion of the high-velocity ("plateau") emission. IRc2 is the only object yet discovered in a region of star formation that has an associated SiO maser. The methyl alcohol (CH_3OH) masers seen toward BN-KL are detected nowhere else in the galaxy. In the near infrared, the intensity of (shock-excited) H_2 emission is much greater than that in any other region in the galaxy, although this could be due, in part at least, to the proximity of BN-KL. Finally, in the far infrared, shock-excited CO is seen in BN-KL but, so far, nowhere else.

Because of limited time, I will confine my remarks to the energetic high-velocity flows. The basic problem remains: What is the underlying physical mechanism that drives the flow? In addition, we are interested in the properties of the flow and its interaction with the surrounding molecular cloud.

Phillips and Beckman[1] and Solomon et al.[2] argue that radiation pressure (on dust grains) can drive the high-velocity winds, provided that the optical depth, τ, in the near infrared (~ 1–$3 \, \mu m$) is large. This is because, for mass loss driven by radiation pressure, the momentum in the wind $\dot{M} \, v_\infty$ is equal to $\tau \, L/c$, where L is the luminosity of the underlying source. In Orion, $\dot{M} \, v_\infty \simeq 100 \, L/c$, so $\tau_{near \, IR}$ must be ~ 100 if this mechanism is to be viable. Although there are no compelling reasons to rule it out in Orion at this time, as discussed by Zuckerman[3] and Beckwith and Zuckerman,[4] such a model almost certainly fails for lower luminosity sources such as T Tauri and Herbig-Haro objects.

0077–8923/82/0395–0162 $1.75/1 © 1982, NYAS

Here $\dot{M} v_\infty$ is still $\sim 100\ L/c$, but $\dot{M}$ is orders of magnitude less than in Orion, which effectively prevents the build-up of large optical depths in the near infrared. Thus, we would be left with the unsatisfying situation of requiring two entirely different mechanisms to explain high and low luminosity objects that have similar values of $\dot{M} v_\infty c/L$.

Although the jury is still out, it seems that the stars that are responsible for the energetic winds are probably almost on the main sequence and are not cocoon stars. In Orion, for example, the most likely candidates are BN and IRc2, as discussed by Scoville and Werner in this volume. Both of these objects are rather hot: BN has a little H II region and, if IRc2 is really extinguished by five magnitudes at 3.8 μm, as argued by Werner, then it, too, must be fairly hot ($\gtrsim 1000$ K). IRc4, on the other hand, which looks more like a deeply enshrouded star (if it is not merely a density enhancement in the molecular cloud), seems to be a less promising candidate for driving the high-velocity winds. In a much lower luminosity system, T Tauri, Dyck et al. have discovered a companion star that radiates only in the infrared and also appears to be deeply enshrouded in a cocoon of warm dust.[5] Indirect arguments, not yet completely compelling, suggest that the energetic wind at T Tauri is driven by the visible star, not the infrared companion.

Concerning Werner's model, which requires a very opaque doughnut of dust and gas around IRc2, it is not obvious that this simple geometry is consistent with the distribution of H_2, H_2O maser, OH maser, and high-velocity CO and SO emission that seem to be elongated along various, often orthogonal, axes.

What about the other end of the flow, where it impinges on the surrounding molecular cloud? Does the 2-μm H_2 emission originate in the outflowing gas itself or in the molecular cloud gas after it has been excited by the wind? Arguments exist on both sides. For example, Nadeau et al. believe that, because it would be difficult to accelerate clumps in the molecular cloud to high velocities with multiple nondissociating shocks, the really high-velocity emission ($V \gtrsim 25$ km s^{-1}) is produced in the mildly shocked molecular wind itself.[6] However, the existence of magnetic fields in the gas may alter this simple picture, as discussed by Hollenbach and by Draine in this volume. Kuiper et al. suggest that the wind must be interacting strongly with the molecular cloud gas, since high radial velocity H_2O maser emission is observed at large angular distances from BN-KL.[7] If the H_2O was merely gas excited in the flow itself, then one would expect this high-velocity emission to appear more nearly along the direct line of sight to BN-KL.

Suppose that the emission originates, at least in part, from the wind itself. Then, except for the H_2 and far-infrared CO emission, which are hot, it is not obvious that shock excitation is required for other molecules observed at millimeter wavelengths. The temperature of these molecules is quite modest and comparable to that of the spike emission (~ 70 K). In addition, relative abundances predicted by shock models don't agree quantitatively with observed abundances in the high-velocity gas.[7] One case in point is HCO^+, which is observed to be substantially enhanced in abundance where the supernova remnant IC443 shocks a nearby molecular cloud. But chemical models predict decreased HCO^+ in such a shocked region. In Orion, high-velocity HCO^+ emission appears[7,8] near a strong peak in the 2-μm H_2 emission but offset in position from BN-KL and from the high-velocity emission apparent from other molecules. This could be interpreted as implying that the HCO^+ is produced in gas just behind the shock front that has excited the 2-μm emission but not in the nonshocked gas in the outflow itself.

Many researchers have considered the chemistry in shocks. Perhaps it is time to carefully consider the chemistry in a high-density flow that originates in a hot region near a star, as suggested above, and then cools and recombines (if it is atomic at the base of the flow) as it flows outward. A possible way to check this possibility would be to search for C I emission toward IC443. C I emission is not seen in the high-velocity gas in Orion (Phillips, this volume), but might be seen in IC443 if strong shocks excite the latter, but not the former, gas.

More generally, although many groups are considering the chemistry of outflowing material around evolved giant stars, none, to my knowledge, are considering pre-main-sequence stars. For the latter stars, an oxygen-rich (O/C > 1) environment is appropriate (certainly in the plateau gas in Orion[9]). Goldreich and Scoville[10] and Scalo and Slavsky[11] have considered envelopes around oxygen-rich evolved stars. The latter authors predict large abundances of H_2O, OH, SO_2, SiO, and SO, all of which are observed in the plateau around Orion but not around evolved stars![12]

So a calculation beginning at high temperature, even, perhaps, inside a chromosphere or H II region, and following an expanding, cooling outflow seems appropriate at this time.

REFERENCES

1. PHILLIPS, J. P. & J. E. BECKMAN. 1980. Mon. Not. R. Astr. Soc. **193:** 245–60.
2. SOLOMON, P. M., G. R. HUGUENIN & N. Z. SCOVILLE. 1981. Astrophys. J. **245:** L19–22.
3. ZUCKERMAN, B. 1981. *In* Infrared Astronomy. D.P. Cruikshank and C.G. Wynn-Williams, Eds: 275–79. Reidel. Dordrecht.
4. BECKWITH, S. & B. ZUCKERMAN. 1982. Astrophys. J. **255:** 536–40.
5. DYCK, H. M., T. SIMON & B. ZUCKERMAN. 1982. Astrophys. J. **255:** L103–6.
6. NADEAU, D., T. R. GEBALLE & G. NEUGEBAUER. 1982. Astrophys. J. **253:** 154–66.
7. KUIPER, T. B. H., B. ZUCKERMAN & E. N. R. KUIPER. 1981. Astrophys. J. **251:** 88–102.
8. RYDBECK, O. E. H., A. HJALMARSON, G. RYDBECK, J. ELLDER, H. OLOFSSON & A. SUME. 1981. Astrophys. J. **243:** L41–45.
9. ZUCKERMAN, B. & P. PALMER. 1975. Astrophys. J. **199:** L35–38.
10. GOLDREICH, P. & N. SCOVILLE. 1976. Astrophys. J. **205:** 144–54.
11. SCALO, J. M. & D. B. SLAVSKY. 1980. Astrophys. J. **239:** L73–77.
12. ZUCKERMAN, B. 1981. Astron. J. **86:** 84–86.

DISCUSSION OF THE PAPER

T. MOUSCHOVIAS (*University of Illinois, Urbana, Ill.*): Since yours was a summarizing talk, please allow me to ask a summarizing question. All the observational and theoretical results presented today unquestionably shed light on the physical condition in Orion. Is it not fair to state, however, that virtually all the objects and phenomena discussed refer to events that follow, rather than precede, star formation? And if we want to understand star formation itself, shouldn't we be observing the less exciting (and less excited) dense clouds that have not yet given birth to stars?

DISTRIBUTION OF STARS IN THE ORION REGION

Syuzo Isobe

Tokyo Astronomical Observatory
University of Tokyo
2-21-1, Osawa, Mitaka, Tokyo, Japan

INTRODUCTION

It is well understood at the moment that most young stars (those younger than 10^8 y) were formed in large molecular clouds. However, the mean densities of molecular clouds are not high enough to form stars. Some kind of compressing mechanism is needed to produce a star-formation process. It is generally believed that there are three triggering mechanisms for star formation: (1) compression by density wave, (2) compression by the expanding gas of supernovae, and (3) compression by the radiation of O- and B-type stars. To find out which mechanism is dominant, one should study those regions where the stars are very young. The Orion region is one of the youngest. Moreover, since it is at high galactic latitude and is near the Sun, we are able to obtain many kinds of good data.

In this paper, we will discuss results obtained from photographic observations of the stars in the Orion region.

OBSERVATIONAL RESULTS

The Orion constellation contains many bright stars. Moreover, on the Palomar Sky Atlas prints, we can count over one million stars in this field and most of these stars are members of the Orion association. Isobe made a star count in this region with 26′ × 26′ mesh regions and drew a contour map of stellar distribution (FIGURE 1).[1] Stars are distributed throughout the region except in the dense molecular cloud discovered later by Kutner et al.,[2] but the stellar density is a little higher in the Barnard Loop than in the other parts. This result suggests that the enhancement of stellar density in the Barnard Loop has some relation to the explosive event(s) that made the Barnard Loop itself.

Isobe and Sasaki[3] extended the observation by Penston et al.[4] to the fainter stars and confirmed that the ages of the stars in the Orion Nebula are in the range from 10^4 to 3×10^7 y. This indicates that low-mass stars in the Orion Nebula are as old as the oldest stars in the whole Orion association.

More recently, Isobe and Nishino measured U, B, and V magnitudes of stars in the region of the eastern part of the Barnard Loop and found that the number of the bluer stars is larger in the Loop than outside the Loop, but that the redder stars are distributed rather uniformly.[5] This result confirms Isobe's conclusion that the stars in the Loop were formed during the same explosive event that made the Loop.[1]

0077–8923/82/0395–0165 $1.75/1 © 1982, NYAS

DISCUSSION

About a decade ago, Blaauw showed that subgroups of stars in the Orion association are sequentially ordered by age. This kind of sequential formation of massive stars can be interpreted by the effect of radiation pressure from the newborn exciting stars and the effect of gas pressure from H II regions in the Orion Molecular Cloud.[7] The youngest visible stars in the Orion association are the Trapezium stars in the Orion Nebula, which have ages of about 10^4–10^5 y. In the molecular cloud just behind the Orion Nebula, some new massive stars, which were observed in the infrared wavelengths as the BN object and KL objects, were formed about 10^3 y ago through the effect of the Trapezium stars. Therefore, star formation actually does proceed by means of the third effect mentioned above.

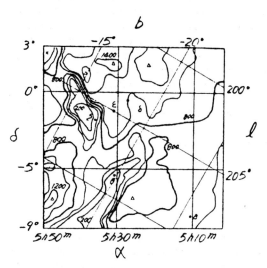

FIGURE 1. A contour map of the number densities of stars in the Orion association on the blueprint of the Palomar Sky Survey. The number of stars in regions 26' × 26' are shown. Triangles and crosses show the peak and bottom positions, respectively.

During sequential star formation, massive stars evolve quickly and finally produce supernova explosions. Since massive stars are formed sequentially, supernova explosions follow sequentially. The first supernova expands into the surrounding interstellar space, piles up interstellar gas, making a central hole, and slows down. The second one expands freely in the central hole without decreasing the expansion velocity, reaches the expanding shell of the first supernova, and they then expand together into interstellar space. The following supernovae develop in the same way but add the expansion energy to the shell.

The expansion energy of the Barnard Loop is very large; it is the energy of ~20–40 supernova explosions. This figure matches well with the idea mentioned in the previous paragraph. Actually, Reynolds and Ogden provided some evidence that the Barnard Loop was formed by supernova events, the latest one of which might have exploded about a million years ago. The bluer stars on the loop observed by Isobe and Nishino[5] are strongly expected to be formed by the second effect, that due to these supernovae.

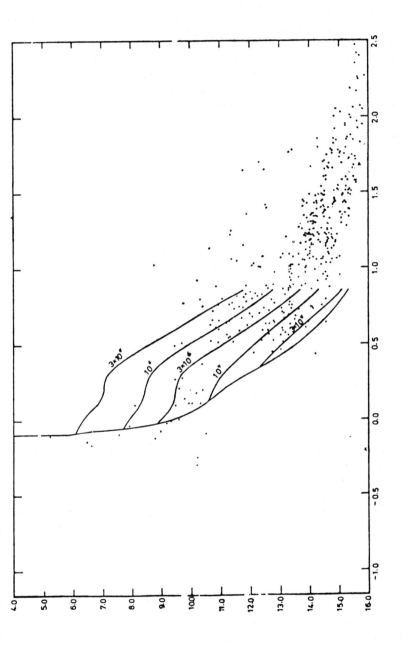

FIGURE 2. R and $(R - I)$ diagrams. Theoretical equal-age lines are given in the figure.

However, the redder stars, which are probably older than the bluer stars, are smoothly distributed throughout the loop. Moreover, even in the Orion Nebula, which includes the youngest massive stars in the Orion association, the ages of stars range up to 3×10^7 y, which is the age of the oldest subgroup, the Orion association Ia.

Few and Booth[9] made a H_2CO molecular observation in the Orion Molecular Cloud and found many clumpy structures with densities higher than 10^6 cm^{-3} even in the region further away from the Trapezium stars by which such new stars as the BN object and KL objects were formed. This suggests that compression of molecular clouds by some effect took place in these regions before the effect of the radiation pressure of O- and B-type stars reached the regions.

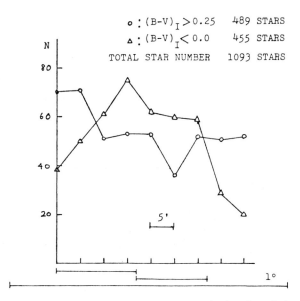

FIGURE 3. The distribution of star number in each striped region of the eastern part of the Barnard Loop. Since absolute calibrations were not made, $(B - V)$ values have meaning only in relative terms.

It is said that a density wave passed through the Orion association ~1–3 $\times$ 10^7 y ago. According to this, the first stage of massive stars were formed in a small area of the whole molecular cloud at that time. Moreover, the density wave compressed the whole molecular cloud and this compression was followed by the formation of low-mass stars.

CONCLUSION

Although the data are not complete enough to give a definite conclusion, it is very possible that three types of triggering mechanisms for star formation are at work in the Orion association.

SUMMARY

The Orion region contains stars with ages ranging from 10^4 to $2-4 \times 10^7$ y. Those stars began to be formed by the different triggering mechanisms. To find out which mechanism was dominant, we examined the characteritics of the distribution of stars in the different Orion regions. From the previous results, (1) the age of the stars in the Orion Nebula ranges from 10^4 to 3×10^7 y and (2) stellar components in regions in the Barnard Loop and outside the Loop are different, it is concluded that three types of triggering mechanisms are at work in the Orion region.

REFERENCES

1. ISOBE, S. 1973. Dust, stars, and hydrogen distribution in the Orion association. *In* Interstellar Dust and Related Topics. J. M. Greenberg and H. C. van de Hulst: 433–44. D. Reidel. Dordrecht.
2. KUTNER, M. L., K. D. TUCKER, G. CHIN & P. THADDEUS. 1977. The molecular complex in Orion. Astrophys. J. **215:** 521–28.
3. ISOBE, S. & G. SASAKI. 1982. Globules in the Orion Nebula. III. Range of ages of the Orion Nebular stars. Publ. Astron. Soc. Jpn. **33:** 241–47.
4. PENSTON, M. V., M. F. J. MAN & M. J. WARD. 1976. Emission free photographic photometry of stars in the Orion Nebula cluster. Mon. Not. R. Astron. Soc. **174:** 449–54.
5. ISOBE, S. & Y. NISHINO. 1981. Distribution of stars in the Orion Barnard Loop. *In* Proc. 2nd Asian Pacific Regional Meeting IAU. B. Hydayat, Ed. In press.
6. BLAAUW, A. 1964. Annu. Rev. Astron. Astrophys. **2:** 213–46.
7. ELMEGREEN, B. G. & C. J. LADA. 1977. Sequential formation of subgroups in OB associations. Astrophys. J. **214:** 725–41.
8. REYNOLDS, R. J. & P. M. OGDEN. 1979. Optical evidence for a very large expanding shell associated with the I Orion OB Association, Barnard's Loop, and high galactic latitude Hα filament in Eidanus. Astrophys. J. **229:** 942–53.
9. FEW, R. W. & R. S. BOOTH. 1979. A survey of formaldehyde in the Orion dust cloud L1641. Mon. Not. R. Astron. Soc. **188:** 181–90.

OBSERVATIONS OF [C I] EMISSION
IN ORION A (M42) AND ORION B (NGC 2024)

G. Münch and H. Hippelein

Max-Planck-Institut für Astronomie
D-6900 Heidelberg-Königstuhl, Federal Republic of Germany

INTRODUCTION

We have already reported the detection of [C I] emission in the spectrum of M42 and presented the results of preliminary measurements of intensities and radial velocities of the lines $^1D_2 \rightarrow {}^3P_0$, at $\lambda 9850.26$, and $^1S_0 \rightarrow {}^1D_2$, at $\lambda 8727.13$.[1] Since the [C I] emission was observed with maximum strength near the location where Hα appears to be strongest, we attempted to explain its excitation through collisions with free electrons from the H$^+$ gas present in the transition layers between the H II region and the surrounding neutral medium. We could account for the column density of C^0 required by the measured line intensities only by further postulating that the process of charge exchange between C$^+$ ions and H^0 atoms took place at the high rate proposed by Steigman[2] and Péquignot et al.[3] Recently, however, Butler and Dalgarno have shown on first principles that the charge transfer process involved actually has a rate some six orders of magnitude smaller than the one we assumed, and therefore it cannot play any role in determining the degree of ionization of C^0.[4] Before the publication of Butler and Dalgarno, we fully realized that extensive additional observations of [C I] emission would be required to study the systematics of its occurrence and to provide firm empirical evidence in regard to its place of origin. The observational program undertaken towards this aim has not yet been completed. Nevertheless, the data collected to date suffice to serve as a basis for a revision of our earlier analysis, evidently made necessary by the arguments of Butler and Dalgarno.[4] In this paper, we shall present the results of recent observations in [C I] of M42 and NGC 2024.

OBSERVATIONS

All observations reported here have been made with a pressure-scanned Fabry-Perot spectrometer in a twin etalon mode, attached to the 1.2-m Zeiss telescope of the Calar Alto Observatory in Spain. A description of the instrument has been published by Hippelein and Münch.[5] A variety of spacer combinations has been employed in order to obtain resolutions up to 4.2 km s^{-1}. The detection limit for the [C I] $\lambda 9850$ line, however, is not set by its photon statistics, but rather by the background set by airglow lines, such as OH(3,0)P$_2$(2) $\lambda 9848.5$, two weaker unclassified lines at $\lambda\lambda 9850.9$ and 9863, and the very strong OH(3,0)P$_1$(2)$\lambda 9877$. Considering that the OH $\lambda 9848$ line has an emission rate of 150 Ray, it would appear to be clear that the measurement of a nebular line with an emission rate of 20 Ray or less requires a painstaking subtraction of the night sky spectrum. The problem of doing photometry in the [C I] $\lambda 9850$ line is further complicated by the nearby presence of H$_2$O telluric absorption at $\lambda\lambda 9848.89$ and 9849.12 Å.

170

0077-8923/82/0395-0170 $1.75/1 © 1982, NYAS

RESULTS

Map of M42 in [C I] λ9850

We have mapped anew the Orion Nebula with a circular field stop 80 arcsec in diameter positioned over the grid of 64 points shown in FIGURE 1. The emission rates given therein for each field have been derived from Gaussian least-square fits to the observed profiles. The uncertainty for all the fields is nearly constant and amounts to 25 Ray. Compared with earlier measurements, the emission rates given in FIGURE 1 are higher by approximately 30% as a consequence of the use of better-tuned prefilters and of the [S III] λ9532 line as an intensity calibrator. The more extended survey now available allow one to establish significant differences between the surface brightness distribution in [C I] and in Hα or [S III] λ9532. In particular, it is noticed that the brightness distribution in [C I] is flatter, less peaked towards the center, than in the other nebular lines. The [C I] map also shows features not found in the [S III] map,

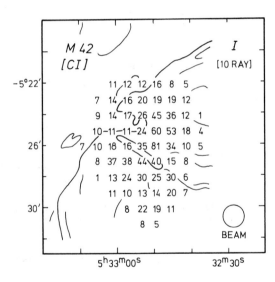

FIGURE 1. The emission rate in the [C I] λ9850 line, expressed in Rayleighs (Ray; omnidirectional 10^6 photons cm^{-2} s^{-1}), over the Orion Nebula.

such as the secondary maximum at RA = 5^{h}32^{m}52^s and Dec = −5°30′ (1950). Contradicting our earlier opinion, we can now say that the [C I] brightness distribution does not follow that of typical nebular lines.

Width of the [C I] λ9850 Line in M42

Bearing in mind that the regions where the [C I] emission arises have not been proved or disproved to be coincident with the C II regions, which originate the radio recombination lines, we have obtained a [C I] λ9850 profile with a resolution of 4.2 km s^{-1}, as measured by the full width at half maximum (FWHM) of the Kr I 9856.24 calibrator line. The observed profile, together with its deconvolution from the instrumental profile, is shown in FIGURE 2. It can thus be seen that the deconvolved profile has a FWHM = 6.1 km s^{-1}, considerably smaller than that of any other proper nebular line. For example, the minimum width observed for [O II] 3726–29 is FWHM = 10.1 km s^{-1}. For the [O I] λ6300 nebular line, we have—at the same position (40 arcsec west, 10 arcsec south of θ^1Ori C) to which the profile of [C I] of FIGURE 2 refers—measured a FWHM nearly equal to that of [O II] given above. The FWHM for the [C I] line should be further corrected for the effects of the finite entrance aperture, but this correction is probably less than 1 km s^{-1}, considering the radial velocity map of the [C I] line given below. While the [C I] line is definitely narrower than any other nebular line, it is still broader than the radio C II recombination lines. Only in two of the fields studied by Jaffe and Pankonin are the C II radio recombination lines broader or just as wide as the [C I] line.[6]

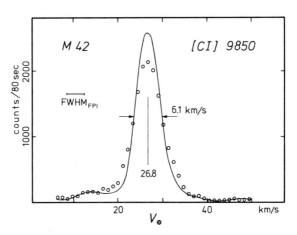

FIGURE 2. The profile of the [C I] λ9850 emission line for a 80 arcsec diameter field centered 40 arcsec west and 10 arcsec south of θ^1Ori C, obtained with a resolution of 4.2 km s^{-1}.

Radial Velocities of [C I] λ9850 in M42

The rest wavelength of the [C I] line, unlike those of other forbidden nebular lines, is well determined by laboratory measurements of numerous intersystem uv lines, and, therefore, accurate radial velocities can be determined. From the Gaussian fits to the observed profiles, we have derived radial velocities for each field, from which the map of FIGURE 3 has been constructed. The curves of constant radial velocities for the [C I] line follow the general shape shown by the radio recombination lines[6] and the radial velocities we measure for the [C I] line agree almost exactly with those of the radio lines. For example, the value of the radial velocity of the radio lines near θ^1Ori C, in the heliocentric frame, is 28.5 km s^{-1}, within 1 km s^{-1} of our value. On the radio map, the maximum radial velocity is 28.9 km s^{-1}, near the position where the [C I] line reaches its maximum of 30 km s^{-1}. In comparison, the H II radio or optical recombination and forbidden lines are blue-shifted by about 8 km s^{-1} with respect to the C II radio or [C I] lines.

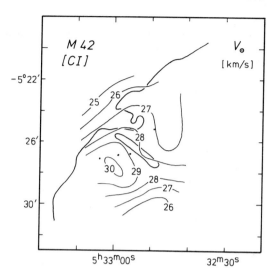

FIGURE 3. Contours of constant heliocentric radial velocity in the [C I] λ9850 line over the Orion Nebula.

Map of NGC 2024 in [C I] λ9850

Visual imagery of NCG 2024 shows a conspicuous dark lane running through its middle. Imagery at [S III] λ9532 shows the same lane, with somewhat reduced contrast relative to the bright areas. The map we have obtained in [C I], shown in FIGURE 4, in contradistinction, only faintly shows a diminished [C I] line intensity at the position of this lane with respect to its surroundings. In particular, the area at RA = $5^h39^m10^s$ and Dec = $-1°57'.0$ (1950), with an emission rate of 28 ± 6 Ray in [C I] λ9850, is at the center of the dark lane and also nearly coincides with the location of maximum radio continuum emission at 1.95 cm, maximum IR continuum at 2.4 μm, and maximum strength radio C II recombination lines. The measured [C I] radial velocity at that

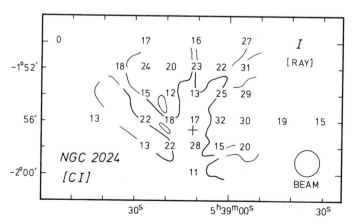

FIGURE 4. The emission rate in the [C I] λ9850 line, in Rayleighs, over NGC 2024.

position, 28 km s^{-1}, also agrees well with that of the C II radio lines arising in the foreground of the main radio continuum source.[7]

The metastable levels $^1D_2$ and $^1S_0$ of the C^0 atom can be excited through the following processes within or near an H II region. (1) Recombination of C^+ with thermal free electrons in excited levels with subsequent radiative cascade. (2) Collision with thermal electrons within the H II region. (3) Resonance absorption of uv stellar continuum radiation and subsequent cascade. (4) Collisionally in nonsteady conditions, as in a shock. Excitation through electron collision can be ruled out, because the concentration of C^0 within the bounds of the H II region is far too small, given that charge exchange between C^+ and H^0 is ineffective.

Process (3) can take place either by absorption in a resonance line of the triplet system or by direct absorption in an intersystem transition. Whether the process is effective or not depends strongly on the model distribution of dust and gas adopted. On the one hand, dust is needed to shield C^0 from the uv ionizing stellar radiation, but, on the other hand, it should have an optical depth not too high in order to make radiation in the region of the lines accessible to C^0 atoms. A powerful observational constraint would be set by an estimate of the C^0 column density from the strength of interstellar absorption lines in the spectrum of the stars imbedded in the H II region. For θ^1Ori C and θ^2Ori A, imbedded in M42, Copernicus observations, unfortunately not of the highest quality, exist, but do not show particularly strong C I absorption lines in the 1200 Å region. The column density of C^0 implied would be insufficient to populate the metastable levels by resonance absorption. In this context, it would be particularly useful to analyze the $2p^3P \rightarrow 3s^3P^0$ (uv2) and $2p^3P \rightarrow 2p^3D^0$ (uv3) multiplets of C I at $\lambda1657$ and 1561 Å, which must be present in existing high-resolution IUE spectra.

The possibility of accounting for the population of the C^0 metastable levels by direct recombination is strongly model dependent. That is to say, it is always possible to build an *ad hoc* model distribution of gas and dust that is sufficiently shielded from the uv radiation of the exciting stars to avoid a too high photoionization rate. Following a more empirical procedure, we shall show that the models constructed to interpret the C II radio recombination lines can also explain the observed intensity of the [C I] line. Let us take, for example, the parameters

$$N_e = N(C^+) = 30 \text{ cm}^{-3}, \qquad T = 200\,^0K, \qquad \text{and} \qquad L = 0.05 \text{ pc},$$

given by Pankonin *et al.* for the electron and ionic densities N_e and $N(C^+)$, for the temperature T, and for the thickness L of the C II region observed in the foreground of NGC 2024.[7] The net recombination coefficient of C^+ at 200 K is $\beta = 5 \times 10^{-12}$ cm^3 s^{-1}[3] and the number of recombinations to the singlet and triplet levels are approximately in ratio of their multiplicities. The rate at which the $^1D_2$ level is being populated is then

$$\tfrac{1}{4} N_e \beta(t) = 3.8 \times 10^{-11} \text{ s}^{-1} \ll A(^1D_2 \rightarrow {}^3P),$$

since the transition probability A for the production of a [C I] photon in either of the $\lambda9850$ or 9824 lines is 2.5×10^{-4} s^{-1}. Since collisional deactivation of ^1D and upward radiative transitions are negligible, every recombination to the singlet system of C^0 leads to a $\lambda9850$ or a $\lambda9824$ photon. The emission rate of $\lambda9850$ through the column would then be

$$E(9850) = \frac{1}{4} N_e N(C^+) \, \beta(T) \, L \, \frac{A(9850)}{A(9850) + A(9824)} = 130 \text{ Ray},$$

an amount that accounts well for the observed strength of the line and could be further reduced to take extinction due to dust into consideration.

Based on the observational evidence presented above, we now think that the metastable levels of C^0, which lead to the emission of the observed [C I] lines, are populated by the recombination of C^+ in high-density regions, predominantly neutral, beyond but near the outer boundaries of H II regions.

SUMMARY

A map of Orion A in radiation of [C I] $\lambda9850$, with a resolution of 80 arcsec, down to a level of intensity 20 Ray is presented. The brightness distribution in [C I] appears to be flatter than that in Hα, with the apparent intensity ratio [C I/Hα] varying between 6.0 and 0.4 $\times$ 10^{-3}. A similar map of Orion B, with a resolution of 2 arcmin, down to a level of 7 Ray, is shown to demonstrate that the visually conspicuous central obscuration lane, behind which the maximum continuum radio emission arises, appears to be bright in [C I]. A high-resolution profile of [C I] $\lambda9850$ in Orion A shows that its FWHM is 6.1 km s^{-1}, significantly smaller than that of any other nonradio nebular line. With the proven inefficiency of charge exchange in building up the concentration of C^0 in the H I–H II transition layers, where electron collisional excitation may occur, on the basis of the observational results, we consider that, most likely, the [C I] emission arises either by recombination in cool C II regions or by being excited by shocks.

REFERENCES

1. HIPPELEIN, H. & G. MÜNCH. 1978. Astron. Astrophys. **68:** L7.
2. STEIGMAN. 1975.
3. PÉQUIGNOT, AD., S. M. V. ALDROVANDI & G. STASINSKA. 1978. Astron. Astrophys. **63:** 313.
4. BUTLER, S. E. & A. DALGARNO. 1980. Astron. Astrophys. **85:** 144.
5. HIPPELEIN, H. & G. MÜNCH. 1981. Mitt. Astron. Ges. **54:** 193.
6. JAFFE, D. T. & V. PANKONIN. 1978. Astrophys. J. **226:** 869.
7. PANKONIN, V., C. M. WALMSEY, T. L. WILSON & P. THOMASSON. 1977. Astron. Astrophys. **57:** 341.

QUASI-STELLAR BUT NEBULAR CONDENSATIONS
IN THE CORE OF THE ORION NEBULA

Jean-Louis Vidal

Observatoires du Pic-du-Midi et de Toulouse
65200 Bagnères-de-Bigorre, France

INTRODUCTION

A new kind of nebular condensation has been recently discovered by Laques and Vidal in the core of the Orion Nebula.[1,2] First, I shall describe the present state of observational knowledge about them, mainly deduced from bidimensional photometry in some emission lines. Second, I shall look for the various possible explanations for such phenomena. Finally, I shall point out the usefulness of high spatial and spectral resolution observations, from ground and space, in order to specify the physical conditions in such condensations and to describe their future evolution.

THE PRESENT STATE OF OBSERVATIONAL KNOWLEDGE
ABOUT THE CONDENSATIONS

Thanks to a set of electrographic plates, made with the 106-cm reflector of the Pic-du-Midi observatory, taken with narrowband ($\sim$10 Å FWHM) in various emission lines (Hα,Hβ, [O III] λ5007 Å, [N II] λ6584 Å, [S II] λ6717 Å, and λ6731 Å) and the continuum, six quasi-stellar condensations have been identified in the core of the Orion Nebula.[1,2] For comprehensive information we refer the reader to the original paper;[2] here, we shall give only a brief summary of this work, *i.e.,* that necessary to fully understand the problem.

These condensations, which are located very close (4 to 8″) to the brightest star of the Orion trapezium (*i.e., θ_1*C Ori, $m_v \simeq$ 4.7) have been clearly differentiated from the faintest stars in the same region; as a matter of fact, they are quite visible on the Hα, Hβ, and [O III] plates, but invisible on the [N II], [S II], and continuum plates. Moreover, thanks to good seeing and the narrowness of the band, the condensations are well isolated from the wings of the spread function of stellar images (and especially those of θ_1C Ori); see FIGURE 1.

These objects do not look like anything seen before, such as Bok's globules, Herbig-Haro objects, or T Tauri stars.

Among other characteristics, their dimensions (ϕ = 1.4″ FWHM on the best plates till 1978, then 1″.0 on more recent [O III] plates taken at Pic-du-Midi and at the CFHT, in Hawaii, *i.e.,* a diameter not significantly different from those of stellar images) make them look like stars. On the other hand, except for one case (at 20 μm), which seems to be a projection effect, no coincidence with an infrared source has been found.

The fluxes of the condensations in the Hα, Hβ, and [O III] λ5007 Å lines have been measured and calibrated thanks to previous spectrophotometric work by Peimbert and

176

0077-8923/82/0395-0176 $1.75/1 © 1982, NYAS

Costero in a neighboring nebular spot.[3] A maximum value has been determined for the brightness in the [N II] and [S II] lines ([N II]/Hα < 4.10^{-2} and [S II]/Hα < 5.10^{-3} for each of the two sulphur lines).

After correction for absorption (from the theoretical Hα/Hβ ratio and the interstellar reddening law for the Trapezium stars), we get the emitted intensities in the three lines.

The electron density, calculated from the Hα intensity and the measured diameter, assuming a standard 8500 K temperature, ranges from 1.5×10^5 cm^{-3} to 4.1×10^5 cm^{-3}, figures that must be considered minimum values.

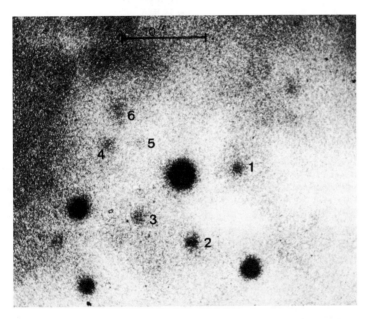

FIGURE 1a. An enlargement of an electrograph of the Trapezium region in Orion in the Hα line, a 3-min exposure. The condensations numbered 1–6 are clearly visible (condensation 6, which was suspected to be double on the basis of this plate, was found to be actually double on more recent plates).

From the [O III]/Hα ratio, which is much less (0.5–0.75) in the condensations than in the general nebular medium (1.23), and after a discussion of the ionization degree and the temperature of these objects, we find much higher values for N_e, from 2.2×10^6 cm^{-3} to 3.5×10^6 cm^{-3}.

At this point, we must say that the previous assumptions concerning the temperature of the condensations (T_e at least equal to T_e of the general medium, due to the high density that enables collisional deexcitation, thus reducing the cooling rate due to the forbidden lines) seem to be correct. Very recent spectrographic observations I have done at ESO in Chile (with the image dissector scanner at the 152-cm telescope), which are still rough and not yet thoroughly reduced, give a preliminary indication of

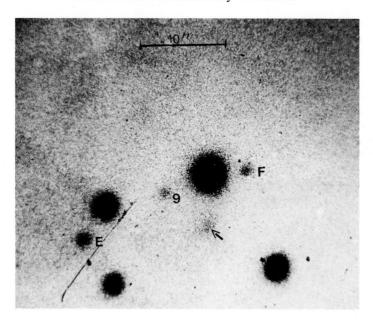

FIGURE 1b. An enlargement of an electrograph of the Trapezium region in Orion in the [N II] λ6584 Å line, a 4-min exposure. The condensations are invisible. On the other hand, the well-known stars E and F (the second being at the limit of detection on the original Hα plate) and a fainter star, numbered 9, are clearly visible. (The arrow points to a ghost of θ_1C due to the interference filter.)

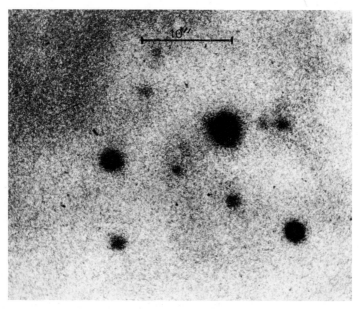

FIGURE 1c. An enlargement of an electrograph of the Trapezium region in Orion in the [O III] λ5007 Å line, a 1-min exposure. Condensation 1 and star F are clearly separated by 2″.

n electron temperature in the condensations 1500 K to 2000 K higher than in the general medium. From these observations, I hope I will be able, also, to directly derive the electron density from the [Cl III] doublet λ5518 Å and λ5538 Å.

Finally, the ionization and recombination analysis leads us to suggest a model for a globule that would be partially ionized, the mean characteristics of which can be deduced; they are those of condensation 1, according to FIGURE 2.

A neutral globule is progressively eaten away by an entering ionization front, while the ionized and ejected matter becomes progressively diluted in the general nebular medium, the density of which is much lower ($N_e \simeq 10^4 \, cm^{-3}$). The values of the electron density, the Hα intensity, and the width that can be ionized allow us to calculate the radii of the neutral globule and its ionized envelope. From a dynamical point of view, such heterogeneities, had they been entirely ionized, would not have had a long lifetime and the probability of observing them would have been nearly zero.

Recently, Franco and Savage have explained the velocity blueshifts ($\sim$25 km s^{-1}) of the high ionization lines (C IV and Si IV) in the uv spectrum of θ_1C Ori; they theorize

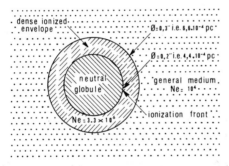

FIGURE 2. A schematic representation of the proposed model for condensation 1. We must emphasize that the outer circle is not, of course, a physical boundary. Rather, the ionized gas becomes progressively more diluted from $N_e = 3 \times 10^6 \, cm^{-3}$ (or more) to $N_e = 10^4 \, cm^{-3}$, *i.e.*, the density of the general nebular medium.

that they are produced through the acceleration of high-density globules in the core of the nebula by the stellar wind of θ_1C.[4]

VARIOUS POSSIBLE EXPLANATIONS

As discussed by Collin-Souffrin, the real nature of such globules is still completely unknown:[5] Are they clumps in the process of condensation by thermal or gravitational instability or slowly evolving relics of the primeval H I medium?

Stable (or Slowly Evolving) Remnants of the Primeval H I Medium

This is the hypothesis we had stated first, grounded upon theoretical work by Dyson, who looked—at a time when molecular clouds were almost unknown—for the explanation of the permanence of supersonic turbulence in the nebular medium, turbulence that would be damped very rapidly without any energy source.[6]

Even though such structures have not been necessary to explain the observed dynamical phenomena since the observation of molecular clouds near H II regions stimulated the calculation of new dynamical models, such as the "Champagne model,"[7-9] such globules, as conceived by Dyson, may nevertheless exist. The figures given by Dyson for the radius are from 10^{15} cm to 10^{17} cm (our diameters are in the range of 10^{15} cm) and for the density of the ionized matter are from 10^5 to 10^6 cm^{-3} (we have deduced densities in the range of 10^6 cm^{-3}).

For gravitationally stable globules, the evolution time (time at the end of which the mass loss by ionization of the neutral globule is 50%) is 2–8×10^4 y, consistent with an age of the Orion nebula in the range of 10^5 y.

Unstable Condensations

Gravitational Instability

Dyson himself has shown that such globules may be gravitationally unstable. However, with his working hypothesis (D-critical front, neutral globule described as an isothermic sphere of gas), the calculated masses (from $0.1\ M_\odot$ to $0.5\ M_\odot$) collapse in very short times, something like $10^3\ M/M_\odot$ y. Under these circumstances, if the condensations pre-existed in the neutral medium, before the formation of the Trapezium stars, it is quite unlikely that those globules which are observed today are stars in formation. Nevertheless, it is not unreasonable to think that these stars, situated in the core of the nebula, were born from similar globules, but with a greater initial mass, in order to let the collapse begin.

However, if another type of instability (see below) appeared later in the medium, the age of these globules would be, perhaps, much less than that of the nebula and then it is not impossible that a gravitational collapse would be in process.

Thermal Instability

Another type of instability, first conceived by Field to explain the two phases in the interstellar medium,[10] could be mentioned here, but with different dimensions as well as temperatures, densities, and durations.

As we know, the physical origin comes from the shape of the $\log p - \log \rho$ curve, in which there appears an unstable region and comparatively rapid transition toward stable states.

This mechanism would have to be studied in detail, in the case presented here, to see if it is, actually, possible.

Layer Instability

Around an H II region of low density, which is ionized by a massive young star, stands the denser H I region, which is compressed and pushed outward by the radiation pressure. The acceleration of a dense layer above a lighter one is an unstable situation.[11] All the small perturbations grow into the so-called "elephant trunks," as seen in the Rosette Nebula. These trunks are unstable themselves and must disintegrate into several round blobs or globules of dense H I, which then contract further under their own gravity plus the pressure of the surrounding H II.[12-14]

USEFULNESS OF HIGH SPATIAL AND SPECTRAL RESOLUTION OBSERVATIONS IN THE VISIBLE AND THE UV

To improve the physical description of such condensations, one has to gather more information (we have seen above that we have, now, relatively few data) by bidimensional and spectrographic high-quality observations. First, an independent determination of diameters, temperatures, and densities should be looked for. Second, knowledge of the velocity field in the ionized part of the globule would be very valuable.

Diameters

Above, we said that, on our best plates, the diameters of the condensations are 1.0″ (FWHM), not significantly different from star diameters, and, as a matter of fact, condensation 6 appears to be double on these new electrographs.

From ground-based observations, it is possible to do better. At Pic-du-Midi, Aurière succeeded in getting plates of central regions of globular clusters (M3 and M15) with a 0.5″ resolution.[15] Recent observations with the new 2-m telescope at Pic-du-Midi and with the 3.6-m in Hawaii are still more promising. With the image photon counting system and recentering techniques, it seems possible to get 0.3″ resolution and maybe better.

With the space telescope the aim is 0.1″.

Temperatures and Densities

The intensity ratios of some well-known emission lines, determined with good spectrophotometric equipment at a good site (*i.e.,* it separates, as far as it is possible, the light coming from the condensation itself from the light coming from the nearby nebular general background) will be the basis of a complete physical analysis. In the visible range, the [O III] lines at $\lambda4363$ Å and $\lambda5007$ Å are the most interesting for temperature determinations, and the [Cl III] lines at $\lambda5518$ Å and $\lambda5538$ Å (since the [O II] ($\lambda3726-3729$ Å) and [S II] (at 6717-6731 Å) lines are too faint or perhaps, absent) for density determinations.

As far as the uv is concerned, the ionization structure and physical conditions can be examined through a number of rather strong collisionally excited emission lines, some of them from ions unobservable in the visible: C II, C III, C IV, N III, Ne III, Mg II, Si II, Si III, S III, and S IV.[16] Particularly, the temperature-sensitive [O III] ratio $[I(\lambda 5007) + I(\lambda 4959)]/I(\lambda 4363)$ can be supplemented by [O III] $I(\lambda 1661) + I(\lambda 1666)$, which arise from the ^5S level. Other ions with temperature-sensitive ratios in or including the ultraviolet are [Ne III] $[I(\lambda 3697) + I(\lambda 3869)]/I(\lambda 3342)$ (much better observed from space than from ground) and Si II $I(\lambda 2335)/I(\lambda 1817)$. Some ratios are sensitive to both temperature and density, such as C III] $\lambda 1907/\lambda 1909$, which needs a good spectral resolution for detection.

Velocity Field Determination

The profile of the brightest emission lines and their shift should be determined with a high resolution (at least $R = 1.2 \times 10^5$, *i.e.*, 2.5 km s^{-1}, which is the highest resolution of the high-resolution spectrograph of the space telescope), in order to determine the velocity field, the Doppler effect being the most important phenomenon causing the broadening of nebular lines. This velocity field analysis of the ionized gas in the condensations must allow one to discriminate between the two kinds of globules mentioned above, that is, the stable ones and the unstable ones. For this point, too, high spatial resolution is absolutely necessary in order to get rid of the nebular background as much as is possible.

REFERENCES

1. LAQUES, P. & J.-L. VIDAL. 1978. Uppsala Astron. Obs. Rep. **12**: 11.
2. LAQUES, P. & J.-L. VIDAL. 1979. Astron. Astrophys. **73**: 97.
3. PEIMBERT, M. & R. COSTERO. 1965. Bol. Obs. Tonantzintla Tacubaya **5**: 3.
4. FRANCO, J. & B. D. SAVAGE. 1981. Preprint.
5. COLLIN-SOUFFRIN, S. 1979. Int. Astron. Union Symp. 54, discussion of a paper by D. E. Osterbrock. M. S. Longair, Ed.: 116. NASA.
6. DYSON, J. E. 1968. Astrophys. Space Sci. **1**: 388.
7. TENORIO-TAGLE, G. 1979. Astron. Astrophys. **71**: 59.
8. BODENHEIMER, P., G. TENORIO-TAGLE & H. W. YORKE. 1979. Astrophys. J. **233**: 85.
9. TENORIO-TAGLE, G. H. W. YORKE & P. BODENHEIMER 1979. Astron. Astrophys. **80**: 110.
10. FIELD, G. B. 1965. Astrophys. J. **142**: 531.
11. SPITZER, L. 1974. Astrophys. J. **120**: 1.
12. EBERT, R. 1955. Z. Astrophys. **37**: 217.
13. BONNOR, W. B. 1956. Mon. Not. R. Astron. Soc. **116**: 351.
14. VON HOERNER, S. 1975. *In* Lectures Notes in Physics 41, H II Regions and Related Topics. T. L. Wilson and D. Downes, Eds.: 53. Springer-Verlag. New York.
15. AURIERE, M. 1979. *In* ESA/ESO Workshop, Geneva, Astronomical Uses of the Space Telescope. F. Maccheto, F. Pacini, and M. Tarenghi, Eds.: 177.
16. OSTERBROCK, D. E. 1979. Int. Astron. Union Symp. 54, M. S. Longair, Ed.: 99.

MEASUREMENTS OF MAGNETIC FIELD STRENGTHS
IN THE VICINITY OF ORION

Carl Heiles

Department of Astronomy
University of California, Berkeley
Berkeley, California 94720

Thomas H. Troland

Department of Physics and Astronomy
University of Kentucky
Lexington, Kentucky 40506

INTRODUCTION

The Orion Nebula and its very dense molecular clouds are associated with two much larger, less dense CO clouds that extend to the north and south (discussed in detail in References 1 and 2) and an even larger H I envelope (discussed in detail in Reference 3). The very dense molecular clouds and young stars responsible for the interest in Orion presumably contain matter that was originally part of these larger, less dense CO and H I clouds.

During the gravitational contraction that forms denser objects like these there are three significant forces: gravitation, gas pressure, and magnetic pressure. We know very little about the last force because magnetic fields are notoriously difficult to observe. The results reported in this paper are some of the first results of a long and arduous attempt to perform such measurements.

OBSERVATIONS

We used the Hat Creek 85-ft telescope to detect the Zeeman splitting of the 21-cm line seen in emission at two positions near the Orion Nebula. An overview of our original experimental technique and equipment is given in a separate paper.[4] However, after that paper was written, the electronics on the telescope were very greatly improved by the addition of dual-channel cooled FET amplifiers with system temperatures of 50 K, a motor-driven zero-loss polarization switch, and a 1024-channel, 3-level digital correlator. Without these improvements, the observations reported herein could not have been done.

We observed two closely-spaced positions located at $(l, b) = (209.6, -20)$ and $(210.35, -20)$. The two positions both lie within 1.5° of the Orion Nebula itself, and close to or within the southern CO cloud discovered by Kutner *et al.*[2] The latter position lies within the 1 K ^{13}CO contour of Kutner *et al.*,[2] the former lies about $\frac{1}{4}$ degree outside this contour.

FIGURE 1 shows the results for both positions. The results consist of two types of spectrum, a polarization-switched spectrum and a conventional frequency-switched spectrum. The latter is simply the ordinary 21-cm line profile. The former is the

183

0077-8923/82/0395-0183 $1.75/1 © 1982, NYAS

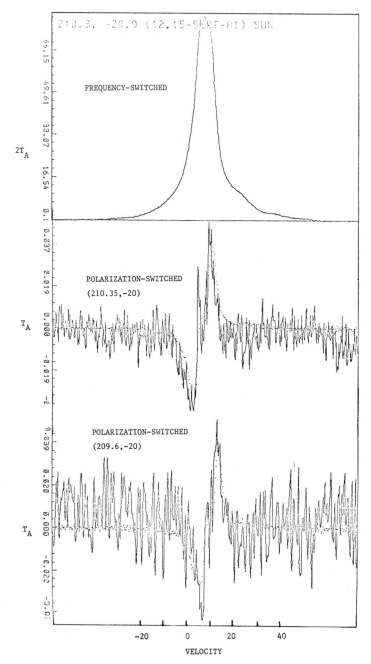

FIGURE 1. Frequency- and polarization-switched profiles for the two positions. Only one of the frequency-switched profiles is shown because they are nearly identical.

difference between right- and left-hand circular polarizations. In the presence of a magnetic field, the former spectrum shows a "Zeeman pattern," that is, the derivative of the ordinary line profile with the amplitude proportional to the line-of-sight component of the field; the sign of the Zeeman pattern depends on the direction of the field.

FIGURE 1 shows clear detections of the magnetic field. In both positions the field strength is $+10$ μG; the positive sign indicates that the field points away from the observer.

In the Zeeman pattern for (210.35, -20), there is an additional narrow component. There is no obvious counterpart of this component in the line profile. To search for this counterpart, we obtained a position-switched spectrum by subtracting the frequency-switched spectra of the two positions. This is shown in FIGURE 2. The counterpart is clear in the position-switched spectrum, showing up as a narrow bump.

We have derived the properties of the narrow component by hand-fitting a baseline and Gaussian parameters to the position-switched spectrum. The derived properties of this assumed Gaussian are as follows: amplitude, 2.8 K; FWHM, 1.9 km s^{-1}; velocity, 7.4 km s^{-1}. The magnetic field strength is -11 μG. There is a large uncertainty in this result because of the position-switched spectrum; it was derived from only these two positions, which are relatively far apart. We will soon derive a better position-switched spectrum by obtaining more data; the present result should be considered provisional in the absence of a better position-switched spectrum.

Note that the direction of the field is opposite that of the field in the wider component that was discussed above.

DISCUSSION

The Field in (209.6, -20) and the Wide Component of (210.35, -20)

The Orion Nebula is surrounded by a large H I cloud that Gordon has discussed in detail.[3] The apparent shape of this cloud is elliptical, with diameters of 30 and 14 degrees parallel and perpendicular to the Galactic plane, respectively. At the 500 pc distance of Orion, these angles amount to about 250 and 125 pc in linear measure. Gordon gives the mass of the cloud, including helium, as 9×10^4 $M_\odot$; the volume density is about 2.5 H I atoms cm^{-3}. (There is additional mass at the center of the cloud in the form of H_2 and H II amounting to about 20×10^4 $M_\odot$.)

With a line-of-sight field of 10 μG, the magnetic energy density is at least 4×10^{-12} erg cm^{-3}. This is two orders of magnitude higher than the gas pressure, which is about 2.8×10^{-14} erg cm^{-3} if the gas temperature is 80 K, as indicated by a few H I absorption measurements of extragalactic sources in the vicinity.[5] The 21-cm line halfwidth is about 11 km s^{-1}, much wider than the thermal width. If this width results from small-scale isotropic turbulence, the kinetic energy density is about 2.7×10^{-12} erg cm^{-3}, which is comparable to but still less than the lower limit on magnetic energy density. The magnetic energy density exceeds the kinetic energy density in this H I cloud.

We now consider the equilibrium of the H I cloud by applying the virial theorem. We use Spitzer's equation 11–26 for a uniform spherical cloud that is uniformly

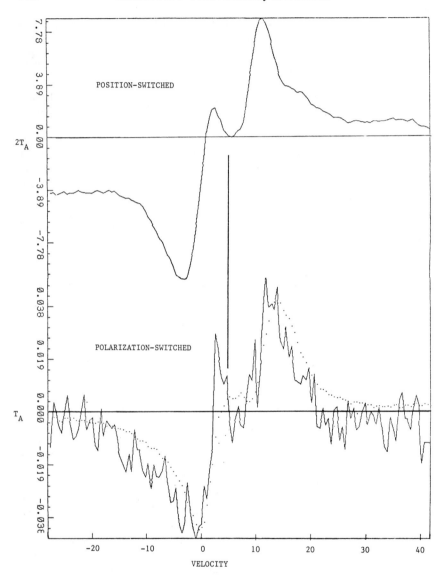

FIGURE 2. Position- and polarization-switched profiles for (210.35, −20). The position-switched profile is reversed: (209.6, −20) − (210.35, −20), instead of the other way around.

magnetized inside and has a dipole field outside.[6] In this model, we assume that the radius of the cloud is 100 pc; this is obtained by assuming that the diameter of the cloud along the line of sight is the same as it is in longitude and by taking the geometric mean of the three diameters. The mass of the H I cloud is thus calculated to be 10^6 $M_\odot$. The discrepancy with Gordon's derived mass of 9×10^4 $M_\odot$, eleven times smaller, results from the fact that Gordon assumed a Gaussian density distribution in the cloud,[3] while we assume a uniform density distribution. The energies involved with the various terms in the virial equation are given in TABLE 1. The gravitational term for the H I cloud includes contributions from both the H I cloud alone and from the H_2 and H II clouds at the center of the H I cloud.

Gravity is insufficient to hold this H I cloud together. Both the macroscopic kinetic energy and the magnetic energy terms are individually larger than the gravitational term, and their sum is at least three (the magnetic term is only a lower limit) times larger than the gravitational term. While, at some time in the distant past, the H I cloud must have collapsed to form the molecular clouds, which subsequently formed the stars we see today, the H I cloud now seems to be unstable to expansion.

We cannot be absolutely certain of the above conclusion, because it neglects the possibility that the cloud is still contracting. If a significant fraction of the 21-cm linewidth results from contracting motions of the cloud, then as the cloud collapses, some of the kinetic energy will be lost to radiation and the net energy balance in the cloud may remain favorable to further contraction.

The Narrow Component of (210.35, −20)

This position is located within the southern CO cloud of Kutner et al.[2] Physical quantities in the ensuing discussion are taken from this paper. This cloud has been mapped in the OH 18-cm main lines by Baud and Wouterloot[7] and has a velocity of about 7 km s^{-1} near the position we observed. Since the velocity of the narrow peak on the position-switched profile of FIGURE 2 is 7.4 km s^{-1}, we identify this peak with the molecular cloud. The column density of H I is about 1.0×10^{19} cm^{-2} and, at our position, the column density of H_2 is about 2×10^{21} cm^{-2}; thus, the ratio (H I/H_2) is about 0.005.

The H_2 volume density in the cloud is about 800 cm^{-3} and the temperature is probably about 30 K, so the gas pressure is about 3.3×10^{-12} erg cm^{-3}. With a line-of-sight field of 11 μG, the magnetic energy density is at least 4.8×10^{-12} erg cm^{-3}, at least 50% larger than the gas pressure. However, the line halfwidth is 1.9 km s^{-1}, which is significantly larger than the thermal width; if this width results from small-scale isotropic turbulence, the kinetic energy density is about 54×10^{-12} erg cm^{-3}, more than an order of magnitude larger than the lower limit on the magnetic energy density.

We apply the same form of the virial theorem to this cloud that we used above. The

TABLE 1

ENERGIES ASSOCIATED WITH TERMS IN THE VIRIAL EQUATION

Term	H I Envelope (10^{50} erg)	CO Cloud (10^{47} erg)
Thermal	0.3	0.4
Macroscopic motions	13	1.6
Magnetic	>12	>0.4
Gravitational	8.5	4.8

CO cloud is highly elongated. Since our form of the virial theorem assumes a spherical cloud, we take the diameter to be equal to the width of the cloud, which we adopt as 7 pc. Our estimate of the macroscopic kinetic energy assumes that the linewidth results from small-scale isotropic turbulence; this contribution dominates rotation by more than an order of magnitude. The energies involved with the various terms in the virial equation are given in TABLE 1.

The gravitational energy term dominates over each of the others individually, and over their sum by a factor of two. This cloud is unstable to gravitational collapse unless the magnetic term is larger than the lower limit by nearly an order of magnitude.

It is quite likely that the magnetic term is, in fact, ten times the lower limit. The magnetic term varies as the square of the magnetic field strength. Thus, if the line-of-sight component of the field is less than one-third the field, then the magnetic term is nine times larger than the lower limit listed in TABLE 1. It is a straightforward and elementary exercise to show that if the ratio of the line-of-sight component to the full field strength is $\leq X$, then the probability that the field is oriented to satisfy this condition is itself $= X$. Thus, with a probability of $1/3$, the field strength in the CO cloud is at least three times the line-of-sight component and strong enough to dominate over gravity and keep the cloud from collapsing. There are other uncertainties, such as the degree to which the field is uniform in the CO cloud and how closely the boundary conditions approximate our assumptions in using the virial theorem. On the whole, it seems quite reasonable to assume that this CO cloud is one case in which the tendency toward gravitational collapse is resisted by magnetic forces.

General Comments

There exists one additional measurement of magnetic field strength in the Orion region: the $-50\ \mu G$ result in the 7 km s^{-1} H I component seen in absorption against the Orion Nebula (see the review by Verschuur[8]). Even though the velocity of this absorption component is nearly the same as that of our detection in the CO cloud, it is unlikely that the two results refer to the same cloud. There are velocity gradients in all the gas clouds in this region and the angular separation of the two measurements is about 1.5°. Furthermore, the aperture synthesis map of Lockhart and Goss shows that the 7 km s^{-1} absorption component occurs in the northeastern part of the Orion Nebula;[9] our position is located toward the south. Nevertheless, Verchuur's absorption result certainly refers to a relatively cold, dense region.

Both Verschuur's absorption result and our result in the CO cloud show the field pointing toward the observer. Both regions are much denser than the surrounding H I cloud, in which the field points in the opposite direction. This reversal is curious, and needs to be investigated by making measurements at more positions. For the CO cloud, our measurement was made in the 21-cm line and we cannot be sure that the H I is uniformly mixed with the molecules; it may instead exist in a sheath on the boundary of the molecular cloud. We intend to make more measurements, both in the 21-cm line and in the 18-cm OH lines, to investigate these points.

According to theoretical calculations, reversals in cloud rotation and field direction can be induced by the transfer of angular momentum by magnetic forces.[10] Such reversals in rotation direction in the centers of two other molecular clouds have already been detected.[11,12] No reversals are detected within the Orion CO cloud, even though the magnetic forces seem to be strong enough to produce them. However, since the existence of such reversals is time dependent, their absence at the present time does not necessarily mean that they did not occur in the past or will not occur in the future. The fact that the magnetic field direction in the CO cloud is opposite that in its surroundings may be indicative of previous or future reversals in rotation direction.

Both the CO cloud and the much larger H I envelope are rotating in roughly the same direction, opposite that expected from differential Galactic rotation. This argues against the possibility that the reversal of the CO cloud rotation, relative to the Galactic rotation, has been caused by magnetic forces. It is also unlikely that the reversal of the H I cloud rotation, relative to the Galactic rotation, has been caused by magnetic forces; the timescale for reversing the rotation of such a large cloud is long, and the interior of the cloud should reverse to a larger extent than the exterior, which does not seem to be the case.

These results have relevance on a Galaxy-wide scale. There are many molecular clouds in the galaxy that are apparently unstable to gravitational collapse. If they were to collapse and form stars, the Galaxy would soon contain many more very young stars and many fewer gravitationally unstable molecular clouds. Either there is an agent that prevents the collapse, or the Galaxy will soon change character in a spectacular way. Magnetic fields have sometimes been invoked as the agent that prevents collapse, but the difficulty of measuring these fields has left this hypothesis untested. Our new capability to measure the field strength in molecular clouds allows us to finally begin the investigation of this question.

Summary

We have measured the magnetic field strength in two regions associated with the Orion Nebula. One region is the large H I cloud that envelops the ionized and molecular clouds; the field strength is 10 μG. The other is the large CO cloud just south of the Orion Nebula discovered by Kutner et al.,[12] the field strength there is 11 μG. The directions of the two fields are opposite. These results are discussed using the virial theorem. In the H I cloud, gravity appears to be weaker than the kinetic and magnetic forces, making this cloud unstable to expansion. In the CO cloud, gravity appears to dominate, unless the magnetic field is at least three times larger than the measured line-of-sight component; the probability for this occurrence is one-third. It is likely, then, that magnetism prevents collapse and star formation in this CO cloud.

[NOTE ADDED IN PROOF: Further measurements taken after the symposium show that the position-switched spectrum presented in FIGURE 2 is erroneous. The only major effect on the discussion is that the field in the molecular cloud points in the same direction as that in the surrounding H I cloud, not in the opposite direction, as stated in the text. There are, in addition, other minor effects, the most important of which are a small change in the magnetic field strength in the molecular cloud and a revision of some of the numbers in TABLE 1. We will present these latest results in a forthcoming paper in the *Astrophysical Journal*.]

REFERENCES

1. KUTNER, M. L., N. J. EVANS & K. D. TUCKER. 1976. Astrophys. J. **209:** 452.
2. KUTNER, M. L., K. D. TUCKER & G. CHIN. 1977. Astrophys. J. **215:** 521.
3. GORDON, C. P. 1970. Astrophys. J. **75:** 914.
4. TROLAND, T. H. & C. HEILES. 1982. Astrophys. J. In press.
5. CROVISIER, J., I. KAZÈS & D. AUBRY. 1978. Astron. Astrophys. Suppl. Ser. **32:** 205.
6. SPITZER, L. 1978. Physical Processes in the Interstellar Medium: 242.
7. BAUD, B. & J. G. A. WATERLOOT. 1980. Astron. Astrophys. **60:** 297.
8. VERSCHUUR, G. L. 1974. *In* Galactic and Extragalactic Radio Astronomy. G. L. Verschuur and K. I. Kellerman, Eds.: 194.
9. LOCKHART, I. A. & W. M. GOSS. 1978. Astron. Astrophys. **67:** 355.
10. MOUSCHOVIAS, T. C. H. & E. V. PALEOLOGOU. 1979. Astrophys. J. **230:** 204.
11. CLARK, F. O. & D. R. JOHNSON. 1981. Astrophys. J. **247:** 104.
12. YOUNG, J. S., W. D. LANGER, P. F. GOLDSMITH & R. W. WILSON. 1981. Preprint.

DISCUSSION OF THE PAPER

P. GOLDSMITH (*University of Massachusetts, Amherst, Mass.*): How does the linewidth in the 21-cm feature you identify with the molecular gas compare with the CO linewidth?

HEILES: The H I half-width is 1.9 km s^{-1}, to be compared with the CO width of about 2 km s^{-1}, so they are quite comparable.

ENERGETIC MOLECULAR FLOWS
IN STAR-FORMING CLOUD CORES

J. Bally

Bell Laboratories
Crawford Hill Laboratory
Holmdel, New Jersey 07733

INTRODUCTION

High-velocity outflows of gas have been discovered in many molecular cloud cores that are undergoing an episode of star formation.[1] This phenomonon was first recognized in observations of the Becklin–Neugebauer–Kleinmann-Low (BN-KL) complex in Orion through the presence of broad ($\Delta V \gtrsim 100$ km s^{-1}) wings on the 115 GHz ^{12}CO emission line profile.[2,3] Subsequent observations of other regions of star formation have indicated that energetic molecular outflows are a common and previously unrecognized phase in the birth and early evolution of stars with important dynamical consequences for the evolution of molecular clouds.

Four distinct methods have been used to identify high-velocity molecular flows in star-forming regions. First, the flows can be recognized directly by the presence of broad ($\Delta V > 30$ km s^{-1}) wings on the mm-wavelength emission lines, especially ^{12}CO. Many of the sources so far discovered appear to be extended when observed with the currently available 1–2 arcmin telescope beams. Second, high-velocity H$_2$O maser emission is often seen in regions of star formation with Doppler shifts up to a few hundred km s^{-1}.[4] In some regions (including Orion A), VLBI observations obtained at different epochs have revealed proper motions of individual maser spots that have velocities consistent with their Doppler shifts. Third, shock-excited molecular hydrogen emission in the 2-μm atmospheric window has been seen in regions exhibiting high-velocity CO and H$_2$O masers.[5-7] These emission lines are believed to be formed in dense gas ($n > 10^5$ cm^{-3}) that has been hit by a shock wave moving with a velocity between 10 and 20 km s^{-1}. Fourth, comparison of photographs of several Herbig-Haro objects taken many years apart has revealed proper motions of individual luminous spots. Velocities on the order of 100 km s^{-1} in a direction away from a central star or infrared source have been observed in HH1–2 and L1551.[8]

These methods have revealed the presence of six distinct outflow sources in the molecular clouds associated with the Orion OB association (TABLE 1). The Orion A outflow has been studied in more detail than any other outflow source.[9] The small angular size and velocity extent of this flow make it unusual when compared with other high-velocity outflow sources. The flow associated with the NGC 2071 complex in the northern part of Orion has properties that are more typical.[10,13]

THE KINEMATICS OF THE FOWS

Most flows that have been resolved in the mm-wavelength emission lines show bipolar structure. In the case of NGC 2071, blueshifted gas is seen on one side of the

191

0077–8923/82/0395–0191 $1.75/1 © 1982, NYAS

TABLE 1

HIGH-VELOCITY GAS FLOWS IN THE ORION MOLECULAR CLOUD

Source	Position α_{1950}	Position δ_{1950}	Detection Method*	Ref.	Flow Parameters Velocity (km s^{-1})	Flow Parameters Angular Size	Flow Parameters Mass ($M_\odot$)
Orion A	05^{h}32^{m}47^s	−5 °24′29″	CO, H$_2$O, H$_2$	9	75	0.8′	5
NGC 2071	05^{h}44^{m}30^s	00°20′40″	CO, H$_2$	6, 10	35	3′ × 5′	20
HH25–26	05^{h}43^{m}40^s	−00°06′36″	CO	11	15	3′ × 4′	3.1
NGC 2024	05^{h}39^{m}14.0^s	−1°55′59″	CO	1	18	1′	?
OMC 2	05^{h}32^{m}59.6^s	−05°11′32″	CO, H$_2$	7	10	0.5′	0.5
HH1–2	05^{h}33^{m}35^s	−6°47′25″	PM, H$_2$	8, 12	300	0.3′ × 2.5′	?

*PM: Optical proper motions.

source while redshifted gas dominates the opposite side. In the source L1551, the red- and blueshifted flow define two oppositely directed streams with an aspect ratio (length/width) of around 7, corresponding to an opening angle of around 17° if the streams are not foreshortened by geometrical projection effects. The morphology and velocity field of HH1–2 also shows bipolar structure in the optical part of the spectrum; in this source, the opening angle is even smaller than that in L1551. The bipolar morphology of the high-velocity gas in star-forming cloud cores has a superficial resemblance to the jets seen in radio galaxies. In the molecular flows, the opening angles are seen to be larger ($10° < \theta < 90°$) and the velocities have Mach numbers in the range $10 < M < 100$.

Not enough is known at present to determine the cause of the bipolar structure. Snell et al. have suggested that the flow in L1551 is constrained by an accretion disk.[14] A large-scale disk has been observed in CS in NGC 2071 by Bally.[10] On the other hand, anisotropic outflow from a central source is not ruled out; Cohen et al. have observed an axially symmetric radio source at the core of L1551 that can be interpreted as an anisotropic stellar wind.[15]

The flow velocity in most sources exceeds the local gravitational escape velocity. The total mass of material inside the region affected by the flow can be estimated from the column density of molecules within the emission line cores of species such as ^{13}CO, $C^{18}O$, and CS. Alternatively, the cloud core mass can be derived using the virial theorem applied to the width (at the half-power point) of the spectral lines. In NGC 2071, both methods indicate that the total mass of the cloud core is, at most, a few hundred solar masses (compared to 20 $M_\odot$ of gas within the high-velocity flow). This core mass is between one and two orders of magnitude too small to gravitationally bind the high-velocity ($v \simeq 35$ km s^{-1}) gas located 1–2 arcmin from the flow center. (See FIGURE 1, which shows a CO map of the high-velocity flow in NGC 2071.) Similar arguments can be made for all sources having $\Delta V \gtrsim 30$ km s^{-1} that have been extensively mapped in CO (Orion A, L1551, AFGL 490, Ceph A, AFGL 961, HH7–11, and NGC 2071). Therefore, the high-velocity gas in molecular cloud cores is outflowing.

The mass, energy, and momentum of the high-velocity gas need to be determined in order to investigate the energetics of these flows. The mm-wavelength observations have been used to determine these parameters from the column density of gas in the high-velocity line wings. The parameters of NGC 2071 are typical: $M_{flow} = 20\ M_\odot$, $P_{flow} = 200\ M_\odot\ \text{km s}^{-1}$, and $E_{flow} \simeq 10^{47}$ erg (see Reference 1).

THE CENTRAL SOURCES

All known high-velocity flows contain near-infrared sources at their centers. Luminosities range from 20 $L_\odot$ for L1551[16] to over $10^5\ L_\odot$ for DR21. The near-infrared spectra of these objects often reveal the Bα or Bγ hydrogen recombination lines.[17] Some of these sources exhibit weak ($S_\nu \simeq$ few mJy) radio continuum emission at 5 GHz.[15,18] NGC 2071 contains a pair of ultracompact 5 GHz continuum sources that coincide in position with the 10-μm infrared sources IRS1 and IRS3 discovered by Persson et al.[19] A comparison of the infrared recombination line fluxes with the 5-GHz radio continuum for the source AFGL 961 gives $I_{B\alpha}(10^{-12}\ \text{erg cm}^{-2}\ \text{s}^{-1})/S_{5GHz}$ (mJy) = 10. Such a large ratio indicates either that the 5-GHz radio emission is very optically thick or that the emission is formed in an ionized stellar wind.[10,11] If the second explanation is correct, stellar wind parameters around $dM/dt \simeq 10^{-5}\ M_\odot\ \text{y}^{-1}$ and $V_w \simeq$ few 10^3 km s^{-1} are deduced from momentum conservation if the wind drives

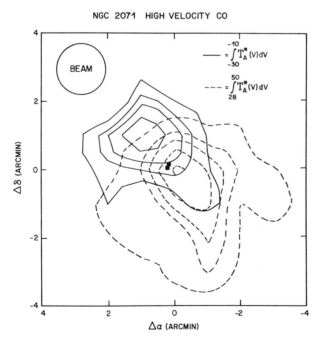

FIGURE 1. A map of the high-velocity ^{12}CO emission in NGC 2071 showing contours of the line intensity integrated over the ranges indicated. Coordinates of the $\Delta\alpha = 0$, $\Delta\delta = 0$ position are $\alpha_{1950} = 05^h44^m30^s$, $\delta_{1950} = 00°20'40''$.

the CO flow. The existence of such powerful winds, especially around low-luminosity pre-main-sequence objects is surprising. The currently available data suggest that the central sources in the flows are extremely young stars that eject mass at a high rate.

BURIED ENERGETIC OUTFLOWS AND THE SUPPORT

OF THE ORION MOLECULAR CLOUDS

Two related problems have plagued our understanding of molecular clouds. (1) What prevents clouds from collapsing in a free-fall time $\tau \simeq 1/(G\rho)^{1/2} \simeq 10^6$ y? (2) How can the observed supersonic mm-wavelength linewidths ($\Delta V \simeq 2$ km s$^{-1} \simeq 10\, C_0$ for $T_{gas} = 10$ K) be maintained? The high-velocity outflows associated with newly formed stars may provide sufficient energy and momentum to prevent gravitational collapse and support the turbulence of molecular clouds. The relevant dynamical parameters of the flows are the force exerted by the flow, $\dot{P}$, the mechanical luminosity of the flows, $\dot{\epsilon}$, and the dynamical lifetime of the flows, τ. These parameters can be determined directly from the observed quantities M (the flow mass), V (the flow velocity), and R (the characteristic flow radius), using the relations

$$\tau = R/V \simeq 10^4 \text{ y},$$

$$\dot{P} = MV^2/R \simeq 0.03 \; M_\odot \, \text{y}^{-1} \, \text{km s}^{-1},$$

$$\dot{\epsilon} = MV^3/2R \simeq 68 \; L_\odot,$$

where the right-hand side corresponds to typical values $M = 10\, M_\odot$, $V = 30$ km s^{-1}, and $R = 10^{18}$ cm. (These numbers are a rough average of those determined individually for Orion A, NGC 2071, and HH24–25.) The total rate of input of momentum and energy into the Orion Molecular Clouds may be about five times larger than the above numbers since there are five flows around luminous buried sources. (HH1–2 is probably associated with a low-luminosity object. It has not been seen in CO and thus may have a lower mass than the other flows.) Assuming that the current situation is typical, a total mean momentum and kinetic energy generation rate for the Orion molecular clouds is

$$\langle \dot{P} \rangle = 0.15 \; M_\odot \, \text{y}^{-1} \, \text{km s}^{-1} \text{ and } \langle \dot{\epsilon} \rangle = 340 \; L_\odot.$$

The total mass of the two Orion molecular clouds is around $M_{\text{Orion}} \simeq 10^5 \, M_\odot$.[22] For a turbulent velocity of 2 km s^{-1}, the total turbulent momentum of the Orion clouds can be generated in 1.3×10^6 y. If there is no dissipation and energy is conserved, then the cloud energy can be generated in 2×10^5 y, less than the free-fall time. A realistic model of cloud support by outflow from young stars needs to consider the swept-up gas flow velocity as a function of distance from the cloud core and include the effects of deceleration by accretion and gravitational forces. For a simplistic picture of a wind-driven bubble in a molecular cloud core of uniform density n_0, the flow velocity is given by $V \propto (\dot{M} V/n_0)^{1/2} r^{-1}$. For the average high-velocity outflow parameters given above, the wind-driven shell decelerates to a velocity of $V = 1$ km s^{-1} at $r \simeq 10$ pc in roughly 10^7 y. It is easy to see how the superposition of many shorter-lived outflow sources can produce the observed motions in clouds.

Norman and Silk have suggested that clouds are supported by T Tauri stellar winds.[23] In their model, the molecular cloud consists of many overlapping wind-generated bubbles. The observations indicate that the energetic outflows that have been observed in molecular cloud cores are the dominant source of bubbles inside molecular clouds. The observed linewidths and support of the molecular clouds are the result of the degraded momentum and of the energy generated by outflows associated with star formation in the past.

SUMMARY

High-velocity molecular gas has been observed in six separate regions in the Orion Molecular Clouds. A detailed mm wavelength study of a region of massive star formation in Orion, NGC 2071, indicates that high-velocity ^{12}CO lines wings are formed in an extended bipolar outflow that originates from the vicinity of a central infrared cluster and a pair of ultracompact 5 GHz continuum sources. The large number of energetic outflows now recognized indicates that an episode of violent mass loss is associated with stellar birth and has important consequences for the early evolution of stars and their associated molecular clouds.

REFERENCES

1. BALLY, J. & C. J. LADA. 1982. Astrophys. J. In Press.
2. ZUCKERMAN, B., T. B. H. KUIPER & E. N. RODRIGUEZ-KUIPER. 1976. Astrophys. J. **209:** L137.
3. KWAN, J. & N. Z. SCOVILLE. 1976. Astrophys. J. **210:** L39.
4. GENZEL, R. & D. DOWNES. 1977. Astron. Astrophys. **61:** 117.
5. FISCHER, J., G. RIGHINI-COHEN & M. SIMON. 1980. Astrophys. J. **238:** L155.
6. BALLY, J. & A. P. LANE. 1982. Astrophys. J. **257:** 612.
7. FISCHER, J. 1981. Ph. D. Thesis. State University of New York, Stony Brook.
8. HERBIG, G. H. & B. F. JONES. 1981. Astron. J. **86:** 1232.
9. KNAPP, G. 1982. This volume.
10. BALLY, J. 1982. Astrophys. J. In press.
11. SNELL, R. L. & S. EDWARDS. 1981. *In* Regions of Recent Star Formation. R. S. Roger and P. E. Dewdney, Eds.
12. ELIAS, J. H. 1980. Astrophys. J. **241:** 728.
13. BALLY, J. 1981. *In* Regions of Recent Star Formation. R. S. Roger and P. E. Dewdney, Eds.
14. SNELL, R. L., R. B. LOREN & R. L. PLAMBECK. 1980. Astrophys. J. **239:** L17.
15. COHEN, M., J. H. BIEGING & P. R. SCHWARTZ. 1982. Astrophys. J. **253:** 707.
16. BEICHMAN, C. A., E. E. BECKLIN & C. G. WYNN-WILLIAMS. 1979. Astrophys. J. **232:** L47.
17. SIMON, T., M. SIMON & R. R. JOYCE. 1979. Astrophys. J. **230:** 127.
18. BALLY, J. & R. PREDMORE. 1982. Astrophys. J. In press.
19. PERSSON, S. E., T. R. GEBALLE, T. SIMON, C. J. LONSDALE & F. BAAS. 1981. Astrophys. J. **251:** L85.
20. PANAGIA, N. & M. FELLI. 1975. Astron. Astrophys. **39:** 1.
21. WRIGHT, A. E. & M. J. BARLOW. 1975. Mon. Not. R. Astron. Soc. **170:** 41.
22. KUTNER, M. L., K. D. TUCKER, G. CHIN & P. THADDEUS. 1977. Astrophys. J. **215:** 521.
23. NORMAN, C. & J. SILK. 1980. Astrophys. J. **238:** 158.

DISCUSSION OF THE PAPER

W. M. IRVINE (*Onsala Space Observatory, Onsala, Sweden*): Although I am not convinced that sulfur-bearing molecules are necessarily chemical tracers of shocks (see the discussion following the paper by J. Welch), it may still be interesting that we (Schloerb, Friberg, Hjalmarson, Höglund, and Irvine, to be published) have detected SO_2 emission from three sources that show one or more of the properties you have mentioned as characterizing these energetic flows: W51, DR21(OH), and W3(OH).

ANISOTROPIC MASS OUTFLOW
IN REGIONS OF STAR FORMATION

L. F. Rodriguez and P. Carral

Instituto de Astronomía
Universidad Nacional Autónoma de México
México 20 D. F., México

P. T. P. Ho

Radio Astronomy Laboratory
University of California, Berkeley
Berkeley, California 94720

J. M. Moran

Harvard-Smithsonian Center for Astrophysics
Cambridge, Massachusetts 02138

INTRODUCTION

Most molecular clouds show carbon monoxide lines that are a few km s^{-1} wide. The detection of very wide CO wings (spreading over ~100 km s^{-1}) near Orion[1,2] indicated that high-velocity gaseous motions could occur in regions of star formation. Recently, mass outflows have been detected in other regions of star formation. Several of the new sources have outflows that are clearly anisotropic, perhaps even antiparallel. The observations show that the blueshifted and redshifted CO emission components are not spatially coincident, but rather arise from two separate regions. Between these two regions there is usually a compact H II region, an H$_2$O maser, or an ir source, which are all indicators of the presence of a recently formed early-type star. Anisotropic mass outflows with these characteristics have been detected in L1551,[3] Cepheus A,[4] and GL 490.[5] Outflows with apparently isotropic distributions have been reported for Orion,[6] GL 961,[7] V645 Cyg,[8] and HL and XZ Tau.[9] However, the apparent isotropy in some of these latter sources may be due to insufficient angular resolution. Finally, for some of these sources—namely, L1551, Cep A, GL 961, V645 Cyg, and HL and XZ Tau—there is evidence at optical wavelengths of shocked gas in the form of Herbig-Haro objects or peculiar nebulosities.

In June, 1979, we made CO observations in the vicinities of three of the nebulosities resembling Herbig-Haro objects reported by Gyulbudaghian *et al.*: GGD4, GGD12–15, and GGD27–28.[10] The cluster GGD12–15 shows, in addition to the normal CO emission from the ambient molecular cloud, the presence of blueshifted and redshifted CO wings, indicating high-velocity gas. The high-velocity lobes come from spatially separated zones located on opposite sides of a water maser source. Another of these objects, GGD27–28, may also show the phenomenon of anisotropic mass outflow, but the signal-to-noise ratio of the data was too low for us to make a detailed analysis.

In order to compare the outflows in various regions, we compiled the observed characteristics of eight sources known to us: Orion, L1551, Cep A, GL490, GGD12–15, V645 Cyg, GL961, and HL and XZ Tau. To be consistent, we reanalyzed the observations to derive such parameters of the outflows as mass, momentum and energy. We used a simple model for the outflow, assuming a power-law distribution of the corrected antenna temperature *versus* radial velocity and a spatial distribution independent of velocity for each emission wing. Other assumptions were (1) optically thin CO emission, (2) a molecular hydrogen to CO number ratio of 10^4, (3) nonradial motions were neglected, and (4) an excitation temperature of 10K was adopted. The resulting masses, momenta, and kinetic energies are considerably smaller than values quoted previously because of the conservative assumptions made. The masses obtained range from ~0.01 to ~10 $M_\odot$, the momenta from ~0.01 to ~40 $M_\odot$ km s^{-1}, and the kinetic energies from ~4 × 10^{41} to ~2 × 10^{45} erg. The high number of sources detected and the wide range of parameters related to them suggest that the outflow phenomenon may occur in the pre-main-sequence life of stars of all spectral types.

Our other major conclusions are the following: (1) Four of the eight outflow sources have associated H_2O maser emission. The total velocity extents in CO and H_2O appear to be correlated, suggesting that the H_2O masers and the high-velocity CO are part of the same outflow. (2) If the momentum in the outflow comes from the radiation field of the star, a special mechanism, such as multiple scattering in the infrared, is required. Alternatively, a substantial mass loss rate on the order of 10^{-4}–10^{-6} $M_\odot y^{-1}$ would be sufficient to drive the outflow. (3) There is a tendency for outflows with higher velocities to be located closer to their origins. This can be explained in terms of a projection effect.

New sources of high-velocity CO emission are rapidly being discovered. Many of them have an asymmetrical spatial distribution for the high-velocity gas. The anisotropy in this outflow may be due to an anisotropic stellar wind,[11] a circumstellar disk,[3] an interstellar doughnut,[12] or an anisotropic density distribution in the ambient cloud.[13] A good part of the future research in this topic will center on clarifying the nature of the focusing mechanism, as well as on identifying the source of energy.

REFERENCES

1. ZUCKERMAN, B., T. B. H. KUIPER & E. N. RODRÍGUEZ KUIPER. 1976. Astrophys. J. **209:** L137.
2. KWAN, J. & N. SCOVILLE. 1976. Astrophys. J. **210:** L39.
3. SNELL, R. L., R. B. LOREN & R. L. PLAMBECK. 1980. Astrophys. J. **239:** L17.
4. RODRÍGUEZ, L. F., P. T. P. HO & J. M. MORAN. 1980. Astrophys. J. **240:** L149.
5. LADA, C. J. & P. M. HARVEY. 1981. Astrophys. J. **245:** 58.
6. SCOVILLE, N. 1980. *In* Int. Astron. Union Symp. 87, Interstellar Molecules. B. Andrew, Ed. D. Reidel. Dordrecht.
7. BLITZ, L. & P. THADDEUS. 1980. Astrophys. J. **241:** 676.
8. RODRÍGUEZ, L. F., J. M. TORRELLES & J. M. MORAN. 1981. Astron. J. **86:** 1245.
9. CALVET, N., J. CANTÓ & L. F. RODRÍGUEZ. 1982. In preparation.
10. GYULBUDAGHIAN, A. L., YU. J. GLUSHKOV & E. K. DESISYUK. 1978. Astrophys. J. **224:** L137.
11. HARTMANN, L. & K. B. MACGREGOR. 1982. Astrophys. J. **257:** 264.
12. CANTÓ, J., & L. F. RODRÍGUEZ, J. F. BARRAL & P. CARRAL. 1981. Astrophys. J. **244:** 102.
13. KONIGL, A. 1982. Astrophys. J. In press..

INFRARED OBSERVATIONS
OF LOW-MASS STAR FORMATION
IN ORION: HH OBJECTS*

Paul M. Harvey and Bruce A. Wilking

Department of Astronomy
University of Texas, Austin
Austin, Texas 78712

Martin Cohen

Ames Research Center
Moffett Field, California 94035

INTRODUCTION

Strom *et al.* reported the first systematic observations of Herbig-Haro objects in the infrared.[1] They found point-like near-infrared (1–3 μm) sources in the vicinity of a number of Herbig-Haro nebulae. In general, the near-infrared sources were not coincident with the HH nebulae. They suggested that the near-infrared sources represented dust-enshrouded exciting stars for the nebulae and that the HH objects were illuminated by reflection along a line of sight of low extinction to the central exciting star. More recent observations have shown the HH objects to be a combination of reflection and emission nebulae[2] where the line emission appears to come from a shocked gas,[3] possibly due to mass outflow from the exciting "star."

The energy distributions of the exciting stars of HH objects have typically shown a steep rise out to the longest observed wavelengths, 3–10 μm.[1,4,5] Therefore, the total luminosities of the exciting stars have been highly uncertain because of the unknown, and presumably significant, contribution at far-infrared ($\lambda \simeq 50$–100 μm) wavelengths. Fridlund *et al.* reported the first observations of an HH object in the far-infrared and showed that the total luminosity of the L1551 infrared source could not be more than a few tens of solar luminosities, ninety percent of which is radiated at wavelengths longer than 20 μm.[6] Because of the large beamsize used for their observations (4.5 arcmin), they were not able to place a significant limit on the size of the emission region or rule out the contribution of some other source to the flux observed. We report here the preliminary results from a program of high-angular-resolution infrared photometry from the NASA Kuiper Airborne Observatory (KAO) and the NASA Infrared Telescope Facility (IRTF) on Mauna Kea, Hawaii. These observations span the wavelength range 1–200 μm and provide the first systematic set of data on broadband energy distribution and total luminosities of HH object–exciting stars. In this paper, we describe the observations and preliminary analysis of the data for the H-H objects observed in the Orion region. In a future paper (Cohen *et al.*, in preparation), we will report on the complete data set and discuss the full implications of the observations.

*This research was supported by a grant from the National Aeronautics and Space Administration, no. NAG 2-67.

199

OBSERVATIONS

Our far-infrared observations were obtained on the Kuiper Airborne Observatory with the photometer described by Harvey.[7] The observations include measurements of the postion of peak emission at both 50 and 100 μm, source size measurements at those wavelengths, and photometry at 40 and 160 μm, all with beamsizes of $d \simeq 40$ arcsec. The absolute flux calibration error is $\pm 25\%$.

Our near-infrared observations were obtained on the NASA IRTF with the Facility Photometer. These observations consist of photometry at 1.2, 1.6, 2.2, 3.5, 4.8, 10, and 20 μm and measurements of source positions at 2.2 and/or 10 μm, with an absolute uncertainty of ± 1.5 arcsec. The beamsize used was 6 or 8 arcsec. The absolute flux calibration uncertainty is $\pm 10\%$ at 1.2–3.5 μm, $\pm 15\%$ at 4.8 and 10 μm, and $\pm 20\%$ at 20 μm.

RESULTS

NGC 1999 (HH1–3)

In addition to the known far-infrared emission from V380 Ori, one far-infrared source was found near HH1, apparently associated with a near-infrared source described by Cohen and Schwarz.[4] Our preliminary position and size measurements at 50 and 100 μm show that the far-infrared emission peaks about 30 arcsec to the west of the near-ir source, with weaker emission extended in the direction of the near-ir object. FIGURE 1 shows the overall energy distribution of this source, assuming that the far-infrared emission is powered by the same star as the shorter wavelength flux. The total luminosity of this object is 30 $L_\odot$, as shown in TABLE 1. The mass of gas required to produce the far-infrared emission is $M \simeq 0.3 M_\odot$, assuming a dust emissivity of 250 cm^2 g^{-1} at 100 μm and a gas-to-dust mass ratio of 200.

In addition to the area around HH1, we searched an area 1 arcmin in radius around

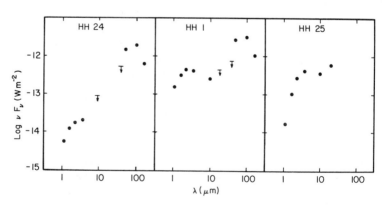

FIGURE 1. The energy distributions are shown for the three HH object–exciting stars observed in Orion. The ordinate is $\nu F_\nu \equiv \lambda F_\lambda$, for which a horizontal line implies equal energy per octave of frequency or wavelength. These energy distributions are similar to those of many BN-type objects, although the total luminosities are a factor of 10–1000 lower.

TABLE 1

TABLE 1

OBSERVED AND DEDUCED PARAMETERS OF HH EXCITING STARS

HH	Position of Exciting Star		Assumed Distance (pc)	Total Luminosity ($L_\odot$)	Far-ir Emitting Gas and Dust Mass ($M_\odot$)
	α_{1950}	δ_{1950}			
1	$05^h33^m55\overset{s}{.}4$	$-06°47'25\overset{''}{.}5$	500	30	0.3
24	05 43 34.4	$-00\ 11\ 06.0$	500	20	0.2
25	05 43 31.1	$-00\ 15\ 21.0$	500	>10	?

HH2 to a sensitivity of 15 Jy (3σ) at 100 μm with negative results and obtained an upper limit to the flux density at the position of HH3 at 100 μm of $F_\nu \leq 4$ Jy (3σ) at 100 μm. This lack of far-infrared emission is consistent with the absence of any reddened 2-μm sources in the vicinity of HH2–3.[5] Therefore, the object near HH1 appears to be the most luminous embedded source near the HH nebulae and, as Cohen and Schwarz point out, is likely to be the exciting star for HH1, 2, and, possibly, 3.[5]

M78 (HH24–25)

Far-infrared emission was found in the HH24 region from a near-infrared source discovered by Strom *et al.*[1] Within the observational uncertainties (± 10 arcsec), the far-infrared emission peaks at the same position as the near-infrared flux, and is extended on a scale of 30–40 arcsec. The energy distribution of this object is shown in FIGURE 1. The total luminosity is 20 $L_\odot$ and the circumstellar gas mass $M \simeq 0.2\ M_\odot$, as shown in TABLE 1.

Searches for far-infrared emission have not yet been made in the HH25 region; however, a probable exciting source has been located in the near-infrared, as shown in FIGURE 1 and TABLE 1. The region around HH25 was searched at 1.6 μm on the IRTF to a flux sensitivity of 4 mJy over an area 1.5 arcmin in diameter. Only one highly reddened object was found, which is coincident with source 59 found by Strom *et al.*[8] The similarity in spectral shape among this object, the other HH exciting stars described here, and those in other areas implies that this is probably the exciting star for HH25 and that it will be found to be a strong source of far-infrared emission.

INTERPRETATION

The two most important observational results of this study are (1) the determination of the total luminosities of the probable exciting stars for the HH objects studied and (2) the measurement of the complete infrared energy distribution of the thermally emitting dust clouds around the exciting stars. These two results lead us to conclude that HH objects are probably the signposts for "low" luminosity analogues to the Becklin-Neugebauer (BN) object phenomenon.[9] Although there is a wide range of observed energy distributions in objects that have been labeled BN-like, there appear to be several common features that, together, distinguish them from other infrared sources in regions of star formation. The most important features are (1) compact near- and far-infrared source size, (2) large visual extinction and steeply rising energy distribution out to wavelengths as long as 50–100 μm, where most of the luminosity is radiated, (3) evidence for high-velocity mass motion in radio-molecular-line spectra, (4) common association with H_2O maser emission at $\lambda \simeq 1.35$ cm, and (5) absorption in the 10-μm silicate band. Of these five characteristics, the first two have been observed in the HH exciting stars in this study. High-velocity gas appears to be the cause of the shock-excited emission lines from the HH nebulae.[3,10] Statistics of water maser occurrence are unreliable because of their typical time variability, but two have already been observed near HH objects.[11,12] The fifth common characteristic of BN objects, 10-μm silicate absorption, has yet to be searched for in HH exciting star spectra. Therefore, within the observational uncertainties and the observed variation of BN object characteristics, the HH stars appear to represent the low-luminosity extension of the BN phenomenon.

Another characteristic common to these HH stars and several BN-type objects is the large rate of mass loss implied by the cloud densities and velocities relative to the momentum available from radiation-pressure-driven mass loss. For instance, in HH1, the estimated mass outflow from the central star is[3] 10^{-5}–10^{-6} $M_\odot$ y^{-1} at $V \simeq 100$ km s^{-1}, or a total momentum rate of $\dot{M}V > 10^{-4}$ $M_\odot$ km s^{-1} y^{-1}. The rate of momentum input that radiation pressure can supply, on the other hand, is $\dot{M}V \simeq \tau(L/c) \lesssim 10^{-5}$ $M_\odot$ km s^{-1} y^{-1} (where τ is the optical depth, L the luminosity, and c the speed of light). This discrepancy also exists for the more luminous BN objects, as described by, *e.g.*, Kwan and Scoville for BN itself[13] and Lada and Harvey for AFGL 490.[14]

A final conclusion from these data is that the dust distributions around at least some HH exciting stars must be asymmetrical, as originally suggested by Strom *et al.*[1] in their reflection nebula hypothesis for the HH nebulae. This follows from the large extinction implied by the near-infrared photometry, assuming that the 1.2 and 1.6 μm fluxes are the reddened continuum of a T Tauri–like photosphere. For instance, in HH24, the 1.2–1.6 μm color implies $A_v \simeq 10$–15 mag, while the optical continuum of the HH nebulae clearly appears to be reflected starlight from the central exciting object. If the central star were surrounded by a spherically symmetric dust distribution with $A_v > 10$ mag, we would not expect to see any reflection nebulae through the dust.

ACKNOWLEDGMENTS

We thank the staffs of the Kuiper Airborne Observatory and the Infrared Telescope Facility for providing and operating the exceptional telescopes that made these observations possible. We also thank M. Joy, R. Levreault, L. Rickard, and H. Thronson for their help in observing these objects.

[NOTE ADDED IN PROOF: Recent observations have, in fact, detected strong far-infrared emission from the exciting star for HH 25.]

REFERENCES

1. STROM, S. E., G. L. GRASDALEN & K. M. STROM. 1974. Astrophys. J. **191**: 111.
2. SCHMIDT, G. O. & J. S. MILLER. 1979. Astrophys. J. **234**: L191.

3. SCHWARTZ, R. D. 1978. Astrophys. J. **223**: 884.
4. COHEN, M. C. & R. D. SCHWARTZ. 1979. Astrophys. J. **233**: L77.
5. COHEN, M. C. & R. D. SCHWARTZ. 1980. Mon. Not. R. Astron. Soc. **191**: 165.
6. FRIDLUND, C. V. M., H. L. NORDH, R. J. VAN DUINEN, J. W. G. AALDERS & A. I. SARGENT. 1980. Astron. Astrophys. **91**: L1.
7. HARVEY, P. M. 1979. Publ. Astron. Soc. Pac. **91**: 143.
8. STROM, K. M., S. E. STROM & F. J. VRBA. 1976. Astron. J. **81**: 308.
9. BECKLIN, E. E. & G. NEUGEBAUER. 1967. Astrophys. J. **147**: 799.
10. ELIAS, J. H. 1980. Astrophys. J. **241**: 728.
11. LO, K. Y., M. MORRIS, J. M. MORAN & A. D. HASCHICK. 1976. Astrophys. J. **204**: L21.
12. LO, K. Y., B. F. BURKE & A. D. HASCHICK. 1975. Astrophys. J. **202**: 81.
13. KWAN, J. & N. J. SCOVILLE. 1976. Astrophys. J. **210**: L39.
14. LADA, C. J. & P. M. HARVEY. 1981. Astrophys. J. **245**: 58.

DISCUSSION OF THE PAPER

B. DRAINE (*Institute for Advanced Study, Princeton, N.J.*): Has there been any spectrophotometry that could resolve a 10-μm silicate absorption feature in HH objects?

HARVEY: Not to my knowledge, but there is some suggestion of the feature in our broad-band data. Higher resolution measurements are indeed possible.

M. HARWIT (*Cornell University, Ithaca, N.Y.*): Have any of the sources you've looked at been observed by the Einstein Observatory?

HARVEY: No.

COMPACT CONTINUUM RADIO SOURCES
IN THE ORION NEBULA

J. M. Moran,* G. Garay,* M. J. Reid,*
R. Genzel,† and P. T. P. Ho‡

*Harvard-Smithsonian Center for Astrophysics
Cambridge, Massachusetts 02138

†Department of Physics
‡Department of Astronomy
University of California, Berkeley
Berkeley, California 94720

INTRODUCTION

We observed the Orion Nebula with the Very Large Array (VLA) of the National Radio Astronomy Observatory,§ searching for compact radio sources that would help establish which of the infrared sources in the Becklin-Neugebauer–Kleinman-Low (BN-KL) region might contain primary energy sources. We detected radio emission from the BN object at 15 and 23 GHz. BN had not been detected at lower frequencies[1] because of its sharply rising spectrum and because of the confusion from extended emission. We found no emission from IRc2, which is thought to be the source of the flow that gives rise to maser emission.[2] If IRc2 has an optically thick compact H II region at 10^4 K, it must be smaller than 3×10^{14} cm in diameter (for a distance of 500 pc). In addition, we found ten compact sources associated with the Trapezium cluster and four other sources near the molecular cloud.

OBSERVATIONS

We observed the Orion Nebula with the VLA at 5 and 15 GHz on 9 August 1981 and at 23 GHz on 16 November 1981. The array was in the B configuration in August and had maximum resolutions of about 0".8 and 0".3 at 5 and 15 GHz. We mapped the region at 15 GHz with a 35 Kλ minimum baseline spacing to remove structures larger than 6". The rms noise in this map was 0.5 mJy and we detected 15 compact sources stronger than 3 mJy. They are listed in TABLE 1. The flux density scale is referenced to 3C147, whose flux density was 2.4 Jy. The phase calibration and position reference source was 0539-057. The 1.3-cm observations were made with the C configuration (a resolution of 0".6), in conjunction with spectral line observations of NH_3 in the molecular cloud. The passband was 9.5 MHz, centered at 22.8 GHz. The contribution of ammonia emission was negligible. The flux density scale is referenced to 3C286, whose flux density was 2.5 Jy.

§NRAO is operated by Associated Universities, Inc., under contract with the National Science Foundation.

0077–8923/82/0395–0204 $1.75/1 © 1982, NYAS

RESULTS

Trapezium Cluster

The ten sources listed in TABLE 1 and shown in FIGURE 1 were detected at both 5 and 15 GHz. The spectra of most of these sources were either flat or rising towards the higher frequency. This is consistent with source models in which the emission comes either from an ionized stellar wind or from a compact H II region that has become optically thin between 5 and 15 GHz. The Lyman continuum emission from an early B star can account for the emission from each of these sources. Nine of the sources are coincident with optically observed stars or condensations.[9-11]

TABLE 1

COMPACT RADIO SOURCES IN THE ORION NEBULA

	α_{1950}*	δ_{1950}	Flux (2 cm)† (mJy)	Diameter (")	Identification‡
		Trapezium Cluster			
1	05^{h}32^{m}48^{s}05	−5°25′30″5	3.1	0.6	
2	48.33	25 19.9	8.6	0.5	LV6
3	48.38	25 15.8	7.1	0.4	θ^1H′, P1866, LV5
4	48.36	25 7.4	11.0	≤0.2	θ^1A, P1865
5	48.61	25 17.4	4.3	≤0.5	LV4
6	48.82	25 9.9	8.5	0.2	LV3
7	49.29	25 9.6	22.4	0.3	θ^1G, P1890, LV2
8	49.38	25 19.4	12.4	0.4	LV1
9	49.52	25 30.1	4.3	0.4	P1894
10	50.22	25 33.9	5.3	≤0.4	H
		Molecular Cloud and Environs			
A	05^{h}32^{m}44^{s}33	−5°23′43″0	8.7	0.2	
B	46.65	24 16.5	9.5	≤0.15	BN object
C	46.70	24 54.4	8.6	0.4	
D	47.42	24 18.8	4.7	<0.4	P1838, IR?
E	50.89	24 30.6	5.1	≤0.4	P1925, X32?

*Positions are accurate to ±0″2 (1σ).
†Flux densities are accurate to ±15%.
‡P refers to stars in the Parenago catalog;[9] LV to [O III] condensations of Laques and Vidal;[10] H to visible stars on plates by Herbig;[11] and X to x-ray sources detected by Ku *et al.*[12] Source D is near an extended minor peak in the 20-μm map of Downes *et al.*[13]

The Becklin-Neugebauer Object

The position of the radio source associated with the BN object (see TABLE 1) agreed with the ir position[13] to within the combined errors of 0^{s}04 in right ascension and 0″3 in declination. The flux densities, plotted in FIGURE 2, were ≤2, 9.5 ± 1.5, and 23 ± 5 mJy at 4.9, 15.0, and 22.8 GHz. Hence, the spectral dependence is $\nu^{2.1\pm0.5}$. At 15 GHz, where the signal-to-noise ratio was the highest, the source was unresolved and smaller than 0″15, or 10^{15} cm. The brightness temperature was greater than 2000 K.

Since the spectral index of the radio emission is very close to two, the radio and infrared continuum emission can be explained by a simple model in which the BN object is a ZAMS star surrounded by an ionized envelope of constant density and a temperature of 10^4 K, which gives rise to the radio emission and Brackett line emission in the infrared. The radio source is optically thick and must have a diameter of $0.''08$ or 6×10^{14} cm (40 AU). The 4-μm Brα flux density, corrected for extinction, is about 6×10^{-11} erg cm^{-2} s^{-1}.[3] The optical depth, assuming a uniform density, a constant temperature, and a size of $0.''08$, is 4×10^{-4}. The line emission requires a Lyman continuum photon rate of 4×10^{46} s^{-1}, which could be supplied by a ZAMS B0.5 star. The volume emission measure is $\sim 2 \times 10^{59}$ cm^{-3}. Using the inferred radio size, the electron density is 4×10^7 cm^{-3} and the emission measure is 3×10^{11} cm^{-6} pc. The opacity is, therefore, 300 at 15 GHz. With this model, the opacity is predicted to be unity at 220 GHz, where the flux density would be ~ 1.5 Jy.

The basic result of our measurements is that the spectrum is essentially that of a blackbody up to 23 GHz and the geometric mean angular size must be about $0.''08 \times (10^4 \text{ K}/T_e)$, where T_e is the electron temperature. If the source were nonthermal, the size could be substantially less than $0.''08$. However, this seems unlikely.

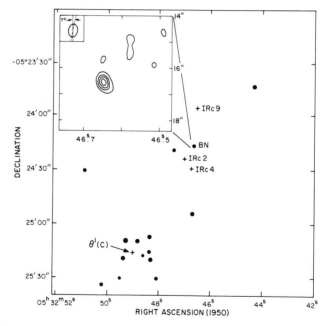

FIGURE 1. A map of the Orion Nebula made with the VLA at 15 GHz. Filled circles denote radio sources detected above the limit of 3 mJy/beam. Also shown are the positions of several infrared sources (IRc) and that of the most luminous Trapezium star, θ^1(C). The diameters of the circles represent the sizes of the sources at the level of 2 mJy/beam. The insert shows a map of the source associated with the Becklin-Neugebauer object. Contours are 20, 40, 60, and 80% of the peak. The synthesized beam shape ($0.''48$ by $0.''42$) is shown in the top left corner of the insert.

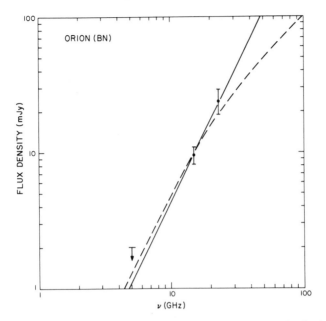

FIGURE 2. The radio spectrum of the BN object. The solid line is the Rayleigh-Jeans approximation to the Planck function. The dashed line is the thermal Bremsstrahlung spectrum expected for an isothermal sphere at 10^4 K with a radial electron density distribution given by $n_e = n_0(r_0/r)^2$, where $r_0 = 3.4 \times 10^{14}$ cm, $n_0 = 10^6$ cm^{-3} for $r < r_0$, and $n_e = 0$ for $r > r_0$.

DISCUSSION

The simple model of a uniform spherical H II region described above is inconsistent with several other observations. First, the Brackett emission lines have broad wings indicative of velocities of ± 100 km s^{-1}.[3] Second, the highly excited CO emission at 2.3 μm, which has been localized to within 1″.4 of BN, is thought to require collisional excitation with densities $> 10^{10}$ cm^{-3}, which probably occur only within a few AU of the star.[4,5] Since the dissociation energy of CO (11.1 eV) is less than the ionization potential of H (13.6 eV), the neutral CO and ionized H cannot be coextensive. It has been widely assumed that the ionized region is inside the neutral gas zone and is, therefore, smaller than a few AU. While the observed Brackett emission could be produced in a source with a diameter as small as 10^{13} cm, the radio emission must come from a much larger source if the temperature is $\sim 10^4$ K. There are two ways to reconcile the microwave continuum and infrared CO data. The inferred neutral gas diameter could be in substantial error, since the CO excitation mechanism is not well understood. Alternatively, the BN object may have a complex structure. Several possibilities involving disk geometry with a bipolar flow along the axis could explain the data.

The ν^2 dependence of the radio emission can be reconciled with the high-velocity outflow suggested by the wings of the Brackett line in a simple way without invoking disk geometry.[6] Consider a spherically symmetric source with an isothermal wind expanding at constant velocity and, hence, an r^{-2} dependence of density, where the ionization ceases abruptly at radius r_0. The flow continues beyond r_0, but in a neutral state. Below a certain frequency, ν_0, such a source will be optically thick for all line-of-sight paths out to a projected distance of $\sim r_0$. Since the observed radius is constant, the radio flux density will vary as ν^2. Above ν_0, the source will be thin at the edge and will look like a wind source with a flux density proportional to $\nu^{0.6}$ and an angular size proportional to $\nu^{-0.7}$.[7,8] Our results constrain the diameter of the ionized source ($2r_0$) to be between 6 and 8×10^{14} cm. The lower limit of 6×10^{14} cm is the diameter of a blackbody at 10^4 K needed to give the measured flux densities and the upper limit of 8×10^{14} cm is the size above which the spectral index becomes too low to fit our measurements. A model that fits the data, shown in FIGURE 2, has a diameter of 6.8×10^{14} cm and a density at its ionization boundary of 10^6 cm^{-3}. For an expansion velocity of 100 km s^{-1}, the mass-loss rate is 3×10^{-7} $M_\odot$ y^{-1}. These parameters can reasonably account for the Brα flux density.

Millimeter observations of the continuum emission from BN will be very important in determining the shape of the continuum spectrum and, therefore, the electron density distribution. VLA observations in the A configuration at a wavelength of 1 cm should make a direct measurement of the source size possible.

REFERENCES

1. SIMON, M., G. RIGHINI-COHEN, M. FELLI & J. FISCHER. 1981. Astrophys. J. **245**: 552–59.
2. GENZEL, R., M. J. REID, J. M. MORAN & D. DOWNES. 1981. Astrophys. J. **244**: 884–902.
3. HALL, D. N. B., S. G. KLEINMANN, S. T. RIDGWAY & F. C. GILLETT. 1978. Astrophys. J. **223**: L47–L50.
4. SCOVILLE, N. Z., D. N. B. HALL, S. G. KLEINMANN & S. T. RIDGWAY. 1979. Astrophys. J. **232**: L121–L124.
5. SCOVILLE, N. Z., R. KROTKOV & D. WANG. 1980. Astrophys. J. **240**: 929–39.
6. SIMON, M., M. FELLI, L. CASSAR, J. FISCHER & M. MASSI. 1982. Astrophys. J. In press.
7. PANAGIA, N. & M. FELLI. 1975. Astron. Astrophys. **39**: 1–5.
8. WRIGHT, A. E. & M. J. BARLOW. 1975. Mon. Not. R. Astron. Soc. **170**: 41–51.
9. PARENAGO, P. P. 1954. Works of the Astronomical Institute at Sternberg: 25.
10. LAQUES, P. & J. L. VIDAL. 1979. Astron. Astrophys. **73**: 97–106.
11. HERBIG, G. H. 1982. This volume.
12. KU, W. H. M., G. RIGHINI-COHEN & M. SIMON. 1981. Science **215**: 61–64.
13. DOWNES, D., R. GENZEL, E. E. BECKLIN & C. G. WYNN-WILLIAMS. 1981. Astrophys. J. **244**: 869–83.

DISCUSSION OF THE PAPER

D. M. WATSON (*University of California, Berkeley, Calif.*): Is there any hope of detecting and resolving the BN source with VLBI?

MORAN: It is not bright enough; the best chance of resolving it is at the K band with the A array of the VLA.

M. SIMON (*State University of New York, Stony Brook, N.Y.*): The ν^2 optically thick spectrum should not be taken to rule out flow in the ionized gas. The $\nu^{0.6}$ radio spectra for winds, as worked out by Wright and Barlow and Panagia and Felli, refer to a situation in which the flow continues to distances far enough away from the star that the density becomes low enough for the flow to have an optically thin envelope in the radio. In this situation, the optically thin part of the flow can contribute ~6 times as much flux in the radio as the optically thick core. The ν^2 spectrum determined for BN must mean that the flow (for which there is evidence from the ir recombination lines) recombines at some distance like 30 AU from the star and is neutral beyond that radius. This size is smaller than the dust thermal radius of BN. Felli, Fischer, and I have determined the radio flow of CRL 490 and S106 IR at 1.3 cm and find a similar situation. The electron density you infer, $n_e \simeq 10^7$ cm^{-3}, is under the assumption that the density is constant within the volume and is a lower bound, since it is the density required to just get the volume optically thick in the radio. I would therefore urge caution in inferring optical depths in Bα because the ir lines are probably formed much deeper within the volume of ionized gas.

P. MEZGER (*Max-Planck-Institute für Radioastronomie, Bonn, F.R.G.*): If you estimate the number of Lyman continuum photons necessary to ionize the compact H II region surrounding the BN object, what spectral type does it correspond to?

MORAN: B0, *i.e.*, 10^{46} ionizing photons per second.

M. W. WERNER (*Ames Research Center, Moffett Field, Calif.*): In what sense is the size inconsistent with models for BN?

MORAN: There is the problem, which also came up in the context of Scoville's discussion yesterday, of getting the dust shell close enough to get it hot enough. We have to move the dust V shell out to 30 AU, and the situation becomes problematical.

THREE GAS CLUMPS NEAR THE EDGE
OF THE VISIBLE ORION NEBULA

D. N. Matsakis

U.S. Naval Observatory
Washington, D.C. 20390

P. Palmer

Department of Physics
University of Chicago
Chicago, Illinois 60637

A. I. Harris

Department of Physics
University of California, Berkeley
Berkeley, California 94720

INTRODUCTION

Ammonia is a useful probe of conditions in molecular clouds. The inversion transitions in the metastable states have lifetimes long enough (10^{-7} s^{-1}) to provide thermal equilibrium over a range of densities and temperatures and the inversion frequencies for the different rotational levels are closely enough the same to allow one to use the same instrument to study a wide range of rotational temperatures. The lower rotational levels have hyperfine "satellite" components of known intensity ratios that are often convenient for the determination of optical depth. Single-dish measurements of ammonia in dark clouds have already shown that the clouds are optically thick and that intensity ratios between different transitions give average rotational temperatures in Orion of 15 K or more. However, the average brightness temperatures of only a few degrees are much lower than one would expect if the clouds uniformly fill the beams. This has led to the expectation that ammonia is clumped on a scale smaller than the single-dish beamwidths. Observations of other molecules have shown clumping to be rather common; the present series of observations were undertaken to confirm the clumping of ammonia and to make an effort towards understanding the role of clumps in the evolution and structure of molecular clouds. While most astronomers believe that these clumps are an interesting but dynamically unimportant feature of giant molecular clouds (GMC), our data allow other possibilities.

OBSERVATIONS

The results reported here are a combination of observations made at the VLA in September 1981 and January 1982. The VLA was used in its D array in September 1981 to observe the central component of the (2, 2) line. This central component of the (2, 2) line and both the central component of the (1, 1) line and one of its hyperfine components were observed in January 1982 after the VLA had been changed to its larger C array. The observing patterns for all three 11.5-hr runs were nearly identical.

210

0077-8923/82/0395-0210 $1.75/1 © 1982, NYAS

The antenna pointing position was the same for all observations, at $\alpha_{1950} = 5^h32^m47^s$, $\delta_{1950} = -5°20'50''$. The complex gain calibrator was 0528 + 134, approximately 18' north of the source, at $\alpha = 5^h28^m07^s$, $\delta = 13°29'42''$.

We observed in the VLA's spectral line mode, with 16 channels for the (2, 2) observations and 32 channels for simultaneous observation of the (1, 1) main and satellite hyperfine lines. Frequency resolution was provided by an autocorrelator system operated to give 24.41 kHz channels (0.3 km s^{-1}). However, since we Hanning-smoothed the channels, the velocity resolution was actually 0.6 km s^{-1}.

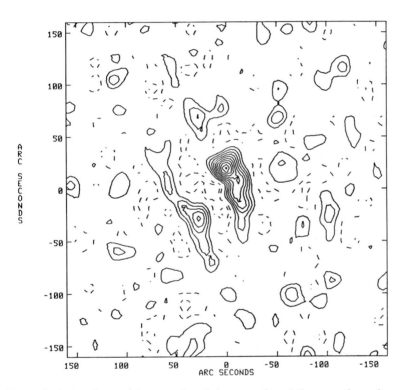

FIGURE 1. A cleaned map of the ammonia emission, tapered to a 14" apparent beam size and centered at a velocity of 9.6 km s^{-1}. The contour levels are spaced by units of 0.0214 Jy/beam and negative contours are dashed.

RESULTS AND DISCUSSION

Lower-resolution single-dish maps of the region near OMC1 and OMC2 show a peak of ammonia emission approximately 3.5 arcmin north of KL. Our first observations of the (2, 2) line in September resolved this peak into the three distinct clouds shown in FIGURE 1. Maps of spectral channels in which there is no NH$_3$ emission indicate that features of apparent brightness equal to or less than two contour levels are probably due to noise.

Each cloud is about 20 arcsec in maximum extent, which corresponds to a linear size of 1.4×10^{17} cm at a distance of 480 pc. There is a systematic shift in the positions of the clumps with velocity, as indicated in maps to be presented elsewhere.[1] Maps of the calibrator across the band show no position shift at all, indicating that the shift is not instrumental. The shift in apparent centers of the clouds over the channels in which they were detected is comparable to their diameters as measured in any one map and corresponds to a 1.2 km s^{-1} gradient across the northern clump and a 1.5 km s^{-1} gradient across the southern one. The center velocity is the same as the single-dish average velocity.

One very important question is, Is all the ammonia emission observed with single-dish antennae at this position accounted for in our VLA data? There could exist extended emission that is resolved out or there could be many smaller clumps that are lost in the noise and limited dynamic range of the maps. Although the comparison with single-dish data is complicated by different resolutions, efficiencies, and calibration strategies, we find a best estimate of 50% for the ratio of integrated VLA line brightness to that seen by single dishes for the two transitions. Since the clumps are significantly larger than our synthesized beam (even in the D-array), we cannot consider the missing flux definite proof of undetected ammonia. The apparent opacity of 7 in the main (1, 1) transition is much larger than the value of about 1 inferred from single-dish data, but this result is extremely sensitive to errors. Inspection of data reveals that our observed main-to-satellite ratio differs from the single-dish value of 0.4 only by 2σ. We can state, however, that if we have resolved out any ammonia emission, it is probably optically thin and contributes up to 50% of the observed single-dish antenna temperatures at this position.

The January observations were made chiefly to determine optical depths and temperature by measuring the ratios of the (2, 2), (1, 1), and one of the (1,1) satellite hyperfine lines. The maps were noisy enough that no unambigious position shifts could be seen between the channels, so each tranisition's maps were averaged together. A comparison of these average maps shows that the (1, 1) main, satellite, and the (2, 2) line emission regions and peaks are coincident within the uncertainty of a few arcsec determined by noise.

The optical depth in the (1, 1) line (τ_{11}) can, in principle, be determined from the relative strengths of the two observed hyperfine components, which have a theoretical opacity ratio of satellite to main of 0.278. Hence, assuming equal T_{ex} for the components,

$$\frac{S_{\text{satellite}}}{S_{\text{main}}} = \frac{1 - e^{-\tau_{\text{satellite}}}}{1 - e^{-\tau_{\text{main}}}} - \frac{1 - e^{-0.278\tau_{11}}}{1 - e^{-\tau_{11}}}.$$

The peak of main line intensity is 243 ± 52 mJy (all errors are 1σ); the peak satellite intensity is 209 ± 48 mJy. Therefore, the ratio of satellite to main is 0.86 ± 0.26, which gives a range for τ_{11} from 3 to infinity, with a center-of-error nominal value of 7.

As pointed out by Cheung et al.,[2] the rotational temperature inferred from τ_{11}/τ_{22} need not equal the excitation temperature of the transitions, even if they are thermalized to the kinetic temperature. At low densities, the population of the (2, 2) state should be less than that inferred from LTE theory because the unpopulated (2, 1) state provides a one-way channel for emptying the (2, 2) state. Numerical calculations with Green's collisional cross sections (preprint) indicate that this can lower the rotational temperature of a 30 K source to 21 K, that of a 20 K source to 16 K, and that of a 15 K source to 13 K. These numbers are, of course, somewhat uncertain, due to uncertainties in the cross sections, but we believe that they indicate the magnitude of the effect. The effect would be quenched in densities exceeding 3×10^8, for, at that density, the (2, 1) state becomes populated. In the presence of trapping, the required density becomes $3 \times 10^8/(\tau_{ir}) \simeq 10^7/\tau_{11}$. This effect poses no problem in the interpretation of the data if the opacity of the (1, 1) transition is taken to be 7. In this case, the excitation temperature is 21 K, the rotational temperature is about 17 K, and the predicted (2, 2) opacity of 1.9 is consistent with the observed brightness. If the opacity of the (1, 1) transition is only 1.0, however, then the excitation temperature is 31 K, the rotational temperature only 21 K, and the predicted (2, 2) opacity of 0.5 implies a (2, 2) brightness of only 11 K, half that observed. To be consistent with observations, we must assume that the Cheung mechanism is quenched by densities on the order of 10^7 in the clumps.

The mass of a clump can be estimated from the velocity gradient, assuming that the clump is gravitationally bound. A balance of centripetal and gravitational acceleration for the measured velocity difference between center and radius of the cloud, 0.75 km s^{-1}, with a radius of 5×10^{16} cm, requires $2M_\odot$ and $n = 2.10^6$. Similar results can be derived from the virial theorem. It should be noted that the clumps are not disc-shaped, as might be expected from the simplest model of a rotating, gravitationally bound cloud. Instead, they appear to be somewhat elongated along the axis of rotation, possibly representing several small clouds along a ridge.

A very crucial parameter of molecular clouds is the degree of inhomegeneity within them. Since ammonia is a tracer of high-density regions, it is not possible to derive parameters of the interclump medium solely from our data. It may be possible to make reasonable estimates from the high-resolution CO maps that are now becoming available; we believe that there are indications in published data that the clumps could contain as much as 50% of the mass within the core region of the cloud. If one assumes that a typical clump has two stellar masses within 10^{17} cm diameters, a C/H ratio of 2×10^{-4}, a CO/C ratio of 0.1, and a CO/C^{18}O ratio of 490, then the column density of C^{18}O in a spherical clump is 1.6×10^{16} cm^{-2}. Since all isotopic forms of CO will be completely thermalized at these densities, we find that the opacity of C^{18}O is given by:

$$T_{C^{18}O} = 0.8 \, \frac{(nC^{18}O)}{(1.6 \times 10^{16} \text{ cm}^{-2})} \frac{(20 \text{ K})^2}{(T_{ex})^2} \frac{(1.0 \text{ km s}^{-1})}{(\Delta V)} .$$

The expected C^{18}O brightness from the clumps would therefore be expected to be a reasonable fraction of the kinetic temperature and the more common isotopic forms would have a brightness equal to that of an opaque blackbody at the kinetic temperature. The emission of the clumps in the CO (^{12}C^{16}O) line would not be detected, of course, since not only the clumps but also the interclump medium must be completely opaque at that frequency. The brightness ratios ^{13}CO/CO and C^{18}O/CO would then equal the clump filling factors plus a contribution from the interclump medium, assuming that the clump and the interclump medium have the same kinetic temperature. Accurate C^{18}O/CO ratios have not been published for this position; however, we note that, in other giant molecular clouds, the C^{18}O/CO ratios that have been measured are roughly consistent with the ammonia filling factors indicated in these same clouds by the low ammonia antenna temperatures. Since the ^{13}CO temperatures are always stronger than the $C^{18}O$ temperatures, we conclude that the integrated column density of the interclump medium (excluding, perhaps, large regions of extremely low densities) must be at least comparable to that of the clumps. If we assume that all the relevant regions are spherical, we find that equal column densities in the two phases require a 10 to 1 density contrast for a 21% filling factor and a 30 to 1 density contrast for a 10% filling factor.

We stress that an interpretation of extreme clumpiness is not required from our data, depends upon many (reasonable) assumptions, and is only appropriate to the inner parts of the GMC, where clumped ammonia is found. The observed clumping could even be due to local chemical or radiative effects. If the clumping is a density effect, observations of molecules such as $C^{18}O$, with the high resolutions now becoming available, will enable us to make a more definite estimate of the density contrast between the clumps and the interclump medium.

SUMMARY

The NH$_3$-containing regions in the Orion Molecular Cloud have been examined at a point 3.5' north of the Kleinmann-Low nebula with the high angular resolution of the VLA. Three small clouds were found, of linear dimensions about 0.05 pc, brightness 15–20 K, displaying a velocity gradient ranging over 1.5 km s[1]. If interpreted as rotation, cloud masses near one solar mass are indicated and the opacity in $C^{18}O$ is near unity.

REFERENCES

1. HARRIS, A. I., C. H. TOWNES, P. PALMER & D. N. MATSAKIS. 1982. In preparation.
2. CHEUNG, A. C., D. M. RANK, C. H. TOWNES, S. H. KNOWLES & W. G. SULLIVAN. 1969. Astrophys. J. 157: L13.

DISCUSSION OF THE PAPER

P. MEZGER (*Max-Planck-Institut für Radioastronmie, Bonn, F.R.G.*): Do you feel that the dense core model of molecular clouds that emerges from your observations, *i.e.,* dense clumps embedded in a lower-density medium, applies to most giant molecular clouds? If so, what are the consequences for mass estimates based on $^{12}C^{16}O$ observations?

MATSAKIS: Masses derived from ^{12}CO have little meaning. We have evidence for clumping in other clouds, *e.g.,* DR 21. The NH$_3$ is only observable within several arcmins of the cloud core, so we have no evidence of similar clumping further out in the clouds.

P. SOLOMON (*State University of New York, Stony Brook, N.Y.*): The model you present has clumps with $n_{H_2} = 10^7$ and virtually empty space in between. You stated that both $^{13}C^{16}O$ and $^{12}C^{18}O$ would be optically thick in these clumps. If this were the case, all three of these isotopic species would show the same antenna temperature, which would be determined only by the filling factor. This is completely contrary to the observations that show the line temperatures to be in the ratio of about 60:12:1.5. Thus, the $^{12}C^{16}O$ to $^{12}C^{18}O$ intensity ratio is observed to be about 40 to 1; your model would suggest 1 to 1.

On a completely separate point: We have measured the masses of molecular clouds on a large scale using ^{13}CO and the virial theorem. These masses agree within a factor of two or better. These comments do not mean that the cores of the molecular clouds are not fragmented; I think that they are fragmented but that the CO is tracing regions of hydrogen density between 10^2 cm^{-3} and 10^4 cm^{-3}. The mass from regions of $n_{H_2} >$ 10^4 cm^{-3} must be added on; this, however, represents a very small fraction of the mass of a giant cloud. Within the core region the total mass is somewhat greater than the CO mass. Outside the cores there are few high-density ($>10^4$ cm^{-3}) fragments and, therefore, very weak or absent ammonia radiation.

R. DICKMAN (*University of Massachusetts, Amherst, Mass.*): The 2-2 NH$_3$ is basically an indicator of clump densities $>10^5$ cm^{-3}. CO is easily excited above $n_H =$ 500 cm^{-3} and, if the interclump medium is as dense as a few 10^3–10^4cm^{-3}, the CO, ^{13}CO, and $C^{18}O$ may be quite uniformly bright over the interclump region. Depending on the density and fractional abundances of the CO isotopic species, even ^{13}CO may be optically thick. The impact of your clumping model on CO observations has to be thought out very carefully.

P. THADDEUS (*Goddard Institute for Space Studies, New York, N.Y.*): If you look at the Orion Nebula, the 10^4K gas is clumped on scales of 1 arcsec. General hydrodynamic considerations suggest that the cool gas will be even more structured. To understand this problem better, we need to make systematic studies of the clumping of the cool gas along the lines you have stated.

IS THE SiO MASER IN ORION ASSOCIATED WITH A LATE-TYPE STAR?

P. R. Schwartz

E. O. Hulburt Center for Space Research
Naval Research Laboratory
Washington, D. C. 20375

The region of the Kliemann-Low (KL) nebula near IRc2 and IRc4 contains not only OH and H_2O masers but also an SiO maser. When observed in the strongest and most familiar SiO maser transitions, $J = 1 \rightarrow 0, v = 1, 2$, and $J = 2 \rightarrow 1, v = 1$, Orion is not dissimilar from the many other known SiO masers in terms of its intensity, spectral shape, or polarization. In one respect, however, Orion is unique: It is the only known SiO maser not positively identified with a late-type long-period variable star. SiO masers are associated with many (and perhaps all) late-type stars that contain OH and H_2O emission but, as illustrated by a recent survey of 27 star-formation regions containing OH and H_2O masers, there are no SiO masers as strong as KL in "classic" OH and H_2O regions within approximately 8 kpc.[1] This remarkable result suggests the possibility that one of the objects within KL, perhaps IRc2, might be a late-type star associated with the region either in a true spatial sense or, perhaps, as a foreground or background object. If this object were a Mira variable and spatially associated with the KL nebula, its probable characteristics—$L \simeq 10^4 \, L_{\odot}$ and a mass loss rate $\dot{M} \simeq 10^{-6} \, M_{\odot} \, y^{-1}$—could explain some of the observations of the region near IRc2.

Before the possibility that IRc2 is a late-type star is considered seriously, the question of whether Orion really resembles other SiO sources must be answered. Recently, Schwartz *et al.* observed a sample of SiO sources in their strong maser transitions in order to make a "snapshot" of this phenomenon.[2] In this experiment, the $J = 3 \rightarrow 2$ and $J = 4 \rightarrow 3, v = 1, 2$, lines were observed in addition to the $J = 1 \rightarrow 0, v = 1, 2$, and $J = 2 \rightarrow 1, v = 1$, lines because of the lack of data on the $J > 2$ lines; the observations were made simultaneously because of the large and rapid time variability of SiO masers. A good example of the results of this experiment, the simultaneous spectra of the S-type Mira variable χ Cyg, is shown in FIGURE 1. Schwartz *et al.*, summarized the observational characteristics of the SiO maser phenomenon as follows.

SiO maser emission is observed in $J = 1 \rightarrow 0$; $v = 1, 2, 3$; $J = 2 \rightarrow 1, v = 1, 2$; $J = 3 \rightarrow 2, v = 1, 2$; and $J = 4 \rightarrow 3, v = 1, 2$ transitions. The $J = 1 \rightarrow 0$ and $J = 2 \rightarrow 1, v = 0$, transitions also probably show some maser emission.

The intensity ratios within the vibrational ladders are such that, in $v = 1, J = 1 \rightarrow 0$ and $J = 2 \rightarrow 1$ are much more intense than $J = 3 \rightarrow 2$ and $J = 4 \rightarrow 3$, and $J = 2 \rightarrow 1$ is probably more intense than $J = 1 \rightarrow 0$, but, conversely, in $v = 2, J = 2 \rightarrow 1$ is much less intense than $J = 1 \rightarrow 0, J = 3 \rightarrow 2$, and $J = 4 \rightarrow 3$, and, in some cases, $J = 1 \rightarrow 0$ is less intense than $J = 3 \rightarrow 2$ and $J = 4 \rightarrow 3$. In $v = 0$ and $v = 3$ it appears as though $J = 1 \rightarrow 0$ maser emission is more intense than any other rotational transition (note: in the broad thermal $v = 0$ features, $J = 3 \rightarrow 2$ is usually more intense than $J = 2 \rightarrow 1$).

The intensity ratios within a rotational state are such that, in $J = 1 \rightarrow 0$; $v = 1/v =$

0077–8923/82/0395–0216 $1.75/1 © 1982, NYAS

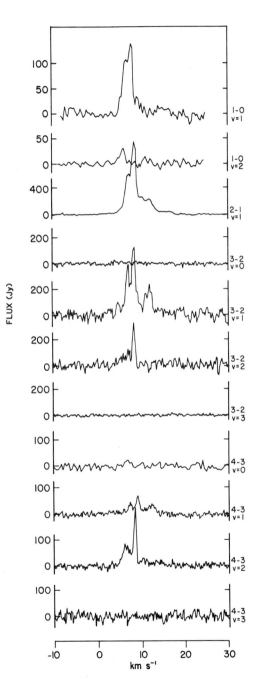

FIGURE 1. SiO maser emission from the S-type long period variable χ Cyg observed in May 1981. The $J = 1 \rightarrow 0$ transitions are near 43 GHz, the $J = 2 \rightarrow 1$ near 86 GHz, the $J = 3 \rightarrow 2$ near 129 GHz, and the $J = 4 \rightarrow 3$ near 172 GHz.[2]

$2 \simeq 1$; in $J = 2 \rightarrow 1$, $v = 1/v = 2 \gtrsim 20$; in $J = 3 \rightarrow 2$, $v = 1/v = 2 \simeq 2$; while, in $J = 4 \rightarrow 3$; $v = 1/v = 2 \simeq 0.3$. In $J = 1 \rightarrow 0$, both $v = 0$ and $v = 3$ are much less intense than $v = 1$ or $v = 2$, with $v = 1/v = 3 \gtrsim 20$.

Maser chains within vibrational states exist with very similar spectra in different rotational transitions. Likewise, spectra within $J = 1 \rightarrow 0$ and $v = 1$, 2 are often similar, also, and it is certainly true that, to within about 1 km s^{-1}, many velocity coincidences exist in all the detected SiO maser transitions.[2]

The SiO maser in Orion KL diverges from this general picture, as FIGURE 2

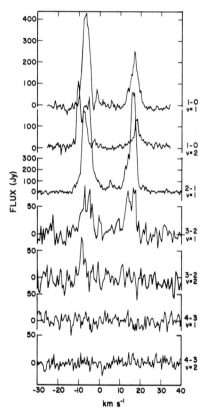

FIGURE 2. The SiO maser spectrum of KL, illustrating the weak $J = 3 \rightarrow 2$ lines and the absence of the $J = 4 \rightarrow 3$ lines.[2]

illustrates. The $J = 3 \rightarrow 2$ transitions are very weak and the $v = 2$ spectrum has a significantly different velocity structure (the positive velocity component is undetectable). The $J = 4 \rightarrow 3$ transitions are undetectable. All objects that Schwartz et al. detected in $J = 1 \rightarrow 0$, $v = 1$, 2, except Orion, were also detected in $J = 4 \rightarrow 3$, $v = 1$, 2.[2]

The picture that thus emerges is that whatever set of conditions excites the SiO $J = 1 \rightarrow 0$ and $J = 2 \rightarrow 1$ masers near late-type long-period variables also exists in KL, but the conditions responsible for excitation of the higher J maser transitions are absent or modified sufficiently to result in a modified $J = 3 \to 2$ velocity structure and not to result in bright $J = 4 \to 3$ masers. KL (or, more particularly, IRc2) appears to the SiO maser region to be very much like but not exactly like the region near a late-type star and, in this respect, appears to be an almost totally unique object completely unlike other star-formation maser regions.

REFERENCES

1. GENZEL, R., D. DOWNES, P. R. SCHWARTZ, J. H. SPENCER, V. PANKONIN & J. W. M. BAARS. 1980. SiO emission in KL—An evolved star in a region of star formation or a unique object in the galaxy. Astrophys. J. **239:** 519–25.
2. SCHWARTZ, P. R., B. ZUCKERMAN, & J. M. BOLOGNA. 1982. Nearly simultaneous observations of vibrationally excited $J = 1 \to 0, J = 2 \to 1, J = 3 \to 2$ and $J = 4 \to 3$ SiO masers. Astrophys. J. **256:** L55–59.

DISCUSSION OF THE PAPER

P. THADDEUS (*Goddard Institute for Space Studies, New York, N.Y.*): Why are SiO masers not seen in other H II regions?

SCHWARTZ: It is a real mystery. Orion must be unique.

AN ANALYSIS OF THE ORION SiO MASER*

Moshe Elitzur

*Department of Physics and Astronomy
University of Kentucky
Lexington, Kentucky 40506*

Introduction

Orion is the source where the original discovery of SiO maser emission was made by Snyder and Buhl.[1] It still stands out as the only SiO maser, so far, found to be associated with a region of star formation rather than an identified late-type star. This led to some debate about the nature of the source and the suggestion that an evolved star may be located in the Kleinmann-Low (KL) nebula, a region of active star formation.

The debate appears to have been settled by observations that identify the SiO maser with the source IRc2.[2] It now appears that IRc2 is the center of activity in the KL region, an activity that manifests itself in the outflow, at a velocity of ~ 18 km s^{-1}, of various molecular species, particularly H_2O masers.[3] IRc2 is apparently a young, luminous ($\sim 10^5 L_\odot$) star with a very high mass-loss rate ($10^{-3} M_\odot$ y^{-1}), which may be responsible for the entire infrared luminosity of the KL nebula.[4]

Attempts to discover SiO masers in other regions of star formation, most recently by Genzel et al.,[5] have all failed up to now and the Orion Nebula stands out as the only star-formation region that supports an SiO maser. In this paper, I shall try to apply the concepts developed for SiO masers in late-type stars[6] in an attempt to understand the Orion SiO maser and its apparent uniqueness. The details of the model will be published elsewhere.

Basic Concepts

SiO maser radiation occurs in low rotation transitions of the $v = 1$, $v = 2$, and $v = 3$ excited vibration states. The high excitation energies imply that the masering material is located very close to the central stars, essentially at the upper parts of their atmospheres. The densities are on the order of 10^{11}–10^{12} cm^{-3} and the temperatures are on the order of 1500–2000 K. The vibration states are excited by collisions from the lower states and then decay radiatively. It can be shown that the excitation rates are J-independent.[7] The radiative decay rates, on the other hand, decrease with J when the vibration transitions are optically thick because of the trapping of photons.[8] The level populations, therefore, increase with J, leading to population inversion and a maser effect.

An operation of the inversion mechanism therefore requires that two basic conditions be met (1) the densities in the maser region cannot be higher than $\sim 10^{12}$ cm^{-3}, because otherwise the vibration decays would be dominated by collisions that

*This research was supported, in part, by a grant from the National Science Foundation, no. AST-80-23712.

220

would thermalize the level populations and quench the maser, and (2) the vibration decay transitions must be optically thick so that their rates become J-dependent. Let us check whether these conditions can be met in the outflow from IRc2. The density n of the outflowing material can be written as

$$n = 1.5 \times 10^{10} \frac{\dot{M}_4}{r_{14}^2 \, v_{10}} \text{ cm}^{-3}, \tag{1}$$

where $\dot{M}_4$ is the mass-loss rate (in units of $10^{-4} \, M_\odot \text{ y}^{-1}$), r_{14} (in 10^{14} cm) is the radius, and v_{10} (in 10 km s^{-1}) the velocity.

As a first-order guess for the parameters of the Orion SiO maser region, we take $\dot{M} = 10^{-3} \, M_\odot \text{ y}^{-1}$, $v = 18$ km s^{-1}, and $r = 2 \times 10^{14}$ cm. The resulting number density is 4×10^{10} cm^{-3}, which is comfortably below the collisional thermalization limit.

Next is the question of the optical thickness of the vibration transitions. The optical depth $\tau_{1,0}$ of the transition $(v = 1, J = 0) \to (v = 0)$ can be written as

$$\tau_{1,0} = 9 \frac{\dot{M}_4}{r_{14} \, v_{10}^2 \, T_3} (1 - \exp(-1.78/T_3))^2, \tag{2}$$

where T_3 is the temperature in units of 1000 K. The temperature dependence arises from the SiO partition function and the induced emission factor. At a temperature of 1500 K and with the other parameters set as above, the optical depth is

$$\tau_{1,0} = 0.45 \, \dot{M}_4. \tag{3}$$

This reult may hold a key to the "uniqueness problem" of Orion because it implies that, all other things being equal, the mass-loss rate must be on the order of 10^{-3} $M_\odot \text{ y}^{-1}$ to activate the maser inversion scheme. The percentage of stars with such a high mass-loss rate may be low. Even with a mass-loss rate of $\sim 10^{-3} \, M_\odot \text{ y}^{-1}$, the optical depth, $\tau_{1,0}$, is only about five in Orion. The optical depth of the $v = 2 \to 1$ transition is about three and that of the $v = 3 \to 2$ transition is only about one. For a temperature that is slightly lower than 1500 K, the $v = 3 \to 2$ transition becomes optically thin. This may explain the lack of maser radiation from $v = 3$ in Orion.

In spite of the relatively small optical depths, the emitted flux can be high. At a temperature of 1500 K, the number of maser photons emitted can be written as

$$\phi_{\text{SiO}} = 1.4 \times 10^{43} \frac{\dot{M}_4^2}{r_{14} \, v_{10}^2} \text{ phot s}^{-1}. \tag{4}$$

Assuming a distance of 400 pc, the observed maser photon flux from Orion is 4×10^{44} phot s^{-1}.[9] With the above-mentioned estimates for r and v, the mass-loss rate would have to be $1.4 \times 10^{-3} \, M_\odot \text{ y}^{-1}$ to explain the observations. This is essentially the same estimate as that derived from the requirement of optical thickness. It is evident that the maser output is large, even though the conditions for maser operation are barely met. Another way of making the same point is by combining (3) and (4) to get

$$\phi_{\text{SiO}} = 1.7 \times 10^{42} \, \tau_{1,0}^2 \, r_{14} \, v_{10}^2 \text{ phot s}^{-1} \tag{5}$$

at $T = 1500$ K. We see that, once the maser becomes operable, namely when $\tau_{1,0}$ is on the order of 1–3, the flux output is immediately appreciable for the parameters of Orion.

We have, therefore, obtained a consistent set of parameters that can explain the SiO maser operation in Orion and, possibly, also provide an explanation for the lack of such action in the other star-formation regions observed up to now. The model also explains why the maser cannot be located further out in the flow. As is evident from (2), the optical depths decrease with radius and the inversion mechanism cannot then operate any more.

At this point, however, the model cannot explain why the maser region is not located at a smaller radius. In fact, the output flux is inversely proportional to the radius (4) and the closest regions of SiO inversion should dominate the maser output. One possible explanation for the lack of inversion at smaller radii could be collisional quenching due to the increase in density (1), but it appears that that would happen only at radii less than 10^{14} cm. Such small radii are ruled out by VLBI observations.[10] An understanding of this problem appears to require a study of the flow itself.

THE OUTFLOW

The material outflowing from IRc2 is traced by various molecules in the region between $r \simeq 10^{14}$ cm and $r \simeq 2 \times 10^{17}$ cm.[4] The flow velocity is remarkably constant at a value of $\simeq 18$ km s^{-1}. Because of the large amount of dust absorption observed in the infrared,[4] radiation pressure on the dust grains seems like a prime suspect for the cause of the flow. The velocity due to radiation pressure is given by

$$v = \frac{L\tau_d}{\dot{M}c}(1 - y),\qquad(6)$$

where

$$y = 4\pi\, GMc/(\kappa L)$$

and τ_d is the optical depth of the dust. In deriving this equation it was assumed that the dust absorption coefficient per unit mass, κ, is constant beyond the dust formation point. Numerically, the equation can be cast in the form

$$v = 2\,\frac{L_5}{\dot{M}_3}\,\tau_d\left(1 - 0.09\,\frac{M_{10}}{\kappa_{15}\,L_5}\right) \text{km s}^{-1},\qquad(7)$$

where L_5 is the luminosity of the central source in $10^5\,L_\odot$, M_{10} is its mass in units of 10 $M_\odot$ and κ_{15} is the dust absorption coefficient in units of 15 cm^2 g^{-1}, which is a standard estimate in the relevant wavelength region.[11] Since all the parameters in (7) are normalized to the most plausible estimates for IRc2, we see that radiation pressure could explain the outflow if the dust optical depth were about 10. If the density decreases as r^{-2}, then the dust optical depth is essentially its value at the base of the flow and can be written as

$$\tau_d = 757\,\frac{\kappa_{15}\,\dot{M}_3}{r_{14}\,v_{10}},\qquad(8)$$

which would lead to expansion velocities too high by almost two orders of magnitude. The most obvious solution to this puzzle is that the dust opacity is much lower than its standard value, presumably because the grains did not condense sufficiently or their dimensions are smaller than average. But when the grain opacity is decreased, the factor y increases and has to be taken into account. The consistent solution of (7) when $v = 18$ km s^{-1} is $\kappa_{15} \simeq 0.11$, so the dust is underabundant by about a factor of 10. Interestingly enough, this is the value that makes $y \simeq 1$; that is, the source is at its Eddington limit. The assumption that the outflow is triggered by radiation pressure on the dust grains is, therefore, self-consistent.

For an absorption coefficient that is $\sim 10\%$ of the standard value, the dust optical depth is on the order of 10. This is also the value required to resolve the "momentum transport problem," according to some estimates,[3,11] so the present model is capable of explaining the 18 km s^{-1} flow. However, the papers by G. Knapp and B. Zuckerman in this volume arrive at higher estimates for the momentum transported by the expanding material. If this is indeed the case, one would have to assume that there is an additional flow in the KL region, for which another explanation would have to be invoked.

Unlike late-type stars, the SiO masering material in Orion is, therefore, mixed with dust that is optically thick in the SiO vibration transitions. This affects two fundamental points of the basic maser model: (1) in calculating the photon trapping factors for the vibration decays, we must take into account the dust optical depth and (2) the molecules are exposed to the dust blackbody radiation field, which tries to thermalize the level populations at the dust temperature.

The first point does not influence the overall picture too much. It is not difficult to show that absorption by the dust will effectively increase the decay rates of the SiO vibration transitions by a factor of $1 + \tau'_d$, where τ'_d is related to the ordinary optical depth of the dust τ_d by

$$\tau'_d = \tau_d \frac{v_{\text{th}}}{v_{\text{ex}}}. \tag{9}$$

Because v_{th}, the gas thermal speed, is about a factor of 10 less than v_{exp}, the expansion velocity, it is evident that $\tau'_d \simeq 1$, so the inversion scheme and all the expressions of the basic maser model are hardly affected.

The second point is of fundamental importance to the model. If the temperatures of the dust and gas were equal, all the level populations would equilibrate with this common temperature and population inversion would be impossible. By solving a simple two-level model for the J and $J - 1$ levels of $v = 1$ and neglecting collisions between the levels themselves, it can be shown that the level populations obey

$$n_J - n_{J-1} = n^{(o)} \frac{\dfrac{C_{0,1}}{\Gamma} \left(1 - \exp\left(-\dfrac{E}{kT_d}\left(1 - \dfrac{T_d}{T} \right) \right) \right)}{\left(1 + \left(J + \dfrac{1}{2} \right) \dfrac{C_{1,0}}{\Gamma} \right) \left(1 + \left(J - \dfrac{1}{2} \right) \dfrac{C_{1,0}}{\Gamma} \right)}, \tag{10}$$

where $n^{(o)}$ is the population per sublevel of the ground vibration state, $C_{v,v'}$ is the collision rate for the $v \rightarrow v'$ transition, E is the energy of the $v = 1$ state, T is the kinetic temperature, T_d is the dust temperature, and Γ measures the vibration decay rate in the presence of the dust radiation field. It is evident that an inversion ($n_J > n_{J-1}$) is possible only when $T > T_d$; that is, the gas must be hotter than the dust.

The dust temperature is obtained by balancing the radiation the grains absorb and emit. If the dust was optically thin, this would lead to the estimate

$$T_d \simeq 1600 \text{ K } (L_5/r_{14}^2)^{1/5}. \qquad (11)$$

The calculations of Scoville and Kwan show that, for the dust optical depth estimated here, its temperature would be reduced by about 50% below the estimate of (11).[12] It is, therefore, reasonable to assume that the dust temperature, T_d, is about 1000 K.

For the maser to operate, the kinetic temperature therefore has to be ~ 1500 K. But this is just the temperature at which grains can begin to condense at these values of pressure, according to Salpeter,[13] and the first substance to condense is corundum (Al_2O_3), which, significantly enough, is not a silicate. The following picture therefore emerges for the maser region: The gas maintains a high temperature ($\gtrsim 1500$ K) out to a distance of about 10^{14} cm. For the densities and temperatures expected, the most significant cooling is through H_2 rotation transitions at a rate[15] of about 2×10^{-20} erg s^{-1} per H_2. To balance this cooling rate, an overall energy input of ~ 6.5 $L_\odot$ is required, which does not seem excessive. It seems most plausible that this energy input is in the form of large gas motions, which are probably turbulent. This would immediately explain the lack of SiO maser action closer to the star; it simply reflects the lack of coherence in the velocity field. If the mean velocity in the turbulent gas is u, then the heating rate H per H_2 molecule is

$$H = 1/2 \, m_{H_2} u^2 \cdot u/r = 2 \times 10^{-20} \, u_{10}^3/r_{14} \text{ erg s}^{-1} \text{ per } H_2, \qquad (12)$$

which can, therefore, balance the H_2 cooling rate at $u \simeq 10$ km s^{-1}.

When the temperature drops below ~ 1500 K, the grains condense and the radiation pressure on the grains begins to drive the material outwards. The motions become ordered and the maser turns on once the flow velocity exceeds the velocity of the random motions. Indeed, the difference between the peak velocities of the SiO is about 22–24 km s^{-1},[14] so the flow velocity, which is half this value, is about 11–12 km s^{-1}. The velocity widths of the maser peaks are about 12–14 km s^{-1}, so the chaotic motions have a mean speed of about 6–7 km s^{-1}, similar to the value estimated above. The SiO maser, therefore, probes the region where grains form and the ordered outflow is established. Indeed, the maser peaks have kept their overall shape and velocities almost unchanged over the past ten years, while exhibiting many variations in the strength and precise location of the components inside the peaks. This agrees rather well with the suggestion that we observe a chaotic velocity field, of ~ 7 km s^{-1}, superposed on an ordered motion of ~ 11 km s^{-1}.

Note that the turbulent energy input must persist throughout the maser region to maintain the gas at a temperature higher than the dust, since, otherwise, the gas will cool by H_2 emission in only about 10^7 s, during which the material will cover a distance of only $\sim 10^{13}$ cm, less than the observed maser dimensions. In addition, once the dust has formed, it provides another efficient cooling mechanism through its collisions with the gas. This cooling rate, C_d, is given by

$$C_d = 1 \times 10^{-20} \frac{\dot{M}_3 T_3^{1/2}}{r_{14}^2 \, v_{10}} \left(\frac{T - T_d}{100 \text{ K}} \right) \text{ erg s}^{-1} \text{ per } H_2, \qquad (13)$$

which is comparable to H_2 cooling. It therefore seems most likely that the maser will turn off because the gas and dust temperatures become comparable before the SiO vibration transitions become optically thin. The nature of the velocity field and the gas-dust temperature difference therefore provide the explanation for the location and dimensions of the SiO maser source.

CONCLUSIONS

It appears that the SiO maser emission from Orion provides us with a rare opportunity to observe the transition from a chaotic to an ordered velocity field due to radiation pressure on grains that have just formed. The reason why star-forming regions do not generally support SiO maser action may be the need for very high mass-loss rates ($\sim 10^{-3}$ $M_\odot$ y^{-1}), as well as the rather close coincidence of parameters, which leads to the gas being warmer than the dust over a sufficiently large region.

SUMMARY

The nature of the source IRc2 in Orion, presumably the center of activity of the KL nebula, is analyzed in light of its maser emission. Questions that receive particular attention are related to the unique nature of this source, the only SiO maser, so far, not associated with a late-type star. It is suggested that the SiO maser emission emanates from a region where, as a result of the formation of dust, the velocity field changes from chaotic to ordered.

REFERENCES

1. SNYDER, L. E. & D. BUHL. 1974. Astrophys. J. **189**: L31.
2. BAUD, B., J. H. BIEGING, R. PLAMBECK, D. THORNTON, W. J. WELCH & M. WRIGHT. 1980. In Interstellar Molecules. B. H. Andrew, Ed. D. Reidel. Dordrecht.
3. GENZEL, R., M. J. REID, J. M. MORAN & D. DOWNES. 1981. Astrophys. J. **244**: 884.
4. DOWNES, D., R. GENZEL, E. E. BECKLIN & C. G. WYNN-WILLIAMS. 1981. Astrophys. J. **244**: 869.
5. GENZEL, R., D. DOWNES, P. R. SCHWARTZ, J. H. SPENCER, V. PANKONIN & J. W. M. BAARS. 1980. Astrophys. J. **239**: 519.
6. ELITZUR, M. 1980. Astrophys. J. **240**: 553.
7. WATSON, W. D., M. ELITZUR & R. J. BIENIEK. 1980. Astrophys. J. **240**: 547.
8. KWAN, J. & N. SCOVILLE. 1974. Astrophys. J. **194**: L97.
9. SNYDER, L. E. & D. BUHL. 1975. Astrophys. J. **197**: 329.
10. GENZEL, R., J. M. MORAN, A. P. LANE, C. R. PREDMORE, P. T. P. HO, S. S. HANSEN & M. J. REID. 1979. Astrophys. J. **231**: L73.
11. SOLOMON, P. M., G. R. HUGUENIN & N. Z. SCOVILLE. 1981. Astrophys. J. **245**: L19.
12. SCOVILLE, N. Z. & J. KWAN. 1976. Astrophys. J. **206**: 718.
13. SALPETER, E. E. 1977. Annu. Rev. Astron. Astrophys. **15**: 267.
14. SNYDER, L. E., D. F. DICKINSON, L. W. BROWN & D. BUHL. 1978. Astrophys. J. **224**: 512.
15. HARTQUIST, T. W., M. OPPENHEIMER & A. DALGARNO. 1980. Astrophys. J. **236**: 182.

PREDICTION OF THE PROTOSTELLAR MASS SPECTRUM IN THE ORION NEAR-INFRARED CLUSTER

Hans Zinnecker

Max-Planck-Institut für Physik und Astrophysik
Institut für extraterrestrische Physik
8046 Garching, Federal Republic of Germany

INTRODUCTION

The nature of the bright ($\sim 10^5 L_\odot$) far-infrared nebula in Orion has been a challenge both for observers and theoreticians ever since its discovery by Kleinmann and Low.[1] Immediately after its discovery, the Kleinmann-Low nebula (KL) was interpreted as a protocluster of several massive stars embedded in an opaque dust cloud.[2] Today it is well known (*e.g.,* Reference 3) that the position of the nebula coincides with the dense core of the Orion A Molecular Cloud, roughly one arcmin to the northwest of the Trapezium stars. The extent of the nebula is about one arcmin, corresponding to a diameter of 0.15 pc at the distance of the Orion Molecular Cloud (500 pc). The central gas density in the KL nebula is believed to be $\sim 10^6$ cm^{-3} and the central gas and dust temperature, ~ 70 K.[4-6] The brightest near-infrared source is the BN object.[7] Its luminosity is $\lesssim 10^4 L_\odot$. Its infrared spectrum has been modelled by Bedijn *et al.*[8] From the diameter and the gas density, given above, the total gas mass of the KL nebula is found to be $\sim 10^2 M_\odot$.*

In the last few years, observations have also revealed that the KL nebula has a very complex structure, owing to an energetic event—most likely, a strong stellar wind that causes high-velocity outflow phenomena (see the review by Scoville[9]). It appears that the infrared source IRc2 (which is near the center of symmetry of the extended contours of the KL nebula) is the primary source of energy for driving the outflow of matter as well as the high flux of far-infrared radiation.[10] In other words, IRc2 is probably the most massive star in the KL nebula. Recently, Lonsdale *et al.* have discovered a near-infrared star cluster embedded in the KL nebula in a region 40″ × 40″ (0.1 pc × 0.1 pc) centered on the BN object.[12] The cluster comprises about 20 members brighter than $K = 11^m.8$, the limiting magnitude of the near-infrared search ($\lambda_{\text{eff}} = 2.2$ μm). This sample of point sources includes the BN object as well as the few sources that had already been detected by Rieke *et al.* at 5, 10, and 20 μm[13] (see also Reference 14). The masses of all those stellar objects are unknown at present, although Lonsdale *et al.* favor the hypothesis that most of the new objects are pre-main-sequence stars of spectral type A (which would correspond to 2–3 $M_\odot$ objects).

The aim of the present theoretical study is to develop a simple model to predict the mass spectrum of the KL cluster. Despite the high stellar number density of the cluster ($\sim 10^4$ pc^{-3}), we favor an accretion model rather than a coagulation model (see References 15–17 for predictions of mass spectra based on coagulation theory).

---

*In the Russian literature, there is—as far as we know—only a single paper on the KL-BN infrared sources,[11] in which polarimetric observations are analyzed to put constraints on the system parameters; yet the mass of the KL nebula is severely overestimated ($10^4 M_\odot$).

0077-8923/82/0395-0226 $1.75/1 © 1982, NYAS

In the following section, we give initial conditions for the model. The main section deals with the fragmentation of the protocluster cloud and contains our analytical model for the protostellar mass spectrum, following which we discuss our main simplifying assumptions as well as related work. Finally, we summarize our results.

INITIAL CONDITIONS

We consider a cloud that is just dense enough for the magnetic field to become dynamically irrelevant. The cloud is assumed to be the central, spherical subsystem of a larger rotating cloud; this subcloud is assumed to represent a closed system so that no matter from the larger external cloud can flow into the central subsystem.

The unperturbed protocluster cloud is taken to be homogeneous with respect to density and temperature; rigid rotation is assumed for simplicity.

TABLE 1 lists the actual set of initial parameters for the protocluster cloud. These data are somewhat arbitrary, though not implausible. For instance, the adopted angular velocity is consistent with the overall velocity gradient of ~ 0.6 km s^{-1} pc^{-1}

TABLE 1

INITIAL CONDITIONS FOR THE KL CLOUD

Density	$n = 3 \times 10^4$ cm^{-3}
Temperature	$T = 10$ K
Angular Velocity	$\Omega = 10^{-13}$ s^{-1}
Diameter	$D = 0.5$ pc
Mass	$M = 125\, M_\odot$

along the Orion A cloud,[18] increased by a factor of five due to some contraction of the central part of the cloud. From the parameters of TABLE 1 we find the following characteristics:

$$\text{freefall time,}\quad t_{ff} = 2 \times 10^5 \text{ y,}$$

$$\text{Jeans mass,}\quad M_J = 1\, M_\odot,$$

$$E_{rot}/E_{grav} = 0.1.$$

The last quantity is the ratio of the rotational energy to the gravitational energy (taken to be positive), which describes how close the system is to centrifugal equilibrium in the beginning. For rigid rotation, centrifugal equilibrium would correspond to a value of this ratio equal to one third;† hence, the system can shrink by a factor of 3.3, in accordance with the present diameter of 0.15 pc.

From M_J and TABLE 1, we infer that our protocluster cloud contains as many as 125 Jeans masses. Such a situation could arise if the magnetic field originally stablized the cloud but "suddenly" decoupled from the bulk of the neutral matter.[19]

---

†Possible axisymmetric (ring-like) and nonaxisymmetric (bar-like) instabilities associated with rotation are not taken into account.

Now, we still have to specify some perturbations on top of the smooth cloud background in order to induce fragmentation during the collapse of the protoculster cloud. To be definite, we imagine 20 density perturbations‡ with a density contrast of 12.5% with respect to the average cloud density. The diameter of these perturbations is taken to be 1.25 times the local Jeans length (which is $L_J = 0.10$ pc, *i.e.,* one-fifth of the initial cloud diameter D from TABLE 1). It follows that the mass of each incipient fragment is about two Jeans masses.

ACCRETION MODEL

Assumptions

The evolution of the protocluster cloud consists of a global collapse, slowed down, however, by centrifugal forces, while, locally, the fragments collapse on their own. The fragments are allowed to reach protostellar densities through quasi-spherical collapse without further fragmentation. This picture implies some mechanism for local transport of angular momentum (*cf.* Reference 20).

The fragments are centrally condensed after a free-fall timescale. This circumstance prevents the fragments from being tidally disrupted again. Moreover, the existence of such centrally condensed fragments does not allow for further fragmentation of the protocluster cloud, since the dynamics of its residual gas is dominated by the gravitational attraction of the protostellar fragments (*cf.* Reference 20). In this situation, the protostellar fragments will try to accrete the residual gas of their parent cloud; in fact, they will compete for the accretion of the remaining gas.

In addition, we make the important assumption that coalescence of protostellar fragments will not occur (conservation of the number of protostellar fragments): It is this assumption that will keep our model simple, *i.e.,* amenable to a *linear* kinetic approach. Justification for this assumption comes from the fact that the protostellar fragments take part in the overall rotation of the protocluster cloud. (The velocity dispersion of the fragments is assumed to be negligible.)

If protostellar fragments gain most of their mass by gas accretion and do not collide with each other, we can easily see that the resulting mass spectrum will be determined by the mass dependence of the accretion rate of the protostellar fragments.

In our case, we shall adopt an accretion rate that is proportional to the square of the instantaneous mass of any protostellar fragment. Thus, the accretion rate is the well-known accretion rate for a point mass surrounded by an infinite nonself-gravitating medium.[21,22]

While this assumption will undoubtedly introduce some oversimplification into the problem, it is unlikely that the model is seriously affected because the point mass approximation may apply as soon as the density contrast $(\Delta\rho/\rho)$ of the initial perturbations has grown to the order of unity. For pressure-free cloud collapse, the initial density contrast (12.5%) grows to 75% by the time the cloud has shrunk by a factor of 3.3, when it reaches centrifugal equilibrium $(\Delta\rho/\rho \simeq D^{-3/2}$, see Reference

---

‡This figure is motivated by the number of near-infrared sources found in the observations of Lonsdale *et al.*[12]

23). The expansion wave–driven accretion mode[24] that predicts a mass-independent accretion rate[25] does not apply in our case because the density profile of the perturbations will join the background density level without exhibiting a dip (no rarefaction wave).

Another assumption of our model is the introduction of some turbulent velocity dispersion of the gas (c_{rms}), which is taken to be constant all over the cloud. We choose $c_{rms} \simeq 1$ km s^{-1}. This value appears to be a reasonable choice; for, from the virial theorem, we would expect $(GM_{cl}/R_{cl})^{1/2} \simeq 2.5$ km s^{-1} for a cloud with mass $M_{cl} = 125$ $M_\odot$ and diameter $2R_{cl} = 0.15$ pc; but, since the cloud is mostly stabilized by rotation, the velocity dispersion has to be less than the virial velocity. (The gas velocity dispersion is, however, supersonic, as is typical for molecular cloud cores; see the list in Reference 26).

So, in an idealized version, we have the following picture: N point masses of nearly equal mass are placed side by side in a reservoir of gas from which each of them can accrete. Due to the common rotation of the gas and the accreting point masses, the latter are at rest with respect to the surrounding gas. Therefore, we do not have to worry about fluctuating accretion rates due to relative motion nor about accretional drag (*e.g.*, Reference 27).

The reservoir of gas is assumed to be infinite, *i.e.*, the gas density is kept constant and time-independent. It turns out that this assumption is not as drastic as it seems at first sight (see the discussion below).

Finally, in spite of the high gas density, the selfgravity of the gas in the accretion process[28] has been neglected. Of course, a more realistic model would have to take all these effects into account. It is hoped, however, that this simplified approach will delineate the essential features of a more complicated model.

Basic Process

From the above assumptions, it follows that the accretion rate for the protostellar objects is given by

$$\dot{M} = \alpha M^2 \qquad (\alpha = \text{const}), \qquad (1)$$

where α is a function of the cloud gas density ρ_{cl} and the turbulent cloud gas velocity dispersion, c_{rms} ($\alpha \simeq 4\pi G^2 \rho_{cl}/c_{rms}^3$, with gravitational constant G). ρ_{cl} is related to the proton number density of the cloud (n) by $\rho_{cl} = \mu m_p n$, where m_p is the proton mass and μ is the mean molecular weight for a cold molecular cloud ($\mu \simeq 2.4$). The solution of (1) yields

$$M(t) = \frac{M_0}{1 - \alpha M_0 t} \qquad (2)$$

for the protostellar mass as a function of time. M_0 is the starting mass $M(t = 0)$ with $t = 0$ corresponding to the moment when the initial free-fall time has passed. Note that $M(t) \to \infty$ for $t \to t_\infty = (\alpha M_0)^{-1}$. The growth in mass of three representative starting masses ($M_0 = 0.5 M_*$, $M_0 = 2 M_*$, and $M_0 = M_*$) accreting from the same gaseous medium ($\alpha = \text{const}$) is shown in FIGURE 1. (Here M_* is an arbitrary stellar mass.) It is

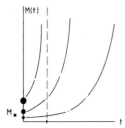

FIGURE 1. Solutions to the accretion equation $\dot{M} = \alpha M^2$ for three different starting masses (see text). The dashed line represents t_∞ for the largest starting mass.

seen from FIGURE 1 that the mass of the first-ranked starting mass diverges before either of the other masses has grown very far. This behavior is called "runaway accretion." It illustrates that small differences in the starting mass are very much accentuated in the course of time (towards $t \rightarrow t_\infty$ of the most massive starting mass). In particular, the difference between the first-ranked and the second-ranked starting mass becomes greater and greater ("gap"). It is interesting to note that such a runaway process is also found in numerical simulations of the growth of the largest protoplanets.[29]

Kinetic Approach

We shall now give a simple kinetic equation according to which the mass spectrum evolves in time due to accretion, starting with a mass spectrum similar to a delta function. Then we shall solve this kinetic equation, employing the accretion rate given in (1). Since the adopted accretion rate is a function of mass only, the kinetic equation will be one-dimensional only.

The kinetic equation is

$$\frac{\partial F}{\partial t} + \frac{\partial}{\partial M}(\dot{M}F) = 0. \tag{3}$$

Here, $F = F(M,t)$ is the number of accreting objects per unit mass (the differential mass spectrum); (3) is nothing else but the continuity equation in "mass-space," which shows that the number of objects is fixed during the accretion process, while the masses change. Equation 3 is solved by the method of characteristics.

We may write

$$F(M)\mathrm{d}M = F_0(M_0)\mathrm{d}M_0, \tag{4a}$$

F_0 being the mass spectrum to start with, *i.e.*, a distribution of masses with just a little spread. (In the case of exactly equal starting masses, M_0, the masses would remain equal for all times, according to our model.)

Alternatively, we may prefer to write

$$\zeta(M)\mathrm{d}\ln M = \zeta_0(M_0)\mathrm{d}\ln M_0, \tag{4b}$$

if we let $\zeta = \zeta(M, t)$ denote the number of accreting objects per unit logarithmic mass interval (the usual notation for the mass spectrum). Correspondingly, ζ_0 should be a narrow distribution in logarithmic mass space.

We proceed by evaluating (4b). First, we have to find M_0 as a function of M. This can be obtained by inverting equation 2 to get

$$M_0 = \frac{M}{1 + \alpha M t}. \tag{5}$$

Second, we have to insert (5) into (4b). We get

$$\zeta(M) = \zeta_0 \left(\frac{M}{1 + \alpha M t}\right) \frac{d}{d \ln M}\left[\ln\left(\frac{M}{1 + \alpha M t}\right)\right] \tag{6}$$

or

$$\zeta(M) = \zeta_0 \left(\frac{M}{1 + \alpha M t}\right)(1 + \alpha M t)^{-1}. \tag{7}$$

Equation 7 is the mass spectrum resulting from the accretion process. For $\alpha M t \gg 1$ [*i.e.*, $M \gg M_0$ and $t \to t_\infty = (\alpha M_0)^{-1}$], ζ_0 becomes a constant, independent of mass, and $(1 + \alpha M t)^{-1} \simeq M^{-1}$. Thus,

$$\zeta(M) \simeq M^{-1} \qquad (M \gg M_0). \tag{8}$$

This is the asymptotic shape of the mass spectrum for large masses that develops during accretion (if $\dot{M} \simeq M^2$). Note that the slope of the mass spectrum does not depend on the distribution of the starting masses. The evolution of the mass spectrum is illustrated in FIGURE 2.

Timescale for the Accretion Process

Up to now, we have not estimated the timescale for the accretion process. This timescale is given by $t_\infty = (\alpha M_0)^{-1}$, *i.e.*,

$$t_\infty \simeq 10^5 \frac{\left(\dfrac{c_{rms}}{1 \text{ km s}^{-1}}\right)^3}{\left(\dfrac{\rho_{cl}}{10^4 M_\odot \text{ pc}^{-3}}\right)\left(\dfrac{M_0}{1\, M_\odot}\right)} \text{y}, \tag{9}$$

where we have normalized to the parameters of our system. ($t = t_\infty$ corresponds to the dashed line in FIGURE 1.) One realizes that the accretion timescale is comparable to the initial free-fall timescale of the cloud.

FIGURE 2. The evolution of the mass spectrum, $\zeta \equiv dN/d\log M$, of gravitating point masses for an accretion rate $\dot{M} = \alpha M^2$ (a narrow symmetric distribution evolves into a broad asymmetric distribution).

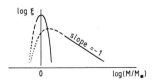

DISCUSSION

Discussion of the Assumptions

Some of the assumptions underlying our model, such as the point mass approximation for the accretion nuclei, have already been discussed. Other assumptions still remain to be discussed, including the assumption of constant gas density and the assumption of noninteracting protostars.

Although the gas density in a finite reservoir drops continuously, while the point masses keep accreting from it, we have neglected the effect of gas exhaustion. We could have replaced the constant cloud density ρ_{cl} by a mean time-dependent cloud density,

$$\rho_{cl}(t) = \rho_{cl}(0) \left(1 - \frac{\overline{M}_*(t)N}{M_{cl}} \right),$$

where N is the number of protostars and $\overline{M}_*(t)$ is the time-dependent mean mass of the mass distribution ($\overline{M}_* N / M_{cl}$ is the fraction of gas that has gone into the protostars at time t). In this case, we would have to solve the problem by an iterative numerical technique. The outcome would be that it takes more and more time for the protostars to

TABLE 2

PREDICTION OF THE MASS SPECTRUM IN THE KL NEBULA

Mass Intervals*	0.5–1	1–2	2–4	4–8	8–16	16–32
Number of Objects	9	6	3	1	0	1

*Equidistant intervals in log-mass (mass in solar units).

accrete a significant amount of matter, until, finally, no more gas is left. However, due to the runaway accretion of the first-ranked starting mass, we expect a violent freezing-in of the mass spectrum rather than a smooth freezing-out. That is, if the first-ranked starting mass grows to a value sufficient to create a hot H II region or a strong stellar wind (bringing accretion to an end) *before* about half the gaseous matter is used up in the protostars, then the mean density cannot have dropped by more than a factor of two. In this case, our assumption of a constant gas density is reasonable. It can be demonstrated that, in fact, our accretion process transforms gas into stars with an efficiency of less than 50% (*cf.* TABLE 2 below). Hence, the assumption of a constant gas density is justified *a posteriori*. Note that, with a star formation efficiency of less than 50%, the resulting stellar cluster will ultimately be unbound as soon as the residual gas will be dispersed or blown off (*e.g.,* Reference 30).

Concerning the assumption of noninteracting protostars, we would like to make the following remark: Even if the velocity dispersion of the protostars were not zero, as assumed in our model, but the same as the turbulent velocity dispersion of the gas (*i.e.,* ~ 1 km s^{-1}), the timescale for inelastic collisions between the protostars (with diameters of ~ 0.01 pc corresponding to the gravitational sphere of influence of a 1 $M_\odot$ point mass in our system) would be $\sim 10^6$ y, *i.e.,* an order of magnitude higher than the accretion timescale. This is why we have stated in the introduction that we prefer an accretion model to a cogulation model.

We realize that the formation of protostars in a protocluster cloud is an extremely complicated problem, since both N-body effects and gas-dynamical effects are important (*e.g.*, Reference 31). It will be a long time yet until numerical codes will be able to provide an exact solution of such a realistic case. In the meantime, we need some approximate models. This may justify our crude approach.

The Results

The prediction for the asymptotic mass spectrum obtained in (**8**) may be presented in a tabular summary (TABLE 2). Attention is drawn to the entry "0," which reflects the "runaway gap." The gap is very pronounced here, because of the small number statistics. We suggest an identification of the entires "1" above and below the gap with IRc2 and the BN object. Thus, the BN object would be the second most massive object in the KL nebula. Considering the number of lower-mass protostars, we notice that the asymptotic mass spectrum (**8**) predicts twice as many objects in the 1–2 $M_\odot$ interval as in the 2–4 $M_\odot$ interval. About one half of the total number of protostars—all those which had originally populated the low-mass wing of the distribution ζ_0—did not gain enough mass during the accretion process to leave the original 0.5–1 $M_\odot$ interval (*cf.* FIGURE 2).

Discussion of Related Work

The OMC1 KL region is usually viewed as the prototype of a molecular cloud core; therefore, our results, scaled appropriately, may apply to other molecular cloud cores as well. For example, Wilking has studied the stellar content of a cluster in the ρ Oph cloud core (distance $\sim$ 150 pc).[32] Interestingly enough, he finds many pre-main-sequence objects ($M < 3M_\odot$) and a few early B stars ($M > 10M_\odot$), but no stars with masses in the range $3M_\odot < M < 10M_\odot$. Could it be that the deficiency of late B stars is due to the runaway accretion of the more massive objects, as outlined in our model?

The results for the mass spectrum from our simple analytical model may be compared with Larson's results from three-dimensional numerical simulations of the fragmentation of a cloud with many Jeans masses[20] (but see Reference 33). These numerical calculations show that the most massive object forms in the densest portion of the cloud (IRc2!), surrounded by several less massive objects, which, in turn, are surrounded by still lower-mass objects. The accretion domain of the most massive objects is the whole cloud, whereas the accretion domain of the less massive objects is smaller, *etc.* The mass spectrum resulting from such an accretion hierarchy is very similar to the mass spectrum derived from our model. So, to some extent, our results are the analytical analogue of Larson's numerical calculations.§

We emphasize that our accretion model is based on coeval star formation (as are

---

§However, our analytical model cannot deal with the formation of binary and multiple systems—the numerical model can.

coagulation models). This means that it applies only to compact clusters, not to open clusters, which are known to have a considerable spread in stellar formation times.

An accretion model for the mass spectrum of an open cluster, in which stellar nuclei form at different times (due to the passage of a shock wave), has been suggested by Bhattacharjee and Williams.[34] These authors also use the $\dot{M} \propto M^2$ accretion rate to derive a mass spectrum starting from unit mass stellar nuclei and a uniform cloud.

Finally, we note that it is possible to study a similar accretion model for protostellar nuclei that are embedded in a clumpy (rather than a smooth) gas cloud. In this case, one can make use of a linear kinetic theory developed by Telford in a different context.[35] It seems that the mass spectrum emerging from a clumpy cloud is asymptotically flatter than the mass spectrum emerging from a smooth cloud, unless the spectrum is somehow frozen before it reaches its self-similar shape.[36]

SUMMARY

We have presented a simple analytical accretion model for the protostellar mass spectrum in the infrared cluster in Orion (OMC1/KL region) in which protostellar cores compete for the accretion of the gas of their parent cloud (cf. Reference 20). Such a model may apply to other molecular cloud cores as well (see Reference 40). The model is a linear model, unlike coagulation models. This is due to the conservation of the number of accretion nuclei, with no collisional mergers occurring. Gas exhaustion effects are neglected, because less than 50% of the cloud gas will be accreted before the most massive star powers the formation of a hot H II region or the formation of an energetic stellar wind, thereby freezing the mass spectrum.

On the basis of an $\dot{M} \propto M^2$ law for the accretion rate of the protostellar cores, we predict a mass spectrum of the form $dN/d \log M \simeq M^{-1}$ for $M \gtrsim 1 \, M_\odot$, independent of the form of the mass spectrum at the beginning of the accretion process. In particular, we predict a runaway growth of the most massive star, with a big gap in mass to the next massive star. Coagulation models would predict an exponential upper cutoff in the mass spectrum. Future observations may be able to distinguish between the predictions of accretion theory and those of coagulation theory.

ACKNOWLEDGMENTS

I should like to thank S. Drapatz and R. Larson for their encouragement and advice. In addition, I am grateful to E. Becklin for providing me with a preprint of the paper by Lonsdale et al. on the discovery of new members of the infrared cluster in the Orion Molecular Cloud.[12]

ADDENDUM

Our model implies that the KL nebula originated through spontaneous condensation in the middle of the OMC1 cloud. However, Ho and Barret, from NH_3 observations, suggest an alternative model in which the KL nebula is the position where two separate clouds are colliding, thereby triggering star formation.[37] Meanwhile, the evidence for two separate clouds has been reinforced by further NH_3

observations[38] and sub-mm observations.[39] So our model may be rivaled by some future model that tries to calculate the theoretical mass spectrum in the KL nebula as a consequence of a cloud-cloud collision.

REFERENCES

1. KEINMANN, D. E. & F. J. LOW. 1967. Astrophys. J. **149**: L1.
2. HARTMANN, W. K. 1967. Astrophys. J. **149**: 287.
3. ZUCKERMAN, B. 1975. In H II Regions and Related Topics. T. Wilson and D. Downes, Eds.: 360. Springer-Verlag. New York.
4. WERNER, M. W., I. GATLEY, D. A. HARPER, E. E. BECKLIN, R. F. LOEWENSTEIN, C. M. TELESCO & H. A. THRONSON. 1976. Astrophys. J. **204**: 420.
5. SCOVILLE, N. Z. & J. KWAN. 1976. Astrophys. J. **206**: 718.
6. GOLDSMITH, P. F., W. D. LANGER, F. P. SCHLOERB & N. Z. SCOVILLE. 1980. Astrophys. J. **240**: 524.
7. BECKLIN, E. E. & G. NEUGEBAUER. 1967. Astrophys. J. **147**: 799.
8. BEDIJN, P. J., H. J. HABING & T. de JONG. 1978. Astron. Astrophys. **69**: 73.
9. SCOVILLE, N. Z. 1981. In Int. Astron. Union Symp. 96, Infrared Astronomy. C. G. Wynn-Williams and D. P. Cruikshank, Eds.: 187. D. Reidel. Dordrecht.
10. DOWNES, D., R. GENZEL, E. E. BECKLIN & C. G. WYNN-WILLIAMS. 1981. Astrophys. J. **244**: 869.
11. GNEDIN, YU. N. & I. G. MITROFANOV. 1976. Sov. Astron. AJ **19**: 673.
12. LONSDALE, C. J., E. E. BECKLIN, T. J. LEE, I. GATLEY & M. STEWART. 1982. Astron. J. Submitted.
13. RIEKE, G. H., F. J. LOW & D. E. KLEINMANN. 1973. Astrophys. J. **186**: L7.
14. BECKLIN, E. E. & C. G. WYNN-WILLIAMS. 1974. Publ. Astron. Soc. Pac. **86**: 5.
15. ARNY, T. & P. WEISSMAN. 1973. Astron. J. **78**: 309.
16. SILK, J. & T. TAKAHASHI. 1979. Astrophys. J. **229**: 242.
17. NAKANO, T. 1966. Prog. Theor. Phys. **36**: 515.
18. LINKE, R. A. & R. G. WANNIER. 1974. Astrophys. J. **193**: L41.
19. MESTEL, L. & L. SPITZER. 1956. Mon. Not. R. Astron. Soc. **116**: 503.
20. LARSON, R. B. 1978. Mon. Not. R. Astron. Soc. **184**: 69.
21. BONDI, H. 1952. Mon. Not. R. Astron. Soc. **112**: 195.
22. McCREA, W. H. 1953. Mon. Not. R. Astron. Soc. **113**: 162.
23. MESTEL, L. 1965. Quart. J. R. Astron. Soc. **6**: 161.
24. SHU, F. H. 1977. Astrophys. J. **214**: 488.
25. STAHLER, S. W., F. H. SHU & R. E. TAAM. 1980. Astrophys. J. **241**: 637.
26. LARSON, R. B. 1981. Mon. Not. R. Astron. Soc. **194**: 809.
27. DODD, K. N. & W. H. McCREA. 1952. Mon. Not. R. Astron. Soc. **112**: 205.
28. CHIA, T. T. 1979. Mon. Not. R. Astron. Soc. **188**: 75.
29. GREENBERG, R., W. K. HARTMANN, C. R. CHAPMAN & J. F. WACKER. 1978. In Protostars and Planets. T. Gehrels, Ed.: 599. University of Arizona Press. Tucson.
30. VON HOERNER, S. 1968. In Interstellar Ionized Hydrogen. Y. Terzian, Ed.: 101. Benjamin. New York.
31. PUMPHREY, W. A. & J. M. SCALO. 1982. Astrophys. J. Submitted.
32. WILKING, B. A. 1981. Ph. D. Diss. University of Arizona.
33. BODENHEIMER, P., J. E. TOHLINE & D. C. BLACK. 1980. Astrophys. J. **242**: 209.
34. BHATTACHARJEE, S. K. & I. P. WILLIAMS. 1980. Astron. Astrophys. **91**: 85.
35. TELFORD, J. M. 1955. J. Meteorol. **12**: 436.
36. ZINNECKER, H. 1981. Ph.D. Diss. Techn. Univ. München. Report MPI-PAE/Extraterrestrische Physik 167.
37. HO, P. T. P. & A. H. BARRETT. 1978. Astrophys. J. **224**: L23.
38. BATRLA, W., P. BASTIEN, T. L. WILSON, K. RUF, T. PAULS & R. N. MARTIN. 1981. Mitt. Astron. Ges. **54**: 278.
39. KEENE, J., R. H. HILDEBRAND & S. E. WHITCOMB. 1982. Astrophys. J. **252**: L11.
40. HO, P. T. P. & A. D. HASCHICK. 1981. Astrophys. J. **248**: 622.

CHEMICAL EVOLUTION OF OB ASSOCIATIONS*

David N. Schramm

Astronomy and Astrophysics Center
University of Chicago
Chicago, Illinois 60637

Keith A. Olive

Theory Group
Conseil Européen pour la Recherche Nucléaire
1211 Geneva 23, Switzerland

It is now well-established that ^{26}Al and ^{107}Pd were present when the solar system formed 4.6 billion years ago (see Reference 1 and references therein). Because these nuclei have half-lives of only a million years, this means that they had to have been synthesized within a few million years of the formation of the solar system. This, in turn, means that the supernovae that produced them probably occurred in the same star-forming region, since the intervals between incidents of star formation in a specific region of the Galaxy are otherwise on the order of 10^8 y. This fact led Reeves to suggest that the Sun had formed as a member of an OB association.[2] Reeves went on to speculate that the other isotopic anomalies observed in carbonaceous chondrites might also have been due to the intermixing of the supernova debris in the association. However, Reeves estimated the magnitude of the effects to be on the order of parts in 10^4 or 10^5 because he assumed that the supernova debris would be intermixed with the entire mass of the molecular cloud out of which the association formed. Olive and Schramm recently re-examined this problem, looking into the detailed chemical evolution of an OB association and, specifically, the intermixing of the debris of the supernovae that occurred within the association.[3] The major difference between Olive and Schramm's results and Reeves' results involves the degree of mixing. Olive and Schramm assume mixing only with the material that goes to make stars, not with the entire $10^5\ M_\odot$ cloud. Thus, the supernova debris is only mixed with $\sim 10^3\ M_\odot$ of material. As a result, the magnitudes of the effects are significantly larger. In particular, they show that the solar system could have been enriched by as much as a factor of two or three over the average interstellar medium (ISM) and the primordial composition of the cloud. They thus show that OB associations probably undergo a significant variation in the heavy element composition between stars that form at the beginning of the association and stars that form just before the association disperses.

The purpose of this paper is to briefly review the results of Olive and Schramm, to present new arguments supporting the mixing assumptions of Olive and Schramm, and to present observational predictions and implications of the arguments.

Let us first go through the new arguments as to why the mixing of the supernova debris in an association should be with $\sim 10^3\ M_\odot$, rather than with the entire $10^5\ M_\odot$. These arguments are crucial to that concerning the magnitude of the effect; it is

*This research was supported, in part, by grants from the National Aeronautics and Space Administration, no. NSG 7212, and the National Science Foundation, no. AST-81-16750, to the University of Chicago.

0077–8923/82/0395–0236 $1.75/1 © 1982, NYAS

interesting that, since the work of Olive and Schramm, the argument has been strengthened considerably, as a result of more detailed examinations of different aspects of the problem. We can now summarize three independent arguments, all of which point in the same basic direction.

(1)   The amount of ^{26}Al in the protosolar nebula is at least as much as is observed, which is at a ratio of 5×10^{-5} Al: ^{27}Al, since every supernova model today only makes a small amount of ^{26}Al, while, at the same time, producing far larger quantities of basic heavy elements like ^{24}Mg, ^{28}Si, etc. It is clear that, if parts in 10^4 of ^{26}Al were present in the early solar system, several orders of magnitude more of the standard heavy elements had to have accompanied it. In addition, with any sort of decay interval between the synthesis of ^{26}Al and the formation of solid objects in the solar system (the minimum estimate for this time is a few million years), it is again clear that even more of the heavy elements must have accompanied ^{26}Al, since the initial absolute amount of ^{26}Al was even higher. Thus, the heavy element enrichment should be at least at the level of 10%, not at the minimum level of parts in 10^4 that is observed in the ^{26}Al anomaly.

(2)   Cox has estimated the amount of material with which supernova debris will mix as it comes to rest in the interstellar medium.[4] He has shown that the debris will come to rest after mixing with approximately 10^3 $M_\odot$ of material. This has been demonstrated by detailed hydrodynamic calculations and strongly supports the mixing assumed by Olive and Schramm. Further refinements of the hydrodynamic calculations involving the explicit situation of a supernova going off near a giant molecular cloud show that the supernova debris will not penetrate the entire cloud, but only impinge on a very thin surface layer. It is, of course, this thin surface layer out of which the new stars form. Thus, the thin surface layer that goes to make the other stars in the association will receive most of the enrichment from the supernova debris, while the bulk of the molecular cloud will remain uncontaminated. In fact, the bulk of the cloud will probably be dispersed before star formation, as additional supernovae blow up at an ever-increasing rate.

(3)   Observations of current star-forming regions in the Galaxy—in particular, the Orion Nebula—show heavy element concentrations that are lower by as much as a factor of two than those in the solar system (see other papers in this volume). Thus, if the heavy element mass fraction, Z, in the solar system is 0.02 and Orion (a region 4.5 billion years younger than the solar system) has a Z of only 0.01, we can deduce (a) that the average heavy element enrichment in the Galaxy is not increasing in a monotonic manner and (b) that the solar system and Orion show abundance differences on the order of a factor of 100%. This is the magnitude predicted by Olive and Schramm.

Now that we have established the assumption of restricted mixing, let us review the results of Olive and Schramm. It was shown there that massive stars produce the bulk of the heavy elements and, in particular, the alpha particle nuclei. Low-mass stars with longer lifetimes produce some of the non–alpha particle nuclei, such as ^{13}C, ^{22}Ne, and the s-process nuclei. The chemical evolution of an OB association with the mixing assumptions given above will differ at different time frames from the average abundances in the Galaxy. In particular, Olive and Schramm did a detailed numerical simulation with state-of-the-art nucleosynthetic models for stars of different masses coupled with stellar lifetimes. They showed that the first few supernovae in the association were primarily enriched in the heavy elements and the alpha particle nuclei and that oxygen was enriched relative to carbon. Thus, if a star formed later in the association, after the first few supernovae had mixed with the cloud, it would have large enrichments in the heavy elements and alpha particle nuclei relative to non–alpha particle nuclei. It would also show enrichments in oxygen relative to carbon. Given the arguments presented earlier with regard to ^{26}Al and ^{107}Pd, it seems clear that our Sun was one of the later stars to form in the association. Therefore, our Sun would have been enriched in heavy elements. If the mixing assumptions described above are correct, then the magnitude of these enrichments would be as much as a factor of two or three over the ISM. This would mean that the typical interstellar medium would be depleted relative to the solar system in those things, not because of evolution since the formation of the solar system, but rather because the solar system was anomalously enriched.

TABLE 1 summarizes certain key materials and their origins with regard to whether they are produced nucleosynthetically in low-mass or high-mass stars. In particular, we note that the ^{12}C, ^{16}O, ^{20}Ne, Mg, Si, S, Ar, and Ca nuclei tend to be produced in

TABLE 1

ORIGIN OF CERTAIN ISOTOPES AND ELEMENTS

Low-Mass Stars $M \lesssim 10\ M_\odot$	High-Mass Stars $M \gtrsim 10\ M_\odot$
^{13}C	^{12}C
$^{17}O\ (^{18}O)$	^{16}O
^{22}Ne	^{20}Ne
O/C < 1	O/C > 1
	Mg, Si, S, Ar, Ca...
s-process $80 < A < 210$	s-process $A < 80$
? F_e ?	? F_e ?
	r-process $80 < A < 210$
? Actinide r-process ?	

massive supernova explosions. The CNO cycle in low- and intermediate-mass stars produces things like ^{13}C, ^{17}O, and some ^{18}O, and the subsequent helium burning of the C and O debris produces ^{22}Ne. It is also known that high-mass stars tend to produce oxygen in greater abundances than carbon; thus, in order to provide the amount of carbon known to be in the Galaxy, the low-mass stars must be producing carbon in greater abundances than oxygen. We know that the bulk of the s-process elements between $A = 80$ and $A = 210$ are produced in the intermediate-mass stars that would not have been the first to blow up in the OB association.[5,6] The high-mass stars, supernova stars, do produce those s-process nuclei with masses less than 80 AMU (see References 7–9 and references therein). The origin of the bulk of the iron is somewhat unclear; it is known that the most massive stars will produce iron, but it is also known that, if the stars in the intermediate-mass range from 4 to 8 $M_\odot$ do blow up *via* carbon detonation supernova models, they, also, could produce a very large amount of iron. The r-process nuclei are another matter. There is no well-established astrophysical site for the r-process (Reference 10 and references therein). It is known that, in the shock processing of a massive star supernova, many neutrons will be released and there will be isotopes produced in the range from iron to lead and bismuth, which are neutron rich and have been ascribed by Burbidge *et al.* to the r-process.[11] However, such neutron capture processing is probably more like the n-process of Blake and Schramm[12] than the classical r-process. It is also clear that this n-processing will not add enough neutrons to produce the actinides. It is interesting that, observationally, we know that the supernovae that made ^{26}Al and ^{107}Pd do not produce actinides. We know this because, while the solar system did have admixtures of ^{26}Al and ^{107}Pd, it did not have any admixture of ^{244}Pu. Since the last nucleosynthetic event that produced ^{244}Pu was 10^8 y before the formation of the solar system, it did not occur within the same OB association. Thus, observationally, we know that the actinide-producing r-process does not occur in every supernova and, in particular, that it did not occur in the supernovae that went off in the OB association just before the formation of the solar system. This leaves open the question of where the actinide-producing events do occur.

From TABLE 1 we can see, then, that the solar system would have been enriched in ^{12}C, ^{16}O, and ^{20}Ne relative to the average interstellar medium and that it would have had higher oxygen-to-carbon ratios than the average interstellar medium. We might also have expected some enhancement in the r-process composition between $A = 80$ and $A = 210$, but not in the actinides. We would have expected some depression in the s-process abundance going to the solar system in the mass range 80–210, which would imply that the ISM might now be more enriched in s-process material than in r-process material in this mass range.

We can understand from these results why the ^{12}C:^{13}C ratio in the interstellar medium appears to be lower than that in the solar system. We can also understand why the cosmic rays might be enriched in ^{22}Ne relative to the solar system, since we expect the cosmic rays to be material that would be closer in composition to the average interstellar medium than the solar system, since the solar system was significantly enriched in supernova debris. Whereas standard models for cosmic ray acceleration seem to indicate that supernovae accelerate the material that surrounds the pre-supernova objects, rather than the freshly synthesized material within the objects, the typical material accelerated in the cosmic rays might be slightly different from the ISM, since supernovae are more likely to occur in OB associations, and, as we have already seen, OB associations are going to be somewhat enriched in supernova material. It is clear that the typical material accelerated by supernova shock waves will not be as enriched as the material behind the shock wave. Also, average supernovae are more likely to be 10 or 12 $M_\odot$ stars than 20 to 30 $M_\odot$ stars, which would be the first ones to blow up in an OB association. It is likely that an average supernova (12 $M_\odot$) will blow up sometime after the OB association has completely dispersed. Thus, the surrounding gas will not be too different from the average interstellar medium.

It is interesting to talk in terms of the oxygen-to-carbon ratio, since the solar system has an oxygen-to-carbon ratio of approximately two and the cosmic rays have an oxygen-to-carbon ratio on the order of unity. There may be regions in the Galaxy that have more carbon than oxygen if the variations of magnitude discussed here are distributed in some manner. Since the chemistry of molecule and grain formation is very sensitive to the oxygen-to-carbon ratio, this could lead to some interesting effects. It is well known that, when there is more oxygen than carbon, all the carbon is locked up in carbon monoxide and there is oxygen left over to make oxides and, therefore, silicates. Whereas, if there were more carbon than oxygen, all the oxygen would be locked up in carbon monoxide and there would be carbon left over to make carbides. This could lead to some very different grain compositions in different regions of the Galaxy.

It should be noted that, in this discussion, we have been concerned solely with abundances and mixing. We have not gone into the question of whether or not the supernovae in an association have a hydrodynamic effect with regard to stimulating the formation of the solar system itself. That is certainly an interesting question, but does not in any way affect the arguments presented here. It does seem clear, regardless of the mechanism, that debris from the first supernovae of the OB association did intermix with the solar system. We have evidence for that from the ^{26}Al and isotopic anomalies. The mechanisms for the mixing are not necessary for the arguments here, although, in other papers, we have presented arguments that seemed to favor grain formation in the supernovae and grain injection into the protosolar nebula as the more probable injection mechanism due to the shrapnel effect.[13-15]

This work, showing that the solar system is probably enriched relative to the typical interstellar medium and that the solar system formed out of an OB association, is particularly relevant with regard to the study of Orion, since Orion is the best-studied example of an OB association, is an example of young interstellar medium material, provides useful comparisons with the solar system, and gives us a stellar nursery of the type out of which our sun once formed to study.

SUMMARY

It is shown that the existence of ^{26}Al and ^{107}Pd in meteorites in the early solar system implies that our solar system probably formed inside an OB association that had been contaminated by the debris of at least one supernova, if not several supernovae. It is shown that, in addition to these radioactive tracers, the contamination of the material out of which the solar system formed by supernovae would have significantly enriched the heavy element composition of the solar system relative to that of the average interstellar medium. In particular, it is shown that the solar system would be enriched in those isotopes which are produced by the more massive stars. These isotopes include ^{16}O, ^{12}C, ^{20}Ne, and some r-process material (although not the actinide r-process). Depending on the amount of mixing that occurred within the OB association, it is conceivable that the magnitude of these enrichment effects could be on the order of a factor of two, which would then mean that our solar system abundances would be anomalous relative to the ISM by factors of about two. Specific isotopic ratios and elemental ratios that might reflect these differences would include the ^{20}Ne:^{22}Ne ratio, which would be higher in the solar system than in the ISM and the cosmic rays, the ^{12}C:^{13}C ratio, which would be higher in the solar system than in the ISM, and the oxygen-to-carbon ratio, which would also be higher in the solar system than in the typical interstellar medium. If OB association chemical evolution does produce effects of this magnitude, it would be anticipated that there would then be variations of these abundances in different associations and regions throughout the Galaxy. Thus, when isotopic and elemental compositions are found that differ from those in the solar system, it may not be the new observations that are anomalous but rather the solar system itself.

ACKNOWLEDGMENTS

We would like to thank Don Cox for illuminating discussions with regard to hydrodynamic mixing effects, Hubert Reeves for discussions on OB associations, John Simpson and John Wefel for discussions on the ^{22}Ne in the cosmic rays, and Manuel Peimbert for a comment with regard to the heavy element content of Orion. We would also like to thank Al Glassgold for encouraging us to write this version for the conference proceedings.

REFERENCES

1. LEE, T. 1979. Rev. Geophys. Space Phys. **17:** 1591.
2. REEVES, H. 1978. *In* Protostars and Planets. T. Gehrels, Ed. University of Arizona Press. Tucson.
3. OLIVE, K. A. & D. N. SCHRAMM. 1982. Astrophys. J. **257:** 276–82.
4. COX, D. P. 1981. Astrophys. J. **245:** 534–51.
5. IBEN, I. & J. W. TRURAN. 1978. Astrophys. J. **220:** 980.
6. ULRICH, R. K. 1973. *In* Explosive Nucleosynthesis. D. N. Schramm and W. D. Arnett, Eds. University of Texas Press. Austin.
7. TRURAN, J. W. & I. IBEN. 1977. Astrophys. J. **216:** 797.
8. COUCH, R. G. & W. D. ARNETT. 1977. Astrophys. J. **194:** 537.
9. WEFEL, J. P., D. N. SCHRAMM, J. B. BLAKE & D. PRIDMORE-BROWN. 1981. Astrophys. J. Suppl. Ser. **45:** 565–84.
10. NORMAN, E. B. & D. N. SCHRAMM. 1979. Astrophys. J. **228:** 881.
11. BURBIDGE, E. M., G. R. BURBIDGE, W. A. FOWLER & F. HOYLE. 1957. Rev. Mod. Phys. **29:** 547.
12. BLAKE, J. B. & D. N. SCHRAMM. 1976. Astrophys. J. **209:** 846.
13. MARGOLIS, S. H. 1979. Astrophys. J. **231:** 236.
14. LATTIMER, J., D. N. SCHRAMM & L. GROSSMAN. 1977. Nature (London) **269:** 116.
15. ELMEGREEN, B. G. 1981. Astrophys. J. **251:** 820–33.

DISCUSSION OF THE PAPER

P. MEZGER (*Max-Planck-Institut für Radioastronomie, Bonn, F.R.G.*): What are your predictions for the solar *versus* the ISM ^{16}O:^{18}O ratio?

SCHRAMM: The ^{18}O in the ISM may appear to be somewhat enriched relative to the solar system since the solar system has extra ^{16}O.

M. PEIMBERT (*Universidad Nacional Autónoma de México, México, D.F.*): The He:O ratio in Orion is larger than that for the Sun. This supports your suggestion. The same seems to be true for the C:O ratio.

S. FEDERMAN (*University of Texas, Austin, Tx.*): Would you expect O:C variations from one subgroup to another in a given association?

SCHRAMM: Yes, it is likely, *e.g.,* by incomplete mixing in the cloud complexes.

SHOCK WAVES IN ORION

David Hollenbach

Ames Research Center
Moffett Field, California 94035

INTRODUCTION

The observations of vibrationally excited molecular hydrogen in the Becklin-Neugebauer–Kleinmann-Low (BN-KL) region of the Orion Molecular Cloud-1 (OMC1) and the early maps of the ~0.1 pc extent of this emission[1] were quickly recognized as the first persuasive evidence for shock waves in interstellar molecular clouds.[2-4] From the extent (FIGURE 1) and intensities of the 2-μm vibrational lines, one could infer that emission originates from thin ($\leq 10^{15}$ cm), hot ($T \simeq 2000$ K) sheets of molecular gas. The cooling time of the molecular gas of ~1 y indicates that a continuous supply of thermal energy is being injected into the gas. Shock waves maintain such thin sheets of hot material, constantly replenishing the cooling gas with newly swept-up hot molecules. Therefore, without any supporting dynamical evidence of supersonic motion, a case could be made for the propagation of shock waves in the Orion molecular gas.

However, at about the same time as the 2-μm observations were being made, microwave spectroscopy of CO in the BN-KL region showed broad wings that extended to nearly ±75 km s^{-1} from the line center.[5,6] Subsequent high spectral resolution measurements of the 1-0S(1) and the S(2) lines of molecular hydrogen, as well as microwave spectroscopy of many other molecules in the same region, showed similar broad wings, which gave direct evidence of the supersonic motions associated with shock waves. The 12-μm S(2) emission of H extends at least ±100 km s^{-1} from the line center.[7]

Shock waves of ~100 km s^{-1} heat newly swept-up material to $T \simeq 10^5$ K and completely dissociate all molecular gas. Recent work on the influence of magnetic fields on interstellar shocks[8-10] has increased theoretical estimates of the critical shock velocity necessary to dissociate molecules from 24 km s^{-1}[3,11] to ~ 50 km s^{-1}; nevertheless, the high velocity of the emission wings eludes a simple shock wave acceleration model. Two models that might explain the 100 km s^{-1} hot H$_2$ are (1) reformation of substantial amounts of H$_2$ in the $T \gtrsim 1000$ K postshock gas behind $v_s \simeq$ 100 km s^{-1} shocks and (2) low velocity, nondissociative shocks in high-velocity ambient gas.[10,12,13]

Observations of the reddening of H$_2$ rotational and vibrational transitions[14,15] indicate that the intrinsic H$_2$ luminosity from the shock waves is on the order of ~200 $L_\odot$ if the average extinction at 2 μm is ~2–2.5 mag, although there is controversy over the magnitude of the extinction.[16-18] The mass of H$_2$ in local thermodynamic equilibrium (LTE) at 2000 K required to produce 200 $L_\odot$ is ~0.03 $M_\odot$, the momentum of the hot gas is ~0.6 $M_\odot$ km s^{-1}, and the kinetic energy of the hot gas is ~6 $M_\odot$ km^2 s^{-2}. (Although emission extends to high velocity and we will demonstrate below that shock waves may travel at ~50 km s^{-1}, the characteristic velocity of the emitting gas is ~20 km s^{-1}.[19]) Shock modeling will provide a measure of the total momentum and energy

242

0077-8923/82/0395-0242 $1.75/1 © 1982, NYAS

in both hot and cool swept-up gas. The short cooling time of the hot gas means that most of the swept-up gas is cool. As shall be shown below, shock modeling indicates a total momentum of $\sim 250\ M_\odot$ km s^{-1} and a total enery of $\sim 6000\ M_\odot$ km^2 s^{-2}.

The inferred momentum and energy must be injected on a timescale on the order of $\Delta t \simeq 0.05$ pc/50 km s^{-1}, or 1000 y. The bolometric luminosities of the embedded sources associated with BN-KL total $L \simeq 10^5\ L_\odot$,[20] the total momentum and energy emitted in 1000 y are $L\,\Delta t/c \simeq 2\ M_\odot$ km s^{-1} and $L\,\Delta t \simeq 6 \times 10^5\ M_\odot$ km^2 s^{-2} in photons and $\dot{M}v_w\Delta t = 100\ \dot{M}_{-3}\,v_{w7}\ M_\odot$ km s^{-1} and $0.5\ \dot{M}v_w^2\Delta t = 5 \times 10^3\ \dot{M}_{-3}\,v_{w7}\ M_\odot$ km^2 s^{-2} in stellar wind, where v_{w7} is the wind velocity in units of 100 km s^{-1} and $\dot{M}_{-3}$ is the mass

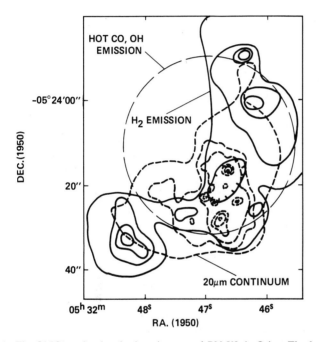

FIGURE 1. The OMC1 molecular shock region around BN-KL in Orion. The 2-μm 1-0S(1) map of hot H$_2$,[37] the 40-arcsec beam position for the peak of the CO[38] and OH[39] observations, and the 20-μm continuum map[40] are superposed.

loss in units of $10^{-3}\ M_\odot$ y^{-1}. Comparison of the energetics of the shocked material in Orion with the energetics of the embedded sources therefore reveals that a hitherto unsuspected mechanism is at work in the protostars; one that efficiently drives high-velocity material from the protostellar vicinity.

Most of the published theoretical work concerning shock modeling of Orion assumes that magnetic fields play a minor role in the shock structure and that the shocks are nondissociative. The purpose of this paper is to describe recent unpublished theoretical work that models the observed H$_2$, CO, and OH emission in Orion (FIGURE 1) with emission from magnetohydrodynamic (MHD) and dissociative shock waves.[9,10,21] The line intensities, but not the line profiles, are well matched with nondissociative MHD shocks of velocity 40–50 km s^{-1} incident upon gas with an H_2 density of 10^5–10^6 cm^{-3} and ambient magnetic fields of 0.5–4 mGs. The low levels of H_2 emission that lie $\gtrsim 50$ km s^{-1} from the line center must be produced either in dissociative shocks or in shocks traversing high-speed ambient gas.

The following section discusses the physical processes dominant in such shock waves in Orion, with special emphasis on MHD shock waves. The third section describes various global assumptions implicit in a simple application of the theoretical results of a one-dimensional steady-state shock wave to the observational data. The fourth section compares the observations of emission lines of H_2, CO, and OH with the shock models. The last section analyzes the implications of these results in terms of the global picture of this star-forming region.

Physical Processes in the Orion Shock Waves

Basic Structure of Magnetohydrodynamic Shock Waves

Drawing on earlier work by Mullan,[22] Draine emphasized the importance of the effects of magnetic fields on shock structure.[8] More recently, several authors have shown that the probable field strengths in Orion almost certainly make the early nondissociative shock models of Orion obsolete.[9,10] For shock speeds in the range $v_s \lesssim$ 40 km s^{-1} in Orion, magnetic fields as low as 50 μGs greatly affect the shock structure.

Magnetic field strengths in Orion are uncertain (the shock model values of ~ 1 mGs presented later in this paper may be the best indirect measure of the field), but are likely to exceed 50 μGs. The observational data[23,24] indicate field strengths of ~ 10 μGs for diffuse clouds, 50 μGs for dense H I clouds, and milligauss fields for dense molecular gas in OH masers. The observed correlation of field strength with density suggests that $B_0 \simeq 0.1$–4 mGs in the ambient gas surrounding BN-KL. These fields are consistent with theoretical estimates[25] and may be sufficient to damp gravitational collapse. Indirect arguments from the observed infrared polarization in the BN-KL region indicates that milligauss fields are needed to align the dust grains.[26–28]

Given a preshock hydrogen nucleus density, n_0, and shock speed, v_s, the crucial observational effect of magnetic fields on shock waves is to lower the peak temperature of the neutral gas, to lower the compression, to heat molecules over a much larger range of velocities, and to spread the heating over a much larger column of gas. From a theoretical standpoint, the inclusion of magnetic fields greatly complicates the solution of shock wave structure. The compression of the component of the magnetic field perpendicular to the shock velocity, $B_\perp$, leads to magnetic forces on the charged particles that drive them through the neutrals. Although the charged grains, ions, and electrons represent a trace component in MHD molecular shock waves, they nevertheless control the dynamics and heating of the neutral gas. These charged particles first experience the increasing magnetic field, $B_\perp$, of the oncoming shock wave and initially accelerate, heat, and compress. The prior compression of the charged particles with respect to the neutrals leads to local enhancements in the ionization fraction, x_i, and the grain:neutral ratio. The final compression, C, for both charged particles and neutrals is

$$C = \left(\frac{8\pi\rho_o v_s^2}{B_{0\perp}}\right)^{1/2} = 19 \left(\frac{v_s}{30 \text{ km s}^{-1}}\right)\left(\frac{n_0}{10^6 \text{cm}^{-3}}\right)^{1/2}\left(\frac{1 \text{ mGs}}{B_{0\perp}}\right). \qquad (1)$$

Frictional (drift) terms between the charged particles and the neutrals dominate their heating. The cooling processes of the various fluids differ, their coupling is weak, and the ion temperature, T_i, the electron temperature, T_e, the neutral temperature, T_n, and the grain temperature, T_g, differ. Often, $T_i > T_e > T_n > T_g$. Therefore, four coupled fluids must be treated in MHD shocks.

Draine discusses and classifies the overall structure of magnetic shocks.[8] He defines a critical preshock magnetic field strength, B_{cr}, which separates two distinct classes of shock wave structure. B_{cr} depends on v_s, n_0, x_{io} (the preshock ionization fraction), and the cooling processes operative in the shock. For low preshock magnetic fields, $B_{0\perp} < B_{cr}$, the dissipation of the neutrals in the shock front is due to neutral-neutral collisions, so the gas is suddenly accelerated and heated in the shock front in a length scale short compared with characteristic cooling lengths. The shock front can therefore be treated as a nonradiative discontinuity, or "jump." Analysis of these shock waves reduces to the problem of following the cooling of the gas behind the shock front. Draine calls such structure "J-type" shocks, and the neutral temperature, density, and flow velocity profiles of these shocks are schematically presented in FIGURE 2a.

When $B_{0\perp}$ exceeds B_{cr}, the dissipation of the neutrals is due to charged particle–neutral gas collisions, which occur over a length scale long compared to the characteristic cooling length if the ionization is low. The shock front is radiative and analysis of the emission from these shock waves involves the calculation of the structure inside the shock front. In the interstellar medium, the neutral acceleration and heating in radiative shock fronts is accomplished by collisions with drifting ions or charged grains. Neutrals heat before they are accelerated or compressed significantly. The length scales for acceleration and heating are given by the mean free path, l_c, for a neutral to be struck by a charged particle; for ions, $l_{ci} \simeq 10^{15}/x_i C n_0$ cm; for grains, $l_{cg} \simeq 10^{21.5}/C n_0$ cm. Inserting the expression for C from (1), we obtain

$$l_{cg} \simeq 3 \times 10^{14} \left(\frac{30 \text{ km s}^{-1}}{v_s}\right)\left(\frac{10^6 \text{cm}^{-3}}{n_0}\right)^{3/2}\left(\frac{B_{0\perp}}{1 \text{ mGs}}\right) \text{cm}, \qquad (2)$$

or a hydrogen column density of heated neutrals

$$N_{hot} \simeq n_0 l_{cg} \simeq 3 \times 10^{20} \left(\frac{30 \text{ km s}^{-1}}{v_s}\right)\left(\frac{10^6 \text{cm}^{-3}}{n_0}\right)^{1/2}\left(\frac{B_0}{1 \text{ mGs}}\right) \text{cm}^{-2}, \qquad (3)$$

when charged grains accelerate the neutrals, as they generally do in molecular MHD shocks (see below). Draine calls these shocks "C-type," since the structure parameters are treated as continuous variables. The shock velocity must be less than the Alfvén velocity in the ions to obtain a C-type structure. The shock structure is schematically shown in FIGURE 2b.

The exact value of B_{cr} is difficult to determine analytically, but numerical modeling indicates that, for $v_s = 30$ km s^{-1}, $n_0 = 2 \times 10^5$ cm^{-3}, and $x_{io} < 3 \times 10^{-7}$, $B_{cr} \simeq 10$ μGs, while these same parameters give $B_{cr} \simeq 30$ μGs for $v_s = 40$ km s^{-1} and $B_{cr} \simeq 300$ μGs for

$v_s = 50$ km s^{-1}.[10] Observations of HCO$^+$ indicate that $x_{io} < 2 \times 10^{-5}$ in Orion[29] and theoretical gas phase chemical modeling of the ambient molecular gas suggests that $x_{io} < 10^{-7}$.[30] Therefore, $B_{0\perp} > B_{cr}$ in Orion when $v_s < 40$–50 km s^{-1}, and C-type solutions apply. Furthermore, C-type shocks are not unique to Orion; C-type solutions are probably required to model all relatively low velocity ($v_s \lesssim 50$ km s^{-1}) shocks in molecular clouds.[9,10] McKee and Hollenbach review evidence for shock waves in other molecular clouds.[31]

Above about 50 km s^{-1}, dissociation and ionization occur, the shock front becomes nonradiative as its length diminishes with x_i^{-1}, and interstellar molecular shock waves

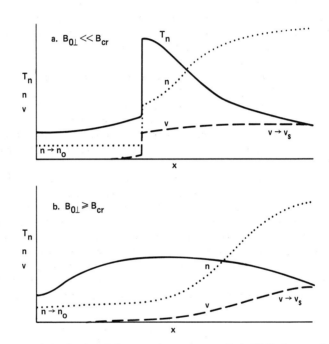

FIGURE 2. A schematic plot of the neutral temperature, T_n (solid line), neutral density, n (dotted line), and neutral flow velocity, v (dashed line), in the frame of the preshock gas. In (a), a J-type shock with a sharp discontinuity and a high T(max) is seen; in (b), a C-type shock is pictured with reduced T(max) but a substantially greater column density of warm gas, emitting over a wider range of v and lower values of n.

assume a strong J-type structure. Here, standard shock jump conditions can be applied across the discontinuity (shock front), and the numerical integration of the conservation equations of the flow can be begun just behind the shock front.[32] The integration becomes a simple time integration of a parcel of neutral gas that cools and compresses as it flows downstream. Neutral heating by ambipolar diffusion is negligible. Physical processes in these fast dissociative shocks are reviewed in Reference 31. The H$_2$ emission in Orion extends to ± 100 km s^{-1}, so the interesting range of parameter space for J-type shocks in Orion is 50 km s$^{-1} < v_s < 100$ km s^{-1}.

The following discussion of physical processes is restricted to Orion-like shock waves in which the maximum neutral gas temperatures T_n (max) exceed ~ 2000 K and the density of the 2000 K gas exceeds either n_{hot} (H$_2$) $\gtrsim 3 \times 10^6$ cm^{-3} or n_{hot} (H) $\gtrsim 3 \times 10^4$ cm^{-3} so that H$_2$ vibrational LTE might be obtained. The shock speed is also restricted to $v_s < 50$ km s^{-1} in order to maintain both C-type structure and the preshock density $n_0 \lesssim 10^7$ cm^{-3} as an upper limit to the Orion ambient density. We shall demonstrate below that v_s must be greater than 20 km s^{-1} in order to produce 2000 K gas in C-type shocks, so the interesting range of parameter space for C-type shocks is approximately 10^5 cm$^{-3} < n_0 < 10^7$ cm^{-3} and 20 km s$^{-1} < v_s < 50$ km s^{-1}.

Heating and Cooling in C-Type Shock Waves

Cool ions streaming through cold neutrals with a relative velocity v_D heat by elastic collisions with the neutrals until equilibrium is achieved at

$$\tfrac{3}{2} k T_i = \tfrac{1}{2} u_n v_D^2,$$

$$T_i = 7 \times 10^4 \left(\frac{v_D}{30 \text{ km s}^{-1}} \right)^2 \text{K}, \tag{4}$$

where u_n is the mean mass of the neutral gas particle. The maximum value of v_D is

$$v_D \text{ (max)} = v_s (1 - 1.9 C^{-2/3}) \tag{5}$$

in C-type interstellar shock waves. Approximately $u_i / u_n \simeq 1$–10 collisions of a streaming ion with H$_2$ molecules is sufficient to thermalize the ion at T_i. Inelastic encounters of ions with neutrals are ineffective in reducing T_i because of their lower cross sections relative to elastic encounters.

The electron equilibrium temperature achieved by elastic encounters with neutrals is close to T_i if electron cooling processes are ignored. However, electrons require $u_n / m_e \simeq 3600$ elastic collisions to reach T_i. Electron heating by collisionless plasma instabilities[33] may dominate the elastic collisional heating by factors of about two.[10] Nevertheless, a single inelastic collision can absorb a significant fraction of the electron energy and cooling of the electrons by their collisional excitation of vibrational and rotational states of H$_2$ makes $T_e \simeq 0.1$–0.2 T_i for 10 km s$^{-1} \lesssim v_D \lesssim 30$ km s^{-1}.[9,10]

Charged grains are also accelerated by electromagnetic forces and, in Orion-like conditions, grains essentially move with the ions and electrons.[9,10] Grains are heated by collisions with the neutrals (radiative heating by the shock wave flux is suppressed by the small grain cross sections for absorption of ir photons), radiatively cool by emission of far-infrared continuum photons, and typically equilibrate at $T_g \lesssim 100$ K.

The neutral H$_2$ molecules are accelerated by collisions with charged grains for $x_i < 3 \times 10^{-7}$ and by collisions with ions for $x_i > 3 \times 10^{-7}$. Most C-type interstellar molecular shocks are mediated by the grain-neutral interaction. These same collisions provide most of the neutral heating, with each collision adding $\sim \tfrac{1}{2} u_n v_D^2$ of thermal energy to the neutral gas. Cooling of hot molecular gas is generally due to collisional excitation, and subsequent radiative decay, of H$_2$, CO, and H$_2$O and collisional dissociation of H$_2$. For densities $n_0 \gtrsim 10^{5.5}$ cm^{-3} and temperatures 500 K $< T_n < 3000$ K, H$_2$O cooling dominates if elemental oxygen is not significantly depleted in the gas,[32] while, for $n_0 \lesssim 10^{5.5}$ cm^{-3} in the same temperature range, H_2 cooling dominates. Heating of neutrals by collisions with streaming grains dominates to $x_i \lesssim 3 \times 10^{-7}$. Equating heating by grains to cooling by H_2O or H_2 results in an equilibrium neutral temperature T_n of approximately

$$T_n = 150 \left(\frac{n_0}{10^4 \text{ cm}^{-3}}\right)^{1/5} C^{2/5} \left(\frac{v_D}{30 \text{ km s}^{-1}}\right)^{6/5} \text{K} \quad \text{if } n_0 \lesssim 10^{5.5} \text{ cm}^{-3},$$

$$T_n = 360 \, C^{2/3} \left(\frac{v_D}{30 \text{ km s}^{-1}}\right)^{2} \text{K} \quad \text{if } n_0 \gtrsim 10^{5.5} \text{ cm}^{-3},$$

(6)

valid for 500 K $< T_n <$ 3000 K and $x_i \lesssim 3 \times 10^{-7}$. The peak neutral temperature can be estimated by inserting (1) and (5) into (6),

$$T_n (\text{max}) = 1300 \left(\frac{n_0}{10^4 \text{ cm}^{-3}}\right)^{2/5} \left(\frac{v_s}{30 \text{ km s}^{-1}}\right)^{8/5} \left(\frac{B_{0\perp \cdot}}{100 \text{ } \mu\text{Gs}}\right)^{-2/5} (1 - 2.3C^{-2/3})$$

$$\text{if } n_0 \lesssim 10^{5.5} \text{ cm}^{-3},$$

$$T_n (\text{max}) = 2600 \left(\frac{n_0}{10^6 \text{ cm}^{-3}}\right)^{1/3} \left(\frac{v_s}{30 \text{ km s}^{-1}}\right)^{8/3} \left(\frac{B_{0\perp}}{1 \text{ mGs}}\right)^{-2/3} (1 - 3.8C^{-2/3}) \quad \text{(7)}$$

$$\text{if } n_0 \gtrsim 10^{5.5} \text{ cm}^{-3},$$

with the same range of applicability as (6). For $T_n >$ 5000 K and $n_0 \gtrsim 10^{5.5}$ cm^{-3}, dissociation of H_2 becomes significant, which leads to diminished cooling, increased ionization, increased neutral heating, and the emergence of a J-type shock front. If $B_{0\perp} = 1$ mGs $(n_0/10^6$ cm$^{-3})$ b, where $b \simeq 1$ (see above) and $C \simeq 20$, then complete dissociation of H_2 by neutral collisions occurs at $v_s \gtrsim 45b^{1/4}$ km s^{-1} for ambient gas of high density, $n_0 \gtrsim 10^{5.5}$ cm^{-3}.

Note that (3) and (7) can be used to calculate the approximate column density N_{max} of gas at $T_n \simeq T_n$ (max). For example, to produce N_{max} (H_2) $\simeq 10^{20}$ cm^{-2} at T_n (max) $\simeq$ 2000 K requires $b \simeq 3$, $v_s \simeq 45$ km s^{-1}, assuming that $N_{\text{max}} \simeq \frac{1}{3} N_{\text{hot}}$, $x_i < 3 \times 10^7$, and $n_0 \gtrsim 10^{5.5}$ cm^{-3}.

A summary of the various dominant heating and cooling processes in C-type shocks appropriate to Orion is presented in TABLE 1.

TABLE 1

HEATING AND COOLING IN C-TYPE SHOCK WAVES IN ORION*

Species	Dominant Heating Process	Dominant Cooling Process
Ions	Elastic scattering with H_2	Elastic scattering with H_2
Electrons	Elastic scattering with H_2 (ion plasma instability?)	Rotational and vibrational excitation of H_2
Grains	Collisions with H_2	Far-infrared continuum emission
Neutrals	Collisions with grains ($x_i \lesssim 10^{-6}$) Collisions with ions ($x_i \gtrsim 10^{-6}$)	Rotational and vibrational excitation of H_2 and H_2O.

*20 km s$^{-1} < v_s <$ 50 km s^{-1}; molecular gas.

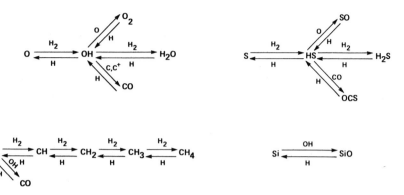

FIGURE 3. The high-temperature (300 K $< T_n <$ 5000 K) neutral chemistry.

Chemistry in the Postshock Gas

The important chemistry of shocks such as those in Orion can be divided into two categories: (1) the neutral chemical reactions (including the dissociation of H_2) at 300 K $< T_n <$ 5000 K, which affect major coolants, and (2) the ion chemistry, which determines the fractional ionization and the average mass of an ion and which leads to important structural effects in C-type shock waves.

Once neutral temperatures rise above about 300 K, endothermic reactions and reactions involving activation energies rapidly dominate the important shock coolants. The essential reactions governing the oxygen, carbon, sulfur, and silicon chemistry are illustrated in FIGURE 3. Reaction rates for the O and C chemistry are summarized in References 10 and 32. In hot molecular gas, the dominant H_2 drives the reactions to the right on FIGURE 3 and the carbon and oxygen are quickly tied up in CO, CH, H_2O, and O_2. Reaction rates for the sulfur and silicon chemistry are summarized by Hartquist *et al.*, who point out that shocks in Orion may produce the relatively high observed abundances of SO, SiO, and OCS.[34]

Dissociation of H_2 proceeds rapidly by any of several channels, depending on the physical state of the gas. Dissociation by collisions with other H_2 molecules proceeds efficiently for $T_n \gtrsim$ 5000 K if $n_0 \gtrsim 10^5$ cm^{-3}, but these rates are diminished by the effects of a non-LTE vibrational population at lower densities.[35] Electron impact dissociation is rarely important.[10] Dissociation of H_2 by collisions with ions and charged grains can never completely dissociate H_2, because a given H_2 molecule is struck only about one time by these species as it traverses the hot streaming portion of the shock wave. Nevertheless, such collisions dominate the production of atomic hydrogen in the Orion shock waves.[10] These H atoms may help maintain the H_2 in vibrational LTE. Similar arguments hold for dissociation of H_2 by collisions with "hot" H_2 molecules formed in charge exchange reactions or from ricochets off grain surfaces.

The ion chemistry in C-type shocks is dominated at low ion temperatures $(T_i \lesssim 10^3 – 10^4$ K) by the reaction chains illustrated in FIGURE 4a.[10] In the first chain, hydrogen molecules are ionized by cosmic rays and quickly converted to H_3O^+ ions, which dissociatively recombine. The second chain schematically diagrams the ionization of metallic atoms by uv photons emanating from the hot portions of the shock wave. In these hot regions, where $T_e \gtrsim 10^4$ K, electronic excitation of Lyman-Werner bands of H_2 leads to the production of uv fluxes that can ionize a significant fraction of metallic atoms such as S, Fe, Mg, and Na. Ultraviolet photons travel upstream from the hot gas a column density of $\sim 10^{21}$–10^{22} cm^{-2}. Depending on the metal atom abundance, this process may dominate the ionization of the cool upstream gas. However, the level of ionization produced by the uv field quickly diminishes once T_i rises above about 1000 K and atomic ions are converted to metallic hydride ions, which rapidly recombine.

At high ion temperatures ($T_i \gtrsim 10^4$ K) dissociation of the ions by collisions with H_2 becomes important and the chemistry proceeds by the chain illustrated in FIGURE 4b.

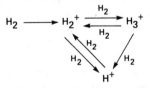

A. LOW ION TEMPERATURE $T_i \lesssim 10^3$ - 10^4 K

B. HIGH ION TEMPERATURE $T_i \gtrsim 10^4$ K

FIGURE 4. Ion chemistry in molecular C-type shocks is schematically diagrammed for two important temperature regimes.

One important effect of this chemistry is the production of H^+ ions. This results in a three-fold reduction in the reduced mass of the ion-neutral encounters. If such encounters mediate shock structure, the result is a lengthening and cooling of the hot neutral region. When the grain-neutral encounters mediate the shock structure, the ion mass indirectly affects the structure because the grain charge—and, hence, the grain coupling to the B field and drifting ions—is dependent upon the ion mass.

The various mechanisms that ionize H_2 include electron, ion, hot atom, and molecule collisions. Neutral collisions are generally insignificant because of their small cross sections and low center-of-mass energies. Electronic collisions dominate if the uncertain rate of plasma instability heating discussed above is high and $T_e > 0.1 T_i$. The ionizaton of H_2 by either ion or electron collisions is a runaway process, since ions beget ions and increased ionization leads to higher neutral temperatures, which ultimately lead to increased ionization rates. The critical ion temperature that leads to runaway ionization in the shock is T_i (cr) $\simeq 10^5$ K, or $v_s \simeq 50$ km s^{-1}. Thus, above $v_s \gtrsim 50$ km s^{-1}, ionization dramatically increases, molecules dissociate, shock fronts appear, and the shock structure becomes J-type.

GLOBAL ASSUMPTIONS OF ORION SHOCK MODELS

One-dimensiomal steady-state shock wave models have been used recently to predict infrared line intensities in Orion.[10,21] Generally, several assumptions are required to apply a simple global model to the observed Orion shock wave region. The basic assumptions in Reference 10 are the following.

(1) The observed shock wave (s) can be described by steady-state shock structures.

(2) The emission intensities can be modeled with a planar shock or shocks traveling at speed v_s relative to an ambient medium of density n_0. The underlying assumption is that a single characteristic n_0 and v_s will provide a model that will match the observed molecular line intensities. The modeling also assumes that v_s can be significantly less than the observed linewidths if the preshock ambient gas is moving at substantial speeds.

(3) All geometrical factors (*e.g.*, seeing shocks edge-on, the number of shocks along various lines of sight), can be incorporated into a multiplicative factor f_A. The parameter f_A is defined as the shock area divided by the projected beam area. For example, a spherical shell with a single shock on its outside face would have $f_A = 4$ if the projected beam area just encompassed the shell.

THEORETICAL MODELS OF INFRARED EMISSION LINES IN ORION

As discussed above, the H_2, CO, and OH emission from the BN-KL region of Orion should be modeled with either (1) a nondissociative C-type shock with $v_s \lesssim 50$ km s^{-1},[10,21] or (2) a dissociative J-type shock with $v_s \gtrsim 50$ km s^{-1}.[10,36] Ambient magnetic fields are almost certainly too strong to allow a third model (3) a nondissociative nonmagnetic (NDNM-type) shock of the sort produced in the original Orion models with $v_s \lesssim 24$ km s^{-1}. However, this section presents models of all three types, since the intensity ratios predicted by model 3 provide independent arguments against their existence in Orion.

TABLE 2 presents the observational data and best or typical fits from Reference 10 for each type of shock. The observational data in TABLE 2 is the approximate average surface brightness in a 1-arcmin beam corrected for 2.5 mag of extinction at 2 μm. The 2-μm H_2 absolute intensities are uncertain to perhaps a factor of two, due primarily to uncertainties and spatial variations in the extinction. The parameter space explored in Reference 10 includes $n_0 = 10^4$–10^7 cm^{-3} and $v_s = 20$–50 km s^{-1} for C-type shocks; $n_0 = 10^4$–10^7 cm^{-3} and $v_s = 50$–100 km s^{-1} for J-type shocks; and $n_0 = 10^4$–10^7 cm^{-3} and $v_s = 10$–24 km s^{-1} for NDNM-type shocks. All models assume preshock abundances of $x(O) = 3 \times 10^{-4}$, $x(OH) = 7 \times 10^{-8}$, $x(H_2O) = 7 \times 10^{-6}$, $x(O_2) = 1 \times 10^{-5}$, and

$x(CO) = 3.7 \times 10^{-4}$. The results are insensitive to the initial distribution of oxygen in O, OH, O_2, and H_2O; the shocks either quickly process the oxygen into H_2O or dissociate oxygen-bearing molecules. All models adopt $f_A = 2$, which corresponds, for example, to the simple case of two planar shocks moving at $\pm v_s$ with respect to the observer into stationary ambient gas.

NDNM shocks cannot be made to match the observations for the following fundamental reason. In order for the shock luminosity to equal that in the H_2 lines, $n_o v_s^3$ must exceed $3 \times 10^{24} \, s^{-3}$. Since $v_s < 24$ km s^{-1}, n_0 must be greater than 2×10^5 cm^{-3}. This high preshock density brings the high-J CO lines produced in the compressed postshock gas close to LTE and makes the 39–29/21–20 ratio ~ 2 instead of the observed 0.25. Furthermore, the shock luminosity is dominated by H_2O unless the H_2O abundance is rather artificially suppressed.

Dissociative J-type shocks also fail to produce the observed absolute intensities. Much of the total emission from the shock is radiated in the optical and the uv as the postshock gas cools from 10^5 K to ~ 5000 K. Molecular hydrogen 2-μm cooling is

TABLE 2

OBSERVATIONAL DATA AND BEST OR TYPICAL FITS FOR EACH TYPE OF SHOCK

	Line	Observed*	C-type†	J-type‡	NDNM-type§
H_2	1–0S(1)	$1(-1)^{37}$	$1(-1)$	$2(-3)$	$2(-3)$
	0–0S(2)	$5(-3)^{41}$	$5(-3)$	$1(-3)$	$8(-6)$
	2–1S(1)	$1(-2)^{37}$	$1(-2)$	$8(-4)$	$4(-4)$
CO	30–29	$2.4(-3)^{38}$	$2.5(-3)$	$3(-4)$	$4(-4)$
	27–26	$6.5(-3)^{38}$	$4.5(-3)$	$5(-4)$	$3(-4)$
	21–20	$1.0(-2)^{38}$	$1.1(-2)$	$1(-3)$	$2(-4)$
	6–5	$1.0(-3)^{42}$	$1.5(-3)$	$4(-4)$	$2(-5)$
OH	$^2\Pi_{3/2}(^5/_2-^3/_2)$	$1(-2)^{39}$	$8(-3)$	$2(-3)$	$6(-6)$

*Surface brightness in erg cm^{-2} s^{-1} ster^{-1} averaged over 1′ beam. The extinction at 2 μm and that at 12 μm were taken to be 2.5 mag and 0.94 mag, respectively.
†Preliminary best fit of Reference 10 with limited exploration of 10^4 cm$^{-3} < n_0 < 10^7$ cm^{-3} and 20 km s$^{-1} < v_s < 50$ km s^{-1}; $f_A = 2$, $n_0 = 2 \times 10^5$ cm^{-3}, $v_s = 50$ km s^{-1}, $x_{io} < 3 \times 10^{-7}$, $B_{0\perp} = 1$mGs, solar abundances.
‡Sample of fit found in the relevant parameter space 50 km s$^{-1} < v_s < 100$ km s^{-1}; $f_A = 2$, $n_0 = 2 \times 10^5$ cm^{-3}, $v_s = 100$ km s^{-1}, $B_{0\perp} = 1$mGs, solar abundances.[10]
§Sample of fit found for $v_s < 24$ km s^{-1}; $B_{0\perp} = 0$; $f_A = 2$, $n_0 = 10^7$ cm^{-3}, $v_s = 10$ km s^{-1}, solar abundances.[36]

suppressed because H_2 does not completely reform at $T_n > 1000$ K. Typically, $x(H_2) \simeq 0.01$ at $T_n = 2000$ K. Artificially forcing H_2 to reform above 1000 K does not increase the absolute 2-μm intensities significantly because H_2O then rapidly cools the gas. However, the intensities produced by the J-type shocks are typically $\sim 10\%$ of the observed intensity, and such fast shocks may contribute to the observed high-velocity wings.

The C-type magnetohydrodynamic shock wave structure results in predicted line intensities and ratios that fit the data remarkably well. The peak neutral temperature in the shock wave is ~ 3000 K, and only $\sim 3\%$ of the H_2 is dissociated. Since the shock velocity is approximately twice that permitted in the NDNM-type shocks, C-type shocks can produce at least eight times as much total shock intensity for the same preshock density (recall that the total intensity is proportional to $n_0 v_s^3$). Furthermore, since the peak temperature is ~3000 K, most of the shock emission emerges in the observed lines. The peak temperature is reached at $n \simeq n_0$ instead of $n \simeq 10 n_0$, typical of J- and NDNM-type shocks. The lower density of the emitting gas provides a good match to the CO line ratios, which require non-LTE populations in the rotational states. The low density also ensures that much of the emergent shock intensity will be H_2 emission, since the relative amount of H_2O cooling increases with density. The hydrogen atom population helps maintain the H_2 in vibrational LTE at ~2000 K. The best fits for C-type shocks occur at $n_0 = 2 \times 10^5$ cm^{-3} (recall that n_0 is the hydrogen nucleus density), $v_s = 50$ km s^{-1}, $B_{0\perp} \simeq 1$ mGs, and $x_{io} < 3 \times 10^7$.[10] To an observer in the preshock ambient gas, the 1–0S (1) H_2 intensity peaks at ~−2 km s^{-1}, has a FWHM of ~10 km s^{-1}, and falls to ~0.01 of the peak intensity at ~−30 km s^{-1} for this model. A range is allowed for these parameters, but n_0 is restricted by the non-LTE line ratios of the high rotational transitions of CO, v_s and $B_{0\perp}$ by the observed 2000 K temperature of the H_2 gas (see (7)), and $n_0 v_s^3$ by the observed intensity in the lines. A crude estimate of the range of parameter values that provide a good fit to the data is obtained by comparing the above values with preliminary results in Reference 21: $n_0 \simeq 2 \times 10^6$ cm^{-3}, $v_s \simeq 40$ km s^{-1}, $B_{0\perp} \simeq 4$ mGs, and $x_{io} < 10^{-7}$. These latter results are matched to Peak 1 of the 2-μm H_2 emission; they only marginally agree with the CO observations if applied to the entire emission region.

<h2>IMPLICATIONS OF THE SHOCK MODELING</h2>

The best-fit shock models indicate a magnetic field in Orion on the order of $B_0 \simeq$ 0.5–4 mGs.[10-21] The magnetic energy density is therefore close to the gravitational energy density in the BN-KL region, and magnetic fields may help support the cloud. Shock modeling holds promise as an independent means of measuring B fields in dense molecular clouds.

The shock modeling independently suggests that the average extinction at 2 μm to the H_2 emission in BN-KL is approximately 2–2.5 mag.[10] Less extinction means a higher ratio of CO intensities to intrinsic H_2 intensities and requires $x(CO)$ to be greater than 3×10^{-4}. More extinction means that n_0 must be increased to provide greater H_2 intrinsic intensity, which, in turn, drives the CO line ratios toward LTE.

A corollary of the above is that the CO abundance must be high, close to the solar abundance of carbon, $x(CO) \simeq 3 \times 10^{-4}$.

The high-velocity line wings of H_2 may be explained by ~100 km s^{-1} dissociative shock waves incident upon stationary ambient gas.[10] Otherwise, nondissociative shocks traversing ambient gas moving at $v > 50$ km s^{-1} are required. Several authors discuss simple models of high-speed ambient gas.[10,12,13] Two groups independently arrive at a similar model, pictured in FIGURE 5.[10,13] A wind of molecules at $v_w \simeq 100$ km s^{-1} impacts upon a shell that is radially expanding into ambient gas at v_{shell}. Two shock waves exist; an inner wind shock with shock speed $v_s = v_w - v_{shell}$ and an outer shock of swept-up ambient gas with $v_s = v_{shell}$. The shocked wind produces the high-speed wings, while the outer shock produces the bulk of the intensity.

Using n_0 and v_s from the best-fit model of Reference 10, the observed extent of the gas provides a measure of the total mass, M_T, momentum P_T, and kinetic energy, E_T in swept-up shock material.

$$M_T = \tfrac{4}{3}\,\pi m_H n_0 R^3 = 5 M_\odot$$

$$P_T = M_T v_s = 250\, M_\odot\, \text{km s}^{-1}, \tag{8}$$

$$E_T = \tfrac{1}{2} M_T v_s^2 = 6000\, M_\odot\, \text{km}^2\, \text{s}^{-2},$$

where $R \simeq 2 \times 10^{17}$ cm is the radius of the region. The total kinetic energy is greater than the observed kinetic energy of hot gas by a factor of 1000. Therefore, an even greater efficiency in converting stellar radiated and wind luminosity into kinetic energy is required than the direct observation of the hot gas implied. The momentum requirements suggest a mass-loss rate of 10^{-2}–$10^{-3}\, M_\odot\, \text{y}^{-1}$ if stellar winds power the shock waves. The instantaneous mass in the wind is comparable to the mass of swept-up material. The swept-up mass in the shell is comparable to the mass estimates of cold high-velocity gas inferred in the CO plateau source.[5,6]

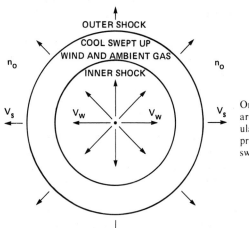

FIGURE 5. The "two-shock" model of Orion. The observed high-velocity wings are produced by the inner shocked molecular wind. The bulk of the intensity is produced by the outer shock, which sweeps up ambient gas.

H_2O is a major coolant in Orion shock waves.[9,10] The latter group predicts an H_2O luminosity about one-half that of H_2. They suggest that high excitation transitions with the lower state at least $E_l/k \gtrsim 2000$ K and with high A_{ul} values (~ 0.1 s^{-1}), such as the $9_{55}-9_{46}$, $10_{56}-10_{47}$, and $11_{57}-11_{48}$ transitions, may be observable from the Kuiper Airborne Observatory with surface brightness on the order of 10^{-3} erg cm^{-2} s^{-1} ster^{-1}.

SUMMARY

Shock modeling of the H_2, CO, and OH infrared emission lines in the BN-KL region of Orion yields several important astrophysical results. The estimates of the strength of the preshock magnetic field, ~ 1 mGs, indicate that magnetic pressure may help support the gas in this region. The determination of the shock speed and the preshock density, along with the measured extent of the gas, reveals that the momentum and energy in the swept-up shocked material is rather high compared with the usual estimates of the available momentum and energy from the embedded sources. The abundance of shock-heated CO is close to the solar abundance of carbon. The extinction to the 2-μm emission region of H_2 is approximately 2–2.5 mag. The high-velocity wings of the H_2 emission lines are most likely produced either behind fast dissociative shock waves or behind slower nondissociative shock waves traversing high-speed material.

REFERENCES

1. GAUTIER, T. N., U. FINK, R. R. TREFFERS & H. P. LARSON. 1976. Astrophys. J. 207: L129–33.
2. HOLLENBACH, D. J. & J. M. SHULL. 1977. Astrophys. J. 216: 419–26.
3. KWAN, J. 1977. Astrophys. J. 216: 713–23.
4. LONDON, R., R. McCRAY & S.-I. CHU. 1977. Astrophys. J. 217: 442–53.
5. ZUCKERMAN, B., T. B. H. KUIPER & E. N. RODRIGUEZ-KUIPER. 1976. Astrophys. J. 209: L137–42.
6. KWAN, J. & N. Z. SCOVILLE. 1976. Astrophys. J. 210: L39–43.
7. BECK, S. C., E. E. BLOEMHOF, E. SERABYN, C. H. TOWNES, A. T. TOKUNAGA, J. H. LACY & H. A. SMITH. 1982. Astrophys. J. 253: L83–87.
8. DRAINE, B. T. 1980. Astrophys. J. 241: 1021–40.
9. DRAINE, B. T., W. ROBERGE & A. DALGARNO. 1982. Astrophys. J. In press.
10. CHERNOFF, D., D. J. HOLLENBACH & C. F. McKEE. 1982. Astrophys. J. In press.
11. HOLLENBACH, D. J. & C. F. McKEE. 1980. Astrophys. J. 241: L47–51.
12. CHEVALIER, R. A. 1980. Astrophys. Lett. 21: 57–61.
13. NADEAU, D., T. R. GEBALLE & G. NEUGEBAUER. 1982. Astrophys. J. 253: 154–71.
14. SCOVILLE, N. Z., D. N. B. HALL, S. G. KLEINMANN & S. T. RIDGWAY. 1982. Astrophys. J. 253: 136–53.
15. BECKWITH, S., N. J. EVANS, I. GATLEY, G. GULL & R. RUSSELL. 1982. Astrophys. J. In press.
16. BECK, S. C. 1982. Ph.D. Thesis. University of California, Berkeley.
17. KNACKE, R. F. & E. T. YOUNG. 1981. Astrophys. J. 249: L65–68.
18. SIMON, M., G. RIGHINI-COHEN, R. R. JOYCE & T. SIMON. 1979. Astrophys. J. 230: L175–78.
19. NADEAU, D. & T. R. GEBALLE. 1979. Astrophys. J. 230: L169–73.
20. GENZEL, R. & D. DOWNES. 1982. In Regions of Recent Star Formation. R. S. Roger and P. E. Dewdney, Eds.: 251–61. D. Reidel. Dordrecht.
21. DRAINE, B. T. & W. ROBERGE. 1982. Astrophys. J. In press.
22. MULLAN, D. J. 1971. Mot. Not. R. Astron. Soc. 153: 145–63.
23. HEILES, C. 1976. Annu. Rev. Astron. Astrophys. 14: 1–28.
24. VERSCHUUR, G. L. 1979. Fundam. Cosmic Phys. 5: 113–34.
25. MOUSCHOVIAS, T. C. 1976. Astrophys. J. 207: 141–56.
26. DENNISON, B. 1977. Astrophys. J. 215: 529–39.
27. DYCK, H. M. & C. A. BEICHMAN. 1974. Astrophys. J. 194: 57–68.
28. JOHNSON, P. E., G. H. RIEKE, M. J. LEBOFSKY & J. C. KEMP. 1981. Astrophys. J. 245: 871–79.
29. ADAMS, N. & D. SMITH. 1981. Astrophys. J. 248: 373–80.
30. UMEBAYASHI, T. & T. NAKANO. 1980. Publ. Astron. Soc. Jpn. 32: 4–20.
31. McKEE, C. F. & D. J. HOLLENBACH. 1980. Annu. Rev. Astron. Astrophys. 18: 219–43.
32. HOLLENBACH, D. J. & C. F. McKEE. 1979. Astrophys. J. Suppl. Ser. 41: 555–89.
33. FORMISANO, V., A. A. GALEEV & R. Z. SAGDEEV. 1981. Preprint.
34. HARTQUIST, T. W., M. OPPENHEIMER & A. DALGARNO. 1980. Astrophys. J. 236: 182–91.

35. ROBERGE, W. & A. DALGARNO. 1982. Astrophys. J. **255**: 176.
36. HOLLENBACH, D. J., D. CHERNOFF & C. F. McKEE. 1980. Bull. Am. Astron. Soc. **12**: 826.
37. BECKWITH, S., S. E. PERSSON, G. NEUBEGAUER & E. E. BECKLIN. 1978. Astrophys. J. **223**: 464–74.
38. WATSON, D. M., J. W. V. STOREY, C. H. TOWNES, E. E. HALLER & W. L. HANSON. 1980. Astrophys. J. **239**: L129–33.
39. STOREY, J. W. V., D. M. WATSON & C. H. TOWNES. 1981. Astrophys. J. **244**: L27–31.
40. DOWNES, D., R. GENZEL, E. E. BECKLIN & C. WYNN-WILLIAMS. 1981. Astrophys. J. **244**: 869–84.
41. BECK, S. C., J. H. LACY & T. R. GEBALLE. 1979. Astrophys. J. **234**: L213–17.
42. GOLDSMITH, P. F., N. R. ERICKSON, H. R. FETTERMAN, B. J. CLIFTON, D. D. PECK, P. E. TANNENWALD, G. A. KOEPF, D. BUHL & N. McAVOY. 1981. Astrophys. J. **243**: L79–82.

DISCUSSION OF THE PAPER

N. SCOVILLE (*University of Massachusetts, Amherst, Mass.*): Our recent high-resolution observations of molecular hydrogen support your proposal for more than one type of shock. Early shock models accounted easily for a temperature of 2000 K through the H_2 dissociation thermostat. How is this done in more recent models?

HOLLENBACH: The velocity is now a more critical parameter in the models.

T. MOUSCHOVIAS (*University of Illinois, Urbana, Ill.*): The calculations described are very nice. But it dangerous to claim predictions for such things as magnetic field strengths. I think that two severe limitations of the calculations should be relaxed before attempting a detailed fit of observations. (1) The assumption of a steady-state shock will be violated by propagation in a density gradient; the jump conditions (including density and temperature jumps) then change significantly. (2) In a real three-dimensional geometry, the expansion perpendicular to the field lines will be prevented on a short timescale and subsequent expansion will be bipolar along field lines. During this phase, a one-dimensional, time-dependent nonmagnetic calculation would make more sense.

G. CARRUTHERS (*Naval Research Laboratory, Washington, D.C.*): Your calculations indicate that the emitting region associated with a C-type shock is wider and more diffuse than that associated with a J-type shock. What angular sizes do these widths correspond to at the distance of Orion, and is there any possibility of spatially resolving them observationally?

HOLLENBACH: The shocks are very thin, on the order of 10^{14} cm.

FRAGMENTATION, PROTOSTELLAR WINDS, AND STAR FORMATION*

Joseph Silk

Department of Astronomy
University of California, Berkeley
Berkeley, California 94720

INTRODUCTION

Molecular clouds are undoubtedly the sites of star formation. Yet, despite the wealth of observational material spanning the microwave, infrared, and optical spectral regions, the detailed processes by which stars form has proved remarkably elusive. What we do possess is strong circumstantial evidence that sheds light on many aspects of molecular cloud evolution and star formation. In the first section of this review, I will summarize some of this evidence, with emphasis on the nature of the highly nonlinear processes by which interstellar clouds containing in excess of 10^4 $M_\odot$ of diffuse matter form stars of much lower mass. The problem of the gravitational fragmentation of a collapsing cloud will be discussed, including the roles of magnetic fields, density fluctuations, anisotropic collapse, and nonlinear interactions between fragments. In cold molecular clouds, where low-mass star formation predominates, and in warm molecular clouds, where massive young stars are present, the effects of protostellar winds are shown to provide an important source of energy and momentum for the diffuse molecular gas.

In outline, the following view of molecular cloud evolution will be presented. Low-mass star formation may be expected to proceed quietly in cold molecular cloud cores. As T Tauri stars form, winds will develop that are likely to drive dense shells of swept-up matter into the cloud. The wind-driven shells eventually interact and break up, to form successive generations of fragments. Accretion and coagulation by fragments that eventually collapse is capable of providing a continuing supply of newly formed low-mass protostars. An external trigger, perhaps provided by collision with a cloud of comparable mass, as would be expected to occur with highest probability in the spiral density wave peaks, will enhance the fragment coalescence rate. This should suffice to initiate massive star formation. Once massive stars form, they are likely to stimulate further massive star formation, which should therefore be self-sustaining and which will, eventually, disrupt the cloud.

OBSERVATIONAL ASPECTS

Stars

Observations of stellar associations provide important clues to the nature of the complex processes by which a molecular cloud produces fragments of protostellar

*This research was supported, in part, by a grant from the National Aeronautics and Space Administration, no. NGR 05-003-578.

257

0077-8923/82/0395-0257 $1.75/1 © 1982, NYAS

mass. One of the more striking results has been obtained by Blaauw, who has used highly accurate proper motions to trace back in time the trajectories of stars in expanding subgroups.[1] He infers that the dimensions and densities of the parent cloud from which the stars formed are similar to those of the giant molecular cloud complexes.

Detection of a pre-main-sequence turn-on in the Pleiades[2,3] quantifies the often-noted discrepancy between the nuclear age derived from massive stars in this open cluster (7×10^7 y) and the contraction age derived from low-mass stars (2×10^8 y). This helps confirm Herbig's suggestion that low-mass stars formed over a substantially longer period than did massive stars.[4] The main sequence turn-off therefore yields the time since the last generation of massive stars formed and the turn-on yields the total time over which star formation has occurred.

Further evidence that supports the hypothesis of noncoeval star formation comes from the dispersion of young cluster stars in the H-R diagram. Iben and Talbot found that the star formation rate increased exponentially with time in a study of two young clusters and that the average mass of newly formed stars increases with time.[5] This result has been confirmed in a recent study of NGC 2264 by Adams *et al.*[6]

One further result that has emerged from statistical studies of young stars may be noted here. Wolf *et al.* have found that the distribution of rotational velocities, $v \sin i$, indicates relatively low rotations for B stars.[7] Moreover, the rotational velocity distribution is consistent with a random distribution of rotation axes, is independent of spectral type (except that the O star distribution in $v \sin i$ is deficient in low rotational velocities), and is independent of the age of the subassociation studied. This result is of considerable importance in understanding how the well-known angular momentum problem has been resolved (the specific angular momentum of the parent cloud being $j \simeq 10^{24}$ cm^2 s^{-1}, whereas that of B stars is $\sim 10^{17}$ cm^2 s^{-1}).

An important sink for angular momentum is in orbital angular momentum, and the study of the distribution of binary periods as a function of secondary mass by Abt and Levy[8] is also of considerable interest here. One relevant result is that binary periods (from 10 h–100 y) form a single-peaked distribution (at $j \simeq 10^{19}$ cm^2 s^{-1}), as would, perhaps, be expected if a single mechanism is responsible for producing these binaries. Moreover, very-long-period binaries (periods greater than ~ 100 y) are found to possess a mass function for the secondaries that resembles that of field stars, whereas that for shorter period binaries is much flatter. This may be indicative of a distinct formation mechanism for the formation of wide binaries with $j > 10^{20}$ cm^2 s^{-1}.

Molecular Clouds

Molecular clouds generally reveal highly supersonic linewidths, typically 1–3 km s^{-1}. Even at high angular resolution, the Orion Molecular Cloud core line profiles remain broad on scales of ~ 0.01 pc. However, studies of cold clouds suggest that the smallest structures, such as Taurus Molecular Cloud-1 (TMC1), may not be far from thermal support on mass-scales of a few $M_\odot$. Dense cold clumps of mass $\gtrsim 1$ $M_\odot$ are seen in NH$_3$ and HC$_5$N. CO maps of warm cloud complexes reveal clumpiness on scales $\gtrsim 100$ $M_\odot$.

In general, cloud lifetimes must exceed the free-fall timescale by a considerable factor. This factor is estimated to be ~10 in the case of TMC1.[9] More generally, one can estimate a mean lifetime for molecular clouds from the observed surface density, Σ, of molecular gas in the galactic plane, the star formation rate, S, and the efficiency, ϵ, of star formation. Using values appropriate to the solar neighborhood, one finds that $\Sigma \simeq 2\, M_\odot \,\mathrm{pc}^{-2}$,[10] $S \simeq 4 \times 10^{-9}\, M_\odot \,\mathrm{y}^{-1}\,\mathrm{pc}^{-2}$,[11] and $\epsilon \simeq 0.01$–0.1.[12] With due allowance for low-mass stars, the higher value for ϵ may be more appropriate. The mean lifetime of a molecular cloud is therefore $\sim5 \times 10^7\,(\epsilon/0.1)$ y. In principle, one could imagine clouds of much shorter mean lifetime with a correspondingly lower value of the efficiency, except that ϵ is constrained by observations. For T associations that range in age from 10^6 to 10^7 y, Cohen and Kuhi find that ϵ generally exceeds 0.01.[12] Hence, a firm lower limit for molecular cloud lifetimes is $\sim5 \times 10^6$ y.

In localized regions there is evidence that molecular gas is dispersing on a much more rapid timescale, consistent with the measured velocity dispersion. One such region is L1551, where a bipolar flow at ±15 km s^{-1} is evidently being generated by a protostellar wind from a T Tauri–like star. Such regions seem to be rare (a recent survey found four such pedestal features in velocity in some 180 dark clouds),[13] but may provide a crucial clue to the energy source in molecular clouds. Spectroscopic studies of T Tauri stars suggest that mass-loss rates as high as $\sim10^{-6}\, M_\odot \,\mathrm{y}^{-1}$ (the value inferred in L1551) are atypical, the more characteristic T Tauri star perhaps having a mass-loss rate $\lesssim10^{-8}\, M_\odot \,\mathrm{y}^{-1}$ (cf. Reference 14). However, high-resolution spectroscopic studies of T Tauri stars reveal evidence for eruptive events that appear to coincide with outbursts in brightness.[15] It is likely that shells of matter are ejected during these outbursts, in a scaled-down analogue of the FU Orionis phenomenon. Hence, a time-averaged mass-loss rate as high as $10^{-7}\, M_\odot \,\mathrm{y}^{-1}$ is inferred from some of these moderately low mass stars. Presumably, these eruptive events, as well as the steadier outflows, occur over periods of $\gtrsim10^6$ y, corresponding to a substantial fraction of the pre-main-sequence convective phase.

More massive pre-main-sequence stars also show evidence for strong mass loss at high velocity. These include the Herbig emission-line stars, believed to be in the range 2–10 $M_\odot$, and extremely luminous infrared stars, such as IRc2 in the core of the Orion Molecular Cloud. The driving source for these winds from luminous protostars is most likely to be due to radiation pressure on circumstellar dust grains.[16] If the optical depth at a few micrometers is sufficiently large (>100 for the Orion source), terminal velocities on the order of 100 km s^{-1} can be attained, comparable with the observed outflows in CO and H$_2$. Indirect evidence that even higher wind velocities are required comes from the proper motion studies of HH1 and HH2, which indicate space velocities for dense knots in these Herbig-Haro objects in excess of 300 km s^{-1}.[4] In this case, the exciting star does not seem to be very luminous. At least for the lower-mass protostars, it is possible that the winds are convectively driven with characteristic velocities of $\gtrsim300$ km s^{-1}.

FRAGMENTATION

Evidence for Fragmentation

The observations of clumpiness in molecular clouds indicate that fragmentation occurs at densities as low as 10^3 cm^{-3}. Indirect support for this conclusion comes from

Blaauw's inference of the dimensions of the prestellar cloud for an expanding subgroup in the Scorpius OB association.[1] The dimensions (45 pc × 15 pc) resemble those of diffuse molecular clouds. This implies that fragmentation into units of stellar mass occurred at densities $\sim 10^2$–10^3 cm^{-3}.

On the other hand, the density of stars in the Trapezium subgroup in Orion exceeds 500 pc^{-3}, equivalent to a gas density of 2×10^4 molecules cm^{-3}. Proper motion studies of Trapezium stars disagree as to whether the group is expanding[17] or contracting,[18] although the velocity dispersion appears to be low (<2 km s^{-1} according to Reference 18). If this subgroup was originally bound by gas, the inferred gas density must be $\sim 10^6$ cm^{-3}. Since the nuclear age is $\sim 10^6$ y, one infers that this subgroup must have become unbound within the past 10^4–10^5 y.

A third piece of evidence is an argument due to Mouschovias, who notes that, if magnetic braking enforces corotation of molecular clouds at densities below $n_{cr} \simeq 10^4$–10^6 cm^{-3} and angular momentum is subsequently conserved in continuing collapse, the residual angular momentum is $\sim 3 \times 10^{18}$ $(n_{cr}/10^6$ cm$^{-3})^{-2/3}$ cm^2 s^{-1} and is characteristic of that in binaries with periods $\lesssim 100$ y.[19] This implies that binaries may have directly formed by fragmentation out of a region with specific angular momentum comparable to that in their orbital motion, provided that the initial fragmentation occurred at densities above n_{cr}. This therefore yields an estimate of the minimum density at which fragmentation must have occurred. One also infers that very wide binaries with periods $\gtrsim 100$ y formed by a different process, most likely by capture, because their mass function resembles that of field stars. Hence, it is unlikely that fragmentation occurs at a density as low as 100 cm^{-3}, although the specific angular momentum in a 1 $M_\odot$ fragment at this density that corotates with the galaxy is appropriate to that of a binary with a period of $\sim 3 \times 10^4$ y.

Evidently, fragmentation occurs at densities characteristic of diffuse molecular clouds. Three-dimensional hydrodynamic simulations of cloud collapse lack sufficient resolution to tackle the fundamental fragmentation problem—How do solar mass fragments form out of a massive cloud? One suggestion has been that fragmentation may occur in successive stages, each stage resulting in fragmentation into two or three fragments, with their specific angular momentum reduced by an order of magnitude.[20] Unfortunately, this process is very inefficient and seems unlikely to be an important mechanism for star formation. Another argument cited against such a hierarchical picture is that the rotation axes of young stars appear to be randomly distributed in space,[7] more suggestive of randomization by gravitational interactions among many fragments than of a binary fragmentation model.

Fragmentation in Spherically Symmetric Collapse

An analytic theory of gravitational fragmentation was first developed by Hoyle,[21] who noted that fluctuations on all scales larger than the initial Jeans length would begin to amplify during collapse. The growth rate for fluctuations is the same as the collapse rate; consequently, the growth is secular rather than exponential and the initial fluctuation level determines the final outcome. For a spherically symmetric uniform system undergoing collapse from rest, the free-fall time is

$$t_{ff} = (3\pi/32G\rho_0)^{1/2},$$

where ρ_0 is the initial density. The perturbation growth rate is

$$\delta \equiv \delta\rho/\rho = \delta_0 \, (t_{ff} - t)^{-1}$$

in the linear regime, for density perturbations of initial density contrast δ_0.[22] Once δ is $\gtrsim 1$, self-gravity becomes important for the fluctuations and rapid growth ensues, as may be demonstrated from an exact nonlinear solution. However, only if the initial amplitude is sufficiently large can we reasonably expect fluctuations to go nonlinear and the collapsing cloud to fragment.

What value is required for δ_0 in order for fragmentation to occur? Since the density increases in uniform spherical collapse as $\rho = \rho_0(t_{ff} - t)^{-2}$, we infer that fluctuations are large when $\delta \simeq 1$, or at a time given by $t_{ff} - t \simeq \delta_0$, just before collapse of the entire cloud at t_{ff}. At this instant, the mean density has increased by a factor $\rho/\rho_0 =$

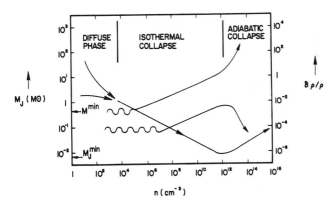

FIGURE 1. The fate of density fluctuations in a collapsing cloud. The Jeans mass (left ordinate) is shown as a function of density for spherical collapse using a silicate grain model. It attains a minimum value of ~0.007 $M_\odot$ at a particle density of ~2×10^{12} cm^{-3}. The evolution of the density contrast (right ordinate) is qualitatively illustrated for a fluctuation containing mass M(min) and for one of mass $<M^{(min)}$ but $>M_J^{(min)}$.

$(t_{ff} - t)^{-2} = \delta_0^{-2}$. Hence, collapse by a factor of 10^4 in density is necessary for perturbations of initial amplitude $\delta_0 \simeq 0.01$.

This estimate assumes that the perturbations are always well above the instantaneous Jeans length. If they are not, growth can be suppressed (FIGURE 1). It is convenient to introduce the instantaneous Jeans mass defined by

$$M_J = \frac{\pi}{6} \rho \, \lambda_J^3 = \left(\frac{\pi^4}{6}\right) v_s^3 \, G^{-3/2} \, \rho^{-1/2}.$$

During the diffuse collapse phase, the cloud will remain approximately isothermal. Hence, as the density increases, M_J will decrease. A fluctuation that is below the Jeans mass at the onset of collapse will eventually begin to grow when it first overtakes the Jeans mass. There is actually a minimum value for the Jeans mass, which effectively occurs when the collapse becomes adiabatic. This inevitably happens at a sufficiently high column density, when radiation trapping occurs and cooling is inhibited.

Consider, then, a fluctuation of wavelength λ that only commences to grow at a time, t_λ, well into the collapse. In other words, at t_λ, the fluctuation mass M_λ first exceeds the Jeans mass. If the density contrast at this time is δ_λ, fragmentation will occur at an epoch t_f, say, when the background density has increased by a factor $(5/\delta_\lambda)^2$, according to a recent analysis of uniform spherical collapse.[23] However, a crucial assumption is that the collapse remains isothermal. In other words, $\rho(t_f)$ must not exceed $\rho(t_{ad})$, where the epoch at which the collapse first becomes adiabatic is denoted by t_{ad}. Fragmentation will only be effective on mass scales larger than M_λ, since smaller scales will not have separated out by t_{ad}. If their density contrast is small at this stage, the fluctuations will not survive into the adiabatic collapse phase as distinct fragments. The minimum mass fragments to form will have just become nonlinear at t_{ad}. If $M_J^{(min)} \propto \rho(t_{ad})^{-1/2}$ denotes the minimum Jeans mass at t_{ad}, we infer that

$$\rho(t_{ad})/\rho(t_\lambda) > \rho(t_f)/\rho(t_\lambda) = (5/\delta_\lambda)^2$$

and the minimum mass fragment is

$$M^{(min)} > (5/\delta_\lambda)\, M_J^{(min)}.$$

This simple result leads to a considerable difficulty in understanding star formation. Not only is collapse by a density enhancement factor of about 10^5 required for fluctuations with an amplitude of ~1% from the instant that they are first Jeans unstable, but the minimum protostar fragment mass becomes uncomfortably large. To see in more detail how this arises, let us briefly review the opacity-limited fragmentation argument that defines $M_J^{(min)}$.

The time evolution of a volume element in the uniform spherically collapsing cloud is defined by a locus in the temperature-density plane. The condition that the volume element be able to freely radiate away its thermal energy as it is compressed and its internal energy increases defines a relation between T and ρ that is initially almost isothermal: $T \propto \rho^{1/8}$ is actually found to apply.[24] Since the Jeans mass can be written

$$M_J = \text{constant } (p/\rho^{4/3})^{3/2}$$

and the equation of state inferred for optically thin uniform spherical collapse is $p \propto \rho^{9/8}$, one sees that the Jeans mass decreases as $M_J \propto \rho^{-5/16}$. For exactly isothermal collapse, one would have $M_J \propto \rho^{-1/2}$.

At sufficiently high density, inhibition of cooling by radiation trapping qualitatively alters this result, since the effective equation of state now resembles $p \propto \rho^{5/3}$. Radiation trapping by fragments comparable in size to the instantaneous Jeans length is inevitable, because the column density across a Jeans mass fragment is proportional to $\rho^{1/2}$ and the optical depth eventually becomes large. The new equation of state in this adiabatic regime is derived by requiring that an isolated fragment be able to radiate away the gravitational energy acquired as it contracts. Since the cooling rate now depends on the fragment size, the evolution track in the (T, ρ) plane is mass dependent. The Jeans mass now rises as $M_J \propto \rho^{1/2}$, and its minimum value occurs when the optical depth across a fragment is on the order of unity. Use of opacities corresponding to conventional grain models (graphite or silicates) and a solar abundance of heavy elements in grains yields a value $M_J^{(min)} \simeq 0.005\ M_\odot$.[24] There is a correction factor that should be incorporated due to the presence of neighboring fragments, which effectively decrease the solid angle over which an individual fragment can radiate freely.[25] This effect raises the minimum fragment mass by a factor $\sim N^{1/6}$, where N is the number of fragments in the cloud.[26]

Even in the absence of any heavy elements, opacity due to H^- formation is important. In this case, the minimum Jeans mass is $\sim 0.3\ M_\odot$.[27] A simple expression for $M_J^{(min)}$ that explicitly demonstrates the role of heavy elements is[27,28]

$$M^{(min)} \simeq 20\ M_c (kT/\mu c^2)^{1/4}\ M_\odot.$$

Here $M_c = (hc/G\mu)^{3/2} \simeq 1\ M_\odot$ is the Chandrasekhar mass and μ is the mean molecular weight. Provided the heavy element abundance remains above $\sim 10^{-3}$ that of the solar value, cooling occurs to below 10 K. However, at lower values, heavy element cooling is unimportant and $T \simeq 10^4$ K is maintained by $Ly\alpha$ cooling.

The problem confronting the fragmentation theory is now very apparent. With $M^{(min)} \gtrsim 10^3\ M_J^{(min)}$ for $\delta \simeq 0.01$, as expected in spherical collapse, it is not at all obvious how fragments of stellar mass can form. Fragments of primordial composition are entirely outside the conventional stellar mass range. Even for solar composition, stars of solar mass are excluded. Indeed, one might more realistically expect that the fragments are only likely to provide lower limits to the actual masses of the protostars that form. The various nonlinear processes that one can imagine, including accretion of uncondensed matter and coagulation of fragments, will tend to increase the masses of fragments. This result led Tohline to conclude that Population III of primordial composition consisted not of stars but of very massive objects.[29] This conclusion, if correct, has important implications for the chemical and dynamical evolution of the protogalaxy. Consequently, it deserves to be examined with some care. In fact, the little evidence one has from nucleosynthetic traces is consistent with the notion that Population III consisted of stars, although practically all must have been considerably more massive than the sun. However at least one halo star has been discovered with essentially zero metallicity ($10^{-4.5}\ Z_\odot$ according to Bessell and Norris[29]) and is presumably a relic of Population III.

Anisotropic Collapse and Fragmentation

The envelope of a collapsing cloud will be more easily supported by the magnetic field, especially if it is somewhat tangled, than the cloud core. This is because the critical mass below which magnetic support is possible for a uniform spherical cloud (the "magnetic Jeans mass") is

$$M_{cr} = 6 \times 10^4\ (B/10^{-6}\ g)^3\ n^{-2}\ \xi\ M_\odot,$$

where $\xi \simeq 0.3$ and $B \propto n^k$, with $1/3 \lesssim k \lesssim 1/2$.[19,30] Consequently, $M_{cr} \propto n^{-3(2/3-k)}$ and is reduced in the cloud core. Evidence was cited above that supports the occurrence of magnetic braking up to densities characteristic of molecular cloud cores. Provided that fragmentation also occurs at densities $>10^3$ cm^{-3}, we have seen that the angular momentum of molecular clouds is likely to reside in the orbital angular momentum of wide binaries.

Accordingly, the collapse of cloud cores is likely to be anisotropic, contracting preferentially along field lines, since field decoupling will only occur gradually. Now the cloud, if cold, is highly Jeans unstable. The characteristic mass for gravitational instability is given by the Bonner-Ebert criterion, which takes the ambient pressure into account:

$$M_{BE} = 1.1(T/10 \text{ K})^{3/2} (n/10^4 \text{ cm}^{-3})^{-1/2} M_\odot.$$

Thus, to decide whether fragmentation occurs, we see that spherical collapse may be an unrealistic assumption. A more plausible assessment of fragmentation may be given as follows.

Consider the collapse from rest of a cloud that is initially uniform, pressure-free, and oblately spheroidal in shape. A similar discussion applies for a prolate collapse. For simplicity, only a small initial deviation from sphericity is assumed. The analysis of the growth of small density perturbations is similar to that for a uniformly collapsing sphere. The collapse of the spheroid is described by two scale factors: $R(t)$ in the directions of the two equal axes and $Z(t)$ in the direction of the smallest axis. The position of any point in the spheroid is then given by $r = r_0 R(t)$, $z = z_0 Z(t)$, where r and z are cylindrical coordinates and r_0 and z_0 refer to the initial position of the point. The density satisfies

$$\rho = \rho_0 (R^2 Z)^{-1},$$

where ρ_0 is the initial density. Now the spheroidal cloud, even if very nearly spherical at the onset of the collapse, becomes progressively more flattened as the collapse continues.[31] In fact, it collapses first along the z-axis into a thin pancake. What this implies is that, in the final stages of the collapse, $R(t)$ changes relatively slowly, while $Z(t) \to 0$ (in practice, the thickness will be finite because the matter will possess a certain amount of thermal energy and pressure).

Recall that, in uniform spherical collapse, a small density perturbation of initial contrast δ_0 amplifies if its scale exceeds the Jeans length and results in fragmentation (that is to say, $\delta\rho/\rho$ becomes large) shortly before the cloud itself has collapsed, in fact within a fraction $1 - \delta_0$ of an initial free-fall time. An interesting difference arises when we study the growth of perturbations in a spheroidal collapse. Density fluctuations that are predominantly aligned with the collapse (z) axis do not become large, whereas fluctuations that are perpendicular to the collapse axis do amplify and separate out before the instant of pancaking. Self-gravity dominates the final evolution of oblate perturbations but is unimportant for prolate perturbations. The rate at which the oblate perturbations grow is found to be[32]

$$\delta \simeq \delta_0 Z^{-1}.$$

An interesting difference is now seen to arise from the one-dimensional nature of oblate spheroidal collapse. Because the density increases as $\rho \propto Z^{-1}$ when $Z \to 0$, we see that the density enhancement achieved by the cloud at fragmentation ($\delta \simeq 1$ in the linear theory) is

$$\rho(t_f)/\rho_0 \simeq Z(t_f)^{-1} \simeq \delta_0^{-1},$$

in marked contrast to the result for spherical collapse. Inclusion of a finite initial pressure that acts to delay fluctuation growth modifies this result, but less severely than in the case of spherical collapse. This is because the retardation means that the entire growth occurs when the collapse is nearly one-dimensional and the geometrical effects dominate the growth rate. If a fluctuation is first Jeans unstable with amplitude δ_λ at an epoch t_λ, one finds that, at fragmentation,

$$\rho(t_f)/\rho(t_\lambda) \lesssim 2\delta_\lambda^{-1}.$$

Adopting the opacity-limited fragmentation result that fragments should have achieved a density contrast on the order of unity before t_{ad}, one now infers that the minimum fragment mass

$$M^{(min)} > (2/\delta_\lambda)^{1/2} M_J^{(min)}$$

for oblate spheroidal collapse. Since this result is found to be valid even for initial flattenings $z_0/r_0 \simeq 0.8$, one infers that it is likely to apply in any realistic situation. The spherical collapse model is too highly idealized to be relevant, given any reasonable range of initial deviations from spherical symmetry. Such deviations would be expected for plausible initial conditions at the onset of the collapse.

The implications for star formation are profound, for one expects the density fluctuation level to be at least $\delta \simeq 0.01$ over a wide range of scales. In primordial clouds, thermal instability associated with H_2 cooling guarantees sizable fluctuations down to mass scales of a few $M_\odot$. In conventional molecular clouds, the complex history of a cloud, involving accumulation of debris from smaller clouds and evolving stars, suggests that fluctuations should be present down to scales of ~ 1 $M_\odot$. Moreover, the violent events inferred to be stirring up the interstellar medium (including supernova explosions and stellar winds) should also generate pressure fluctuations over a wide range of scales. These are able to penetrate $\sim \delta_\lambda^{-1}$ wavelengths into a cloud before dissipating. Consequently, for a cloud of mass M_c, one expects the fluctuation level to be $\delta_\lambda \simeq (M_\lambda/M_c)^{1/3} \gtrsim 0.01$ over any stellar mass scale M_λ.

With $M_J^{(min)} \simeq 0.005$ $M_\odot$ in molecular clouds and 0.3 $M_\odot$ in primordial clouds, the preceding discussion implies that fragmentation is likely to be effective on mass scales as small as 0.05 $M_\odot$ (molecular clouds) to 3 $M_\odot$ (primordial clouds). This does not imply that smaller fragments cannot form, as will happen when quasi-equilibrium disks form around protostars. Moreover, various nonlinear processes can act to increase the fragment mass, and these must be examined if one hopes to be able to compare this theory with the initial mass function.

PROTOSTELLAR WINDS AND STAR FORMATION

Bubbles and Shells

Collisions between fragments must play an important role in spherical collapse; indeed, the concept of individual surviving fragments is questionable.[33] Fragment interactions will result in coagulation and accretion (or possibly in self-destruction). However, a realistic collapse will be nonspherical, and this will diminish the role of collisions. Individual self-gravitating fragments may undergo collapse on a timescale rapid enough that the first protostars are likely to form directly from fragments in the cloud core, where magnetic fields and rotation may be expected to be unimportant in significantly impeding collapse. Once the first stars form, their protostellar winds can have a significant impact on the surrounding gas, perhaps temporarily inhibiting further fragmentation and star formation. Obviously, given the uncertainties in the competing processes of fragmentation and coagulation, there are many possible outcomes and it is necessary to rely on observational evidence for indications of processes that are likely to be important.

One is tempted to try to relate wind input of energy to one of the great mysteries about molecular clouds, namely, the origin of their supersonic line widths. Overall collapse or expansion provides an untenable explanation for the line widths, and one is left with a cloud model that consists of a number of supersonically moving clumps of gas. The outstanding questions are, What drives the clump motions and how are the clumps maintained for periods $\geq 10^7$ y? A similar difficulty is encountered both in warm molecular clouds and in cold dark clouds.

The most natural explanation is that protostellar winds are continuously driving mass motions.[34] Cloud longevity can be understood if the winds are not disruptive, a plausible assumption for low- or intermediate-mass pre-main-sequence such as T Tauri or Herbig emission stars embedded in cold clouds. Now, in a dense molecular cloud, a wind at $\lesssim 200$ km s^{-1} will be radiative and approximately momentum conserving. One may crudely estimate the mean velocity dispersion acquired by an average volume element in a cloud of mass M_c containing a mass M_* in stars that have lost a fraction ΔM_* of their mass at some characteristic wind velocity V_w as

$$\langle \Delta v \rangle \simeq \left(\frac{M_*}{M_c} \right) \left(\frac{\Delta M_*}{M_*} \right) V_w.$$

Evidently, a substantial fraction of the cloud matter can be stirred up with $\langle \Delta v \rangle \simeq 1$ km s^{-1} if $V_w \simeq 200$ km s^{-1}, $M_*/M_c \simeq 0.1$ (as observed in dark clouds), and $\Delta M_*/M_* \simeq 0.1$. For this to persist over 2×10^7 y, a considerable part of the cloud would have to be consumed in star formation: indeed, exhaustion of cloud material may lead to the formation of a T association. On the other hand, intervention of an external trigger, perhaps associated with a nearby supernova or an expanding H II region, may change the cloud evolution in a manner that will now be outlined, and so form an O association.

Let us suppose that the first stars to form are T Tauri stars. These low-mass stars develop winds at the onset of the Kelvin-Helmholtz contraction phase. Low-mass stars that will take 10^8 y or more to reach the main sequence are identified as T Tauri stars in regions as young as Taurus-Auriga.[12] These winds sweep up shells of material. The final radius of such a shell is limited by the ambient cloud pressure to

$$R_{(max)} \simeq \left[\frac{\dot{M} V_w}{4\pi \rho V_*^2} \right]^{1/2} \simeq 0.1 \left[\frac{(\dot{M}/10^{-7} M_\odot \text{ y}^{-1})}{(n/10^4 \text{ cm}^{-3})^{1/2}} \frac{(V_w/300 \text{ km s}^{-1})}{(V_*/1 \text{ km s}^{-1})} \right]^{1/2} \text{pc},$$

where V_* is the stellar velocity dispersion and $\dot{M}$ is the mean protostellar mass-loss rate. The condition for such shells to grossly affect the cloud evolution and inhibit further fragmentation and star formation is that any pair of shells should intersect within the lifetime of the protostellar wind phase. If δM is the amount of mass lost by an individual protostar, this condition can be expressed as

$$\frac{1}{n_* \pi R^2_{(max)} V_*} < \frac{\delta M}{\dot{M}} \quad \text{or}$$

$$n_* \gtrsim 10 \, \frac{(V_*/1 \text{ km s}^{-1}) \, (n/10^3 \text{ cm}^{-3}) \, (0.1 \, M/\delta M)}{(V_w/10^2 \text{ km s}^{-1})} \text{ pc}^{-3}.$$

For comparison, one of the best-studied dark cloud regions in Taurus-Auriga[12,35] contains aggregates of between 4 and 30 T Tauri stars pc^{-3} detectable at $A_v \lesssim 4$. If many T Tauri stars are variable and relatively quiescent for much of their T Tauri phase, as may be indicated by recent observations of emission line and photometric variability, these values of n_* combined with conventional T Tauri mass loss rates could conceivably underestimate the effective mass input due to eruptive outbursts by an order of magnitude. Possible support for this conjecture stems from recent indications that post–T Tauri stars may still be active, at least as x-ray sources, and from the relatively high time-averaged mass-loss rates inferred from the bipolar CO outflows. Eruptive high-velocity outflows in intermediate-mass stars, such as appears to be associated with IRc2, can provide an even more efficient momentum input. This is because the protostellar wind outflow is likely to be at $V_w \simeq 1000$ km s^{-1} and the interaction of the wind with the ambient gas is adiabatic until the swept-up matter shell has entered the momentum-conserving phase when radiative cooling losses become rapid at a velocity $V_{rad} \simeq 200$ km s^{-1}. The momentum acquired by the molecular cloud gas is augmented by a factor of $\sim V_w/V_{rad}$ for a given amount of ejected mass relative to that for an approximately momentum-conserving wind.

In order for the observed molecular cloud linewidths ($v \simeq 1$–3 km s^{-1}) to be maintained over a dynamical timescale, one finds that the rate of mass input from momentum-conserving winds must be $\sim 3 \times 10^{-7} \, (n/10^3 \text{ cm}^{-3})^{3/2} \, M_\odot$ pc^{-3} y^{-1}. Combining this result with the previously derived lower bound on n_*, we find that shell intersections occur, provided that the effective duration of the wind phase satisfies

$$t_w \equiv \delta M/\dot{M} \gtrsim 4(V_*/v)t_{ff}.$$

Since the cloud free-fall time, $t_{ff} = 2 \times 10^6 \, (n/10^3 \text{ cm}^{-3})^{-1/2}$ y, we infer that the mass-loss timescale t_w must at least be comparable to the pre-main-sequence contraction time of a 1 $M_\odot$ star ($\sim 10^7$ y) in order for a single generation of protostars to provide support. If successive generations of protostars form, lifetimes as long as $\sim 10^8$ y may be attainable before the cloud reservoir of gas is likely to become significantly depleted.

These arguments suggest that wind-driven shells are likely to intersect. Shell intersection will result in the formation of supersonically moving clumps, since, in general, intersection occurs before the shells are in pressure balance. The characteristic clump masses are on the order of 0.1 $\bar{\rho}4\pi R^3_{max}/3 \simeq 0.1 \, \bar{\rho}/n_* \simeq 1 \, (\bar{n}/10^4 \text{ cm}^{-3}) \, (10$ pc$^{-3}/n_*) \, M_\odot$, with sizes of ~ 0.1 pc. They will only be weakly confined by ram pressure because of their low Mach numbers and so will continuously replenish the interclump medium.

While individual clumps can be thermally supported, the supersonic clump motions provide an explanation for suprathermal linewidths in molecular clouds. There are already some indications that individual thermally supported fragments with narrow lines have been detected. Until clumps become gravitationally unstable, their cross section for interaction remains large. Accretion of ambient gas and collisions between clumps are likely to occur. The collisions are highly inelastic and should lead to the net growth of clumps by coalescence.

However, this process may require a considerable number of collisions, on the average, per clump. This is due to the oblique nature of the typical collision, an effect that has been recently studied by Scalo and Pumphrey.[36] When clumps of different sizes collide, only the overlapping portions are found to dissipate kinetic energy; the nonoverlapping segments are sheared off. In this way, the timescale for clump kinetic energy to be dissipated by collisions can be increased by an order of magnitude over the mean geometrical collision timescale. Thus, winds from T Tauri stars may suffice to keep clouds stirred up for timescales well in excess of the duration of the wind phase. One must therefore also consider post–T Tauri stars as contributors to the observed linewidths and these stars may greatly exceed the T Tauri stars in number.

Clump Coagulation and Star Formation

The timescales for the clumps to lose kinetic energy and to grow in mass by coalescence must be similar. Thus, within 10^6–10^7 y (depending on the efficiency of the clump collision process), it seems inevitable that the clumps will become Jeans unstable. The Jeans mass is $\sim 3(T/15 \text{ K})^{3/2} (10^5 \text{ cm}^{-3}/n)^{1/2} M_\odot$. Because the clumps are initially ram pressure confined and of low mass, it seems plausible that their coagulation will lead to further low-mass star formation. In fact, one can greatly strengthen this argument by noting that the Jeans mass M_J is reduced as the confining ram pressure due to accumulated winds increases. Since the ram pressure, p, is produced by momentum input from the shells injected at a characteristic scale of $\sim n_*^{-1/3}$, we infer that $p \propto n_*^{2/3}$ and, therefore, $M_J \propto p^{-1/2} \propto n_*^{-1/3}$. This means that, once low-mass star formation results in wind-driven shell intersections, the clumps will coagulate and continue to form low-mass stars. A very considerable enhancement of the coagulation process would be needed to drive the clump mass into the massive protostar regime when it is first gravitationally unstable. The process becomes self-perpetuating: low-mass stars form, develop winds, sweep up dense shells that intersect, and form clumps that, in turn, coalesce to form more low-mass stars.

Coagulation characteristically results in a clump mass spectrum with a power-law tail $N(m, t) \propto m^{-\nu}$, where ν depends on the adopted form of the collision rate and lies in the range $1 \lesssim \nu \lesssim 3$.[37] If $\nu > 2$, the net mass in clumps $\int N(m, t)m \, dm$ is predominantly due to low-mass clumps. Coalescence then dominates over the build-up to the Jeans mass and consequent feedback on the mass of newly forming stars. Consequently, star formation occurs at a relatively steady rate and the mass in clumps remains nearly constant, with an approximate balance between the collapse of clumps to form stars and the injection of new clumps.

On the other hand, if $\nu < 2$, most of the clump mass is in massive clumps of mass near M_J. Thus, as M_J decreases due to the increase in ram pressure as more stars form, the star formation rate rapidly increases. In fact, a simple solution of the coagulation equation in this case indicates that a burst of star formation occurs. The long lifetimes required for cold molecular clouds evidently require $\nu > 2$. Measurement of the clump mass spectrum could perhaps provide a test of the role of clump coagulation in star formation.

Onset of Massive Star Formation

The shock generated by a collision between two clouds of comparable mass should substantially enhance the clump coalescence rate. Such collisions may be expected to occur preferentially in spiral arms; the density wave enhances the cloud-cloud collision rate by up to an order of magnitude.[38] Since the maximum mass of newly formed protostars is approximately given by the product of the mean number of collisions and the mean initial fragment mass, one can also expect to increase the mean protostellar mass by an order of magnitude. This is sufficient to form massive protostars.

Once massive stars begin to form, this process induces further massive star formation. Luminous protostars are capable of providing an important energy input into the surrounding molecular gas through dynamical dissipation of wind kinetic energy and grain heating.[39] The velocity dispersion characteristic of warm molecular clouds (2–3 km s^{-1}) considerably exceeds that in dark clouds and reflects the enhanced stirring that is presumably being produced by massive protostars, as in Orion.

Massive star formation will be self-perpetuating, until the cloud is disrupted. The enhanced coagulation produces fragments of mass far above the Jeans limit and such fragments must collapse, probably undergoing subfragmentation. Now the Jeans mass within a supersonically moving fragment of kinetic temperature T and optical depth τ is proportional to $T^2 \tau^{-1}$. Moreover, the rate at which gravitational energy is being radiated during Kelvin-Helmholtz contraction is proportional to M^3. Once a protostar of appreciable mass forms by subfragmentation, its energy input into the surrounding medium raises the local Jeans mass. Massive stars can therefore induce the formation of stars of even greater mass.

This is obviously an unstable situation: The molecular cloud will rapidly be eroded by winds and H II regions as a burst of massive star formation passes through. Although a single massive star is incapable of disrupting a large molecular cloud, the coherent interaction of a number of stars will ensure its dispersal. It will, however, take some 3×10^7 y before a sufficient number of O and B stars have evolved, if nuclear energy is to be tapped to provide sufficient energy in winds, H II regions, and even supernovae.

COMPARISON WITH OBSERVATIONS

The following model for the Orion Molecular Cloud and other molecular clouds can now be proposed. A massive cloud consists of a magnetically supported relatively diffuse envelope and a cold dense core that is highly Jeans unstable. Systematic core collapse is averted, however, by momentum input in the form of wind-driven shells from low-mass protostars if the mean mass input from winds exceeds $\sim 10^{-6}$ $M_\odot$ pc^{-3} y^{-1}. The critical number density of such stars is about 10 pc^{-3} for the shells to intersect, break up into clumps, and drive supersonic clump motions. The characteristic clump mass is $\sim 1\ M_\odot$, and the characteristic clump separation is ~ 0.1 pc for time-averaged winds of $\sim 10^{-7}\ M_\odot$ y^{-1}. Dissipation of clump kinetic energy in collisions leads to coagulation and accretion over 10^6–10^7 y and further low-mass star formation. Because of the ram pressure confinement of the clumps, the Jeans mass tends to be reduced as more protostars form; thus, low-mass star formation is a self-regulating process. An externally induced shock, perhaps associated with a collision by another cloud or a nearby supernova explosion, is capable of enhancing the clump collision and coalescence rates and therby induces massive star formation. One probably cannot appeal to the Elmegreen-Lada mechanism to initiate star formation in Orion: The required critical surface density swept up by the ionization front–driven shock implies far more extinction than is seen in the direction of the shocked H_2 in the Orion Molecular Cloud. Once massive star formation is initiated, the stirring of the surrounding cold molecular gas by energetic protostellar winds suffices to maintain a high clump coalescence rate, provide a strong heat source for the gas fragments, and ensure that massive stars will continue to form.

As the coherent formation of massive stars acts to drive away the residual gas, the observer would then see a more or less sequential arrangement of subgroups, with the youngest adjacent to the remaining core of the molecular cloud. The oldest subgroups formed in the cloud envelope and their mean density and extent at birth would reflect these conditions. The Trapezium subgroup is the youngest in Orion; it must have formed in the densest part of the molecular cloud core, with $n \simeq 10^6$ cm^{-3} required for it to once have been bound. The spatial arrangement of the subgroups reflects inhomogeneities in the structure of the original molecular cloud and, hence, need not show any great degree of ordering.

Interferometric observations in SO of the inner core of the Orion molecular cloud reveal that the region of suprathermal linewidth emission extends over ~0.1 pc.[40] Now, the high-velocity flow or plateau source seen in CO involves ~30 $M_\odot$ in an outflow at ~50 km s^{-1}. This is capable of randomizing and providing enough momentum to account for the linewidths found in the core source, which contains ~200 $M_\odot$ at a dispersion of ~10 km s^{-1}.

To establish that molecular linewidths throughout the Orion Molecular Cloud and other clouds can be interpreted in terms of momentum input from the randomization of high-velocity wind flows, one must know the locations of the centers of these wind flows. Some indications that these exist with sufficient frequency have recently been obtained.[41] The available data on bipolar wind-driven flows suggest two classes of central object: one with high ($>10^3$ $L_\odot$) and one with low ($\lesssim 100$ $L_\odot$) bolometric luminosity. If we identify these as phases associated with the pre-main-sequence evolution of massive stars (>5 $M_\odot$) and of all lower-mass stars, respectively, we can then estimate the appropriate birth rate from the mean star formation rates given by Miller and Scalo.[11] This leads to estimates of the mean momentum input to molecular clouds that lie within an order of magnitude of the mean momentum input to the interstellar medium from supernovae (TABLE 1). The lower-mass protostars may contribute most of the energy input.

Evidently, protostellar winds may play a role comparable to that of supernovae in the dynamics of interstellar gas: They are the dominant energy sources in molecular clouds, whereas supernova energy input is certainly important for the diffuse interstellar medium. In fact, the possible role of low-mass protostars in energizing interstellar clouds could be even more extensive: The large number of these sources means that, even with low mass loss rates, their collective effect could be significant in the diffuse gas.

MASS LOSS BY PROTOSTARS AND ENERGY INPUT INTO THE INTERSTELLAR MEDIUM

	L ($L_\odot$)	M^* ($M_\odot$)	V^* (km s^{-1})	$MV\left(\dfrac{V_0}{10 \text{ km s}^{-1}}\right)$ (erg)	
Massive protostars					
IRc2	10^4–10^5	10	50	1×10^{47}	
Cep A	2.5×10^4	10	25	5×10^{46}	
AFGL 490	1.4×10^3	30	30	2×10^{47}	
$\langle MV(V_0/10 \text{ km s}^{-1})\rangle$					1×10^{47} erg
Formation rate (>5 $M_\odot$)†					5×10^{-13} pc^{-3} y^{-1}
Mean energy (momentum) input‡ ($\times 10$ km s$^{-1}/V_0$)					6×10^{-29} erg cm^{-3} s^{-1}
Low-mass protostars					
HH 7, 11 IR	70	4	20	1.6×10^{45}	
L1551 IRS-5	25	0.3	15	9×10^{44}	
HH 26 IR	~100	3.1	15	9×10^{45}	
$\langle MV(V_0/10 \text{ km s}^{-1})\rangle$					9×10^{45} erg
Formation rate (>0.5 $M_\odot$)†					1×10^{-11} pc^{-3} y^{-1}
Mean energy input‡ ($\times 10$ km s$^{-1}/V_0$)					1×10^{-28} erg cm^{-3} s^{-1}
Supernovae					
$E(V_0/10$ km s$^{-1})$ (300 km s$^{-1}/V_{rad}$)§					1.7×10^{49} erg
Formation rate (1 per 50 y)†					1×10^{-13} pc^{-3} y^{-1}
Mean energy input‡ ($\times 10$ km s$^{-1}/V_0$)					2×10^{-27} erg cm^{-3} s^{-1}

*Mass and velocity estimates are for high-velocity CO gas flows.
†From Miller and Scalo,[11] with an assumed scale height of 100 pc.
‡The characteristic random motion within a molecular cloud is taken to be V_0 (normalized to 10 km s^{-1}) and momentum conservation is assumed above V_0.
§Supernova remnants with initial kinetic energy $E = 5 \times 10^{50}$ erg are assumed to become radiative, approximately conserving momentum at $V < V_{rad} = 300$ km s^{-1}.

SUMMARY

The problem of the gravitational fragmentation of a collapsing cloud was discussed, with emphasis on the roles of magnetic fields, initial density fluctuations, and deviations from sphericity. Nonlinear interactions between protostellar fragments and the ambient gas are likely to play an important role in cloud evolution. It was argued that the energy input to molecular clouds from protostellar winds provides their dominant energy source and is capable of supporting the clouds against gravitational collapse. Estimates of the observed bipolar flow and protostar formation rates suggest that protostellar winds provide an energy input that is within an order of magnitude of the mean input into the interstellar medium by supernovae. A massive molecular cloud such as that in Orion consists of a magnetically supported relatively diffuse envelope and a cold dense core. Systematic core collapse is inhibited by momentum input in the form of wind-driven shells from low-mass protostars, provided that the mean mass input from protostellar winds exceeds $\sim 10^{-6}\ M_\odot\ \mathrm{pc}^{-3}\ \mathrm{y}^{-1}$. Eruptive mass loss from post–T Tauri stars and high-velocity outflows from intermediate-mass stars may greatly augment the efficiency of this process. Shell intersections lead to fragmentation and drive supersonic clump motions. As more protostars form, the Jeans mass of the ram-pressure confined clumps is reduced. Thus, as coagulation and accretion processes gradually drive the newly formed clumps gravitationally unstable, only low-mass stars continue to form in a self-regulating manner. An externally induced shock enhances the clump collision and coalescence rates and initiates massive star formation. Massive protostars similarly continue to energize the surrounding gas and induce further massive star formation. As the coherent action of massive star formation drives away residual gas, the observer would expect to see a more or less sequential arrangement of subgroups, with the youngest and most compact identified with an origin in the molecular cloud core.

REFERENCES

1. BLAAUW, A. 1978. Problems of Physics and Evolution of the Universe: 101. Armenian Academy of Sciences. Yerevan.
2. LANDOLT, A. 1979. Astrophys. J. **231**: 468.
3. STAUFFER, J. R. 1980. Astron. J. **85**: 341.
4. HERBIG, G. H. & B. F. JONES. 1981. Astron. J. **86**: 1232.
5. IBEN, I. & R. TALBOT. 1966. Astrophys. J. **144**: 968.
6. ADAMS, M., K. STROM & S. STROM. 1981. Private communication.
7. WOLF, S. C., S. EDWARDS & G. W. PRESTON. 1982. Astrophys. J. **252**: 322.
8. ABT, H. A. & S. LEVY. 1976. Astrophys. J. Suppl. Ser. **30**: 273.
9. TOLLE, F., H. UNGRECHTS, C. M. WALMSLEY, G. WINNEWISSER & E. CHURCHWELL. 1981. Astron. Astrophys. **95**: 143.
10. GORDON, M. A. & W. B. BURTON. 1976. Astrophys. J. **208**: 346.
11. MILLER, G. G. & J. M. SCALO. 1979. Astrophys. J. Suppl. Ser. **41**: 513.
12. COHEN, M. & L. V. KUHI. 1979. Astrophys. J. Suppl. Ser. **41**: 743.
13. FRERKING, M. A. & W. D. LANGER. 1982. Astrophys. J. **256**: 523.
14. DECAMPLI, W. M. 1981. Astrophys. J. **244**: 124.
15. MUNDT, R. 1981. Bull. Am. Astron. Soc. **13**: 856.
16. PHILLIPS, J. P. & J. E. BECKMAN. 1980. Mon. Not. R. Astron. Soc. **193**: 245.
17. STRAND, K. AA. 1958. Bull. Am. Astron. Soc. **4**: 5.
18. FALLON, F. W., H. GEROLA & S. SOFIA. 1977. Astrophys. J. **217**: 719.

9. MOUSCHOVIAS, T. 1977. Astrophys. J. **211**: 147.
10. BODENHEIMER, P. 1981. Astrophys. J. **224**: 688.
11. HOYLE, F. 1953. Astrophys. J. **118**: 513.
12. HUNTER, C. 1962. Astrophys. J. **135**: 594.
13. TOHLINE, J. 1980. Astrophys. J. **239**: 417.
14. SILK, J. 1977. Astrophys. J. **214**: 152.
15. SMITH, R. C. 1977. Mon. Not. R. Astron. Soc. **179**: 521.
16. SILK, J. 1980. *In* Star Formation, Tenth Saas Fee Advanced Course of the Swiss Society of Astronomy and Astrophysics. I. Appenzeller, J. Lequeux, and J. Silk, Eds.: 131. Geneva Observatory. Geneva.
17. SILK, J. 1977. Astrophys. J. **211**: 638.
18. REES, M. J. 1977. Mon. Not. R. Astron. Soc. **176**: 483.
19. SESSELL, M. S. & J. NORRIS. 1981. *In* Int. Astron. Union Symp. 68, Astrophysical Parameters for Globular Clusters. A. G. Davis Philip and D. S. Hayes, Eds.: 137. L. Davis Press. Schenectady.
30. MOUSCHOVIAS, T. & L. SPITZER. 1976. Astrophys. J. **210**: 326.
31. LIN, C. C., L. MESTEL & F. SHU. 1965. Astrophys. J. **142**: 1431.
32. SILK, J. 1982. Astrophys. J. **256**: 514.
33. LAYZER, D. 1963. Astrophys. J. **137**: 351.
34. NORMAN, C. & J. SILK. 1980. Astrophys. J. **238**: 158.
35. JONES, B. F. & G. H. HERBIG. 1979. Astron. J. **84**: 1872.
36. SCALO, J. M. & W. A. PUMPHREY. 1982. Preprint.
37. SILK, J. & T. TAKAHASHI. 1979. Astrophys. J. **229**: 242.
38. NORMAN, C. & J. SILK. 1980. *In* Int. Astron. Union Symp. 87, Interstellar Molecules. B. H. Andrew, Ed.: 137. D. Reidel. Dordrecht.
39. SILK, J. 1977. Astrophys. J. **214**: 718.
40. WELCH, W. J. 1982. This volume.
41. BALLY, J. 1982. This volume.

DISCUSSION OF THE PAPER

B. ZUCKERMAN (*University of Maryland, College Park, Md.*): I have the following difficulty in understanding your model. If the timescale for collapse is 10^6–10^7 y and the timescale for the formation of low-mass stars is 10^7–10^8 y, why doesn't the cloud collapse before any stars are formed?

SILK: The stars involved are intermediate-mass, as well as low-mass, stars, and winds begin at the onset of the pre-main-sequence contraction phase.

T. MOUSCHOVIAS (*University of Illinois, Urbana, Ill.*): I have a question on timescales in the coagulation picture. Since the mean density of a molecular cloud is on the order of 10^4 cm^{-3}, a fragment, by definition, will have a density larger than that, say, 10^6 cm^{-3}. Since you ignored magnetic fields, that density implies collapse over a timescale of 1–3×10^4 y. How can fragments collide and coalesce over a timescale shorter than that?

SILK: It is collapse that results in star formation, but the fragments must be gravitationally bound to undergo collapse. So coalescence is initially inevitable. Then the stars generate winds that form the shells that produce more fragments.

ORION AND THEORIES OF STAR FORMATION

Richard B. Larson

Astronomy Department
Yale University
New Haven, Connecticut 06511

INTRODUCTION

Star formation and its various effects are evidently responsible for many of the complex phenomena observed in the Orion Nebula region, and one would therefore like to be able to explain the observations of this region on the basis of theories of star formation. At present, however, any theoretical understanding of star formation is still too rudimentary to bear any detailed comparison with the real world, so one can only discuss some possible processes that may play a role in star formation and then appeal to the observations to try to clarify their importance. In this paper, I shall mention some of the small-scale processes that may be important in the formation of stars in molecular clouds and then describe some data on young stars in Orion and other regions that provide information about star formation in these regions.

As a caution against overly simplified theories of star formation, some large-scale properties of the Orion region are worth keeping in mind. One is that the entire Orion complex of molecular clouds and young stars, whose total mass probably exceeds $2 \times 10^5 \, M_\odot$, is located more than 100 pc below the Galactic plane; thus, whatever process formed the Orion clouds must have assembled them well out of the Galactic plane or, equivalently, must have accelerated the material to a substantial velocity away from the Galactic plane. Moreover, the CO maps that Thaddeus presents in this volume reveal that the Orion clouds are very extensive and complex in structure, with several major concentrations and filamentary extensions, and that they also have a complex velocity field. These observations suggest that the Orion clouds were formed in a more or less violent way by processes that left them in a turbulent state; if so, turbulent motions may play a major role in the subsequent evolution of the clouds and the processes that occur may be much more complex than either a simple contraction process or a regular propagation of star formation through each cloud. In fact, the available data show star formation proceeding simultaneously at several sites in several different molecular concentrations in Orion and do not, when examined closely, clearly follow the widely discussed picture of sequential star formation, but suggest a more complex situation in which star formation can be initiated at many sites.

Ultimately, it is necessary to understand how, on small scales, the molecular cloud material actually becomes condensed into stars and groups of stars. Related problems of long-standing interest are understanding the mass spectrum with which stars form, the time dependence of star formation, and the efficiency of the conversion of gas into stars. While complete theoretical answers cannot yet be given to these questions, it is at least possible to discuss some of the processes that may be involved and some observational constraints on their importance.

0077–8923/82/0395–0274 $1.75/1 © 1982, NYAS

POSSIBLE PROCESSES OF STAR FORMATION

Gravity must clearly play a major role in compressing interstellar matter into stars, and the classical picture of star formation is based on the concept of successive gravitational fragmentation elaborated by Jeans, Hoyle, and others. If a cloud begins to contract under gravity, regions of enhanced density containing more than the Jeans mass can collapse separately to high densities and may fragment again into smaller and smaller units as the density increases and the Jeans mass decreases. Most numerical calculations of collapse and star formation have been motivated by this picture, but the multidimensional calculations have not, so far, supported the concept of successive fragmentation into objects much smaller than the initial Jeans mass. Accordingly, if gravitational fragmentation is the only process that operates, one might expect characteristic stellar masses to be comparable to the Jeans mass in the cloud from which the stars form, being smallest in the regions of highest density.

On the other hand, purely hydrodynamic processes such as shock compression can also play a role in compressing the gas in molecular clouds; shocks should occur often because supersonic internal motions are observed in all but the smallest clouds. The observed motions generally appear to be complex or turbulent, so the shock-compressed regions will probably have an irregular filamentary or clumpy structure; such supersonic turbulent motions may even be responsible for the formation of protostars.[1] If this is the primary mechanism of star formation, one might expect the stars of lowest mass to form in the regions with the highest turbulent velocities. Another purely hydrodynamic process that may become important after dense protostellar clumps have formed, either by gravity or by turbulence, is collisional coalescence and growth of protostellar objects.[2]

In reality, gravity and hydrodynamics probably act together in the formation of stars. For example, once a small dense core or embryonic star has begun to form in a protostellar condensation, its mass can grow by a large factor by gravitationally accreting the more diffuse surrounding gas. The accretion process may involve infall through an accretion shock at the surface of the stellar core if the infalling material has little angular momentum, or viscous inflow in an accretion disk if the material has a large angular momentum. More complex processes involving accretion of lumps or streams of gas may also occur. Numerical simulations[3] and analytical theories[4] of the accretive growth of embryonic stars suggest that this process can account approximately for the form of the observed stellar mass spectrum. If such accumulation processes are important in the formation of stars, it might be expected that the most massive stars would form in the densest regions, *i.e.*, in the dense cores of contracting molecular clouds or forming star clusters where the gas density and accretion rate are highest. Gravitational accretion is a runaway process, since the accretion rate increases with the mass of the accreting object; therefore, stars of extremely large mass may form, unless other effects, such as stellar winds, intervene to cut off accretion before all the available gas has been consumed.

Once stars have begun to form in a molecular cloud, they may exert an important influence on subsequent star formation. For example, the stars will begin to contribute to the gravitational field in the cloud and may exert significant tidal forces on the gas. If the cloud contains a dense embedded cluster of stars, such as appears to exist in Orion, the star cluster may produce an approximately symmetrical, centrally condensed gravitational potential well that acts to concentrate the remaining gas to very high densities at the center, possibly favoring the accretive build up of very massive stars there. Since newly formed stars produce stellar winds with mass-loss rates that increase strongly with stellar mass, the accretive build up of a massive star may be cut off when its wind becomes strong enough to blow away the surrounding gas. This would imply an upper mass limit that increases with increasing ambient gas density, as was suggested previously on the basis of radiation pressure and ionization effects.[5]

Given all the possible processes that have been mentioned, and others, such as magnetohydrodynamic effects, that probably also occur, detailed predictions of how stars form cannot yet be made; to understand more about the way in which stars actually form, it is necessary to refer to observations of regions of star formation. Data on the young stars in these regions are particularly useful because they provide a record of the past history of star formation and can be used to study correlations between stellar properties, such as the mass spectrum, and other characteristics of each region, such as the spatial distribution of gas and young stars.

YOUNG STARS IN ORION AND OTHER REGIONS

Cohen and Kuhi have published an extensive study of T Tauri stars in many regions of star formation,[6] providing quantitative data on temperatures and bolometric luminosities from which the ages and masses of the stars can be estimated. These data have been used to study the mass spectra, spatial distributions, and ages of the T Tauri stars in the three most populous regions, namely, Taurus, Orion, and NGC 2264, and to look both for possible differences in the stellar properties and for correlations with other properties of the associated star-forming molecular complexes.[7]

The spatial distributions of the Cohen-Kuhi stars in Taurus, Orion, and NGC 2264 are plotted on similar linear scales in FIGURES 1-3, coded by stellar mass. In Taurus (FIGURE 1), the youngest stars are dispersed over a relatively large region about 40 pc across and they are mostly in small groups containing from a few up to about ten stars. These small groups are closely associated with a number of Barnard and Lynds dark clouds whose positions are approximately indicated in FIGURE 1. Most of the T Tauri stars have masses less than 1 $M_\odot$; the most massive star in the region is a highly obscured B2 star, located, as indicated, in the B18 cloud near the center of the region.

The distribution of the T Tauri stars in Orion is strikingly different (FIGURE 2), being more compact and centrally concentrated than that in Taurus; most of the Orion stars are in a single large cluster about 10 pc across, centered on the Trapezium. A secondary concentration is associated with the reflection nebula NGC 1999, and the remaining T Tauri stars are scattered along the length of the L1640 dark cloud. Most of the known T Tauri stars in Orion have masses greater than 1 $M_\odot$, in contrast with those in Taurus whose masses are mostly less than 1 $M_\odot$. Another difference is that the Orion stars are somewhat older, having a median age of 1.0×10^6 y, as compared with 6×10^5 y for Taurus. Further evidence that the Orion region is older and more evolved than Taurus is provided by the presence there of large numbers of flare stars, which are generally fainter and older than the T Tauri stars; Orion has 325 known flare stars, whereas Taurus has only 13.[8]

NGC 2264 (FIGURE 3) differs from Taurus in the same sense and perhaps to an even more extreme degree than Orion; the spatial distribution of the T Tauri stars is more compact and their median age is even greater, about 2×10^6 y. As in Orion, most of the known stars have masses greater than 1 $M_\odot$.

The data therefore suggest that, as a star-forming region evolves, the gas becomes more spatially condensed and the stars form in more massive and concentrated clusters, with larger and larger masses. There is, in fact, direct evidence that, in some young associations or clusters, the more massive stars form later than the less massive

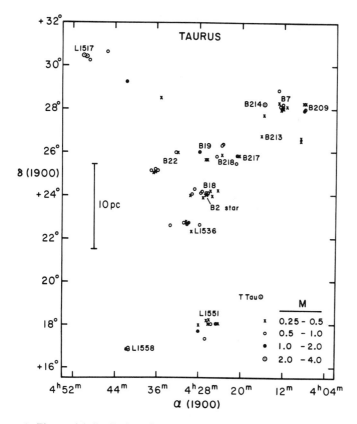

FIGURE 1. The spatial distribution of the Cohen-Kuhi stars in Taurus, coded by mass, as indicated in the lower right. The positions of the associated Barnard and Lynds dark clouds are also indicated, as is the position of the star of earliest spectral type, a heavily obscured B2 star.

stars.[9,10] To determine whether there is also evidence that the more massive stars in Orion have formed with a more compact spatial distribution than the less massive ones, we have shown, in FIGURE 4, only those T Tauri stars that have ages less than 10^6 y and are still seen close to their places of formation; although the statistical significance of the result is not large, the more massive T Tauri stars in FIGURE 4 are, in fact, considerably more spatially concentrated than the less massive ones. It is particularly

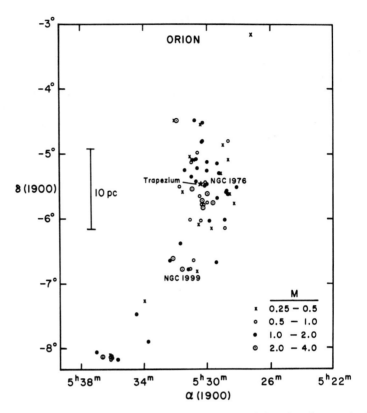

FIGURE 2. The spatial distribution of the Cohen-Kuhi stars in Orion, plotted on nearly the same linear scale as that used in FIGURE 2. The Orion Nebula (NGC 1976) and the associated young cluster are centered on the Trapezium, which contains the most massive star in the region.

FIGURE 3. The spatial distribution of the emission-line stars in NGC 2264. The position of the most massive star, the O7 star S Mon, is also indicated.

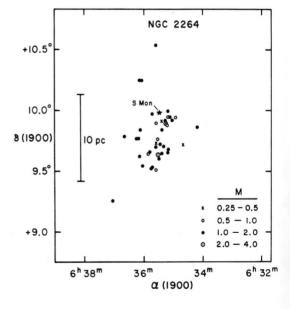

striking that the most massive star in the Orion region, θ^1C Ori in the Trapezium, is right at the center of the Orion cluster.

The formation of massive stars is continuing at present in the dense core of the Orion Molecular Cloud OMC1, whose compact cluster of luminous infrared sources (the BN-KL cluster) is also centrally located and has a projected distance from the Trapezium of only about 0.1 pc. When the stars in the infrared cluster blow away their surrounding gas and become visible, as will probably soon happen, they will appear as a

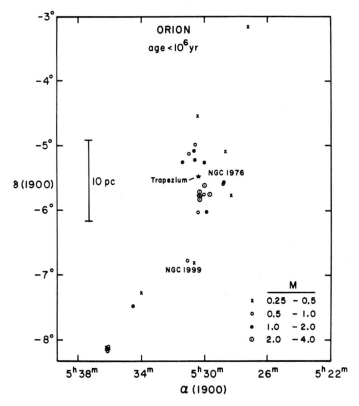

FIGURE 4. The spatial distribution of the emission-line stars in Orion that have ages less than 10^6 y.

new Trapezium-like subgroup of massive stars near the center of the Orion Nebula cluster. The infrared objects in OMC2 apparently represent a group of less massive stars forming further away from the center of the Orion cluster.

MASS SPECTRA OF YOUNG STARS

Because most of the T Tauri stars studied by Cohen and Kuhi lie on nearly vertical Hayashi tracks in the HR diagram,[6] their estimated masses depend primarily on their spectral types and their derived mass spectra are not grossly affected by magnitude-dependent selection effects. There are striking differences in the distribution of spectral types of the T Tauri stars in Taurus, Orion, and NGC 2264. FIGURE 5 shows the ratio of the numbers of stars with spectra earlier or later than K6.5 (corresponding to masses larger or smaller than about 1 $M_\odot$) and brighter than various limiting apparent magnitudes, plotted as if all stars were placed at the distance of Orion. It is evident that, regardless of the limiting magnitude chosen, a large difference persists between Taurus and the other two regions. Even after corrections are made for the different median ages of the stars in these regions,[7] the difference remains significant at about the 3σ level.

The mass spectra of the Cohen-Kuhi stars in Taurus and Orion are shown in FIGURE 6. Here the difference between Taurus and Orion appears as a turndown in the Orion mass spectrum at masses below 1 $M_\odot$. This deficiency of low-mass stars may not apply to the entire population of young stars in Orion, which includes the numerous faint flare stars that are widely dispersed throughout the region, but it is nevertheless characteristic of the youngest stars that have just recently formed in the Orion Nebula cluster. It is noteworthy that the Taurus mass spectrum is very similar to the initial mass spectrum of field stars,[11] while the Orion mass spectrum is similar to that of many open clusters, which often show a deficiency of low-mass stars.[12] Thus, Taurus may be a typical site for the formation of field stars, while in Orion we see the formation of a typical open cluster.

POSSIBLE INFERENCES FOR STAR FORMATION

The data discussed above are all consistent with an evolutionary picture in which the properties of a large star-forming complex change systematically with time while stars continue to form for a period of at least $\sim 10^7$ y. At an early stage, a star-forming region may resemble the Taurus dark clouds, where the average density is relatively

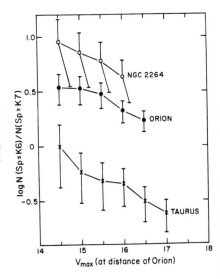

FIGURE 5. The ratio of the number of emission-line stars with spectral types of K6 and earlier to the number with spectral type K7 and later plotted against the limiting magnitude that the stars would have if all were placed at the distance of Orion.

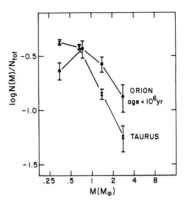

FIGURE 6. The mass spectra of the emission-line stars in Taurus and Orion. Plotted are the numbers of stars in the equal logarithmic mass intervals 0.25–0.5, 0.5–1.0, 1.0–2.0, and 2.0–4.0 $M_\odot$, normalized by the total number of stars in each region. The error bars give $N^{1/2}$ statistical errors.

low and small groups of low-mass stars are forming in scattered small molecular clumps and filaments. With time, the gas may become progressively more condensed into massive, dense clouds or cloud cores like OMC1, forming stars in larger and more centrally concentrated clusters, with a mass spectrum that increasingly favors massive stars. The most massive stars may form as a result of rapid accumulation processes in the dense central cores of these contracting clouds or forming star clusters. The culmination of this type of evolution may be represented by the present Orion Nebula region, where a dense cluster with very massive stars at its center has been forming for the past $\sim 10^7$ y and where the formation of massive stars is continuing near the center.

The fact that the most massive stars appear to form only in a small region of very high density at the center of a forming star cluster is not easily explained by the traditional fragmentation theory of star formation, but suggests that accretive build up processes are involved. Moreover, the massive stars appear to form after, and perhaps as a result of, the previous formation of large groups of less massive stars; a possible explanation is that the gravitational field of a cluster of mostly low-mass stars helps to concentrate the remaining gas to the high densities required to form the more massive stars.

The observations of gas flows associated with obscured young objects in Orion and elsewhere show, however, that no sooner does a massive star form than it begins to produce a strong stellar wind that blows away the gas in its vicinity, probably cutting off further accretion and star formation processes and exposing the newly formed stars to view. The formation of a massive star therefore requires that a large amount of gas be accumulated rapidly into a star before a wind can blow it away; this may explain why massive stars appear to form only in very dense regions, since only there can the required very high accretion rate be attained. Clouds or clumps of lower density can only produce lower accretion rates and, therefore, could only form less massive stars, in agreement with what is observed.

REFERENCES

1. LARSON, R. B. 1981. Mon. Not. R. Astron. Soc. **194:** 809.
2. SILK, J. & T. TAKAHASHI. 1979. Astrophys. J. **229:** 242.

3. LARSON, R. B. 1978. Mon. Not. R. Astron. Soc. **184:** 69.
4. ZINNECKER, H. 1981. Ph.D. Thesis. Max-Planck-Institut Für Extraterrestrische Physik.
5. LARSON, R. B. & S. STARRFIELD. 1971. Astron. Astrophys. **13:** 190.
6. COHEN, M. & L. V. KUHI. 1979. Astrophys. J. Suppl. Ser. **41:** 74.
7. LARSON, R. B. 1982. Mon. Not. R. Astron. Soc. **200:** 159.
8. GURZADYAN, G. A. 1980. Flare Stars. Pergamon Press. Oxford.
9. IBEN, I. & R. J. TALBOT. 1966. Astrophys. J. **144:** 968.
10. WILLIAMS, J. P. & A. W. CREMIN. 1969. Mon. Not. R. Astron. Soc. **144:** 359.
11. MILLER, G. E. & J. M. SCALO. 1979. Astrophys. J. Suppl. Ser. **41:** 513.
12. SCALO, J. M. 1978. *In* Protostars and Planets, Int. Astron. Union Colloq. 52. T. Gehrels, Ed.: 265. University of Arizona Press. Tucson.

DISCUSSION OF THE PAPER

J. PAZMINO (*New York, N.Y.*): Is the difference between the cohesiveness of the Orion and Taurus clouds due to a difference in tidal forces caused by Orion being out of the Galactic plane?

LARSON: It is possible that Taurus is being dispersed by tidal forces and Orion is not, but I don't see what it has to do with Orion's location out of the plane. The real problem is how Orion got there in the first place.

A. STARK (*Bell Telephone Laboratories, Holmdel, N.J.*): The problem of how Orion got so far out of the Galactic plane can be rephrased in terms of velocity: If you accept the age of 12 million years for the complex, how did Orion get such a high velocity ($\sim$15 km s^{-1}) perpendicular to the Galactic plane when it passed through (or originated in) the plane center? This is a remarkably high velocity for such a massive object.

T. MOUSCHOVIAS (*University of Illinois, Urbana, Ill.*): A comment on the spatial separation of low- and high-mass stars: We should be careful in making a direct association between the Jeans mass (as a function of position in a cloud) and the mass of the star that will form out of that fragment. The fact that M_J is larger in the envelope of a cloud does not imply that, once a fragment begins to collapse, a collapsing fragment will not itself fragment further. The three-dimensional numerical calculations do not yet have a good enough resolution grid to address this issue properly.

LARSON: In general, I agree with your remarks. However, the theoretical demonstration of successive fragmentation still has to be made.

A. E. GLASSGOLD (*New York University, New York, N.Y.*): Do you believe that propagating star formation occurs within a cloud?

LARSON: Although the subgroups of the Orion association are considered the prototypes for propagating star formation, I have certain reservations about this picture. First, the geometry is really much more complex than linear propagation along a single cloud. Second, the stars around the Orion Nebula itself have a large range in age. My general opinion is that propagation may be relevant for the formation of massive stars but does not apply to low-mass stars.

M. L. KUTNER (*Rensselaer Polytechnic Institute, Troy, N.Y.*): On the question of sequential star formation, I agree with your statement that the sequence from Ia to Id is a questionable one. Since the Orion association is the prototypical one for which sequential formation of subgroups is argued, the case for coherent sequential formation over a scale of tens of parsecs is weak, at best, for any association.

CRITICAL REMARKS
ON THE ORION STAR-FORMING REGION*

E. E. Salpeter

Department of Physics
Newman Laboratory
Cornell University
Ithaca, New York 14853

INTRODUCTION

Following this paper, George Field will summarize what answers we have been given at this symposium, but I am supposed to criticize and complain about the questions that should have been asked and those that were not answered. I will give a short theorist's "wish list" of observations that I would like to see made in the future, but most of my complaints have already been voiced (or dealt with or both) in previous papers. I therefore revert to two time-honored customs—I will plagiarize other people's comments and propagandize for a forthcoming paper of my own. I will concentrate particularly on the relative importance of low-mass and high-mass stars, so I start by discussing the initial mass function.

THE INITIAL MASS FUNCTION

The initial mass function, ξ, for main-sequence stars (or birthrate function) refers to the number of stars in different mass ranges that are born per unit of time in the Galactic disk. To derive ξ from the observed luminosity function (in mass units), ϕ, for main-sequence stars, we need to know the main-sequence lifetime t_{MS} (ξ is roughly proportional to ϕ/t_{MS} when t_{MS} is less than the age of the Galaxy). Unless there are some unexpected surprises, the initial mass function, ξ, is now reasonably well known.[1,2] On a log-log plot, the quantity $M \times \xi$, which determines what fraction of the mass of the interstellar medium is processed into different types of stars, increases slowly with decreasing mass, but this quantity has "two drooping ends," *i.e.*, the mass ranges below 0.2 $M_\odot$ and above 20 $M_\odot$ are less important.

On the high-mass side, there might be such an unexpected surprise: O stars are known to suffer mass loss in the form of stellar winds. The known O stars outside molecular clouds have mean mass-loss rates of $\lesssim 10^6$ $M_\odot$ y^{-1}, which is of interest, but does not affect the main-sequence lifetime, t_{MS} much. However, in principle at least, there could be a substantial class of O stars with a mass-loss rate so much larger that their lifetime, t'_{MS}, is much shorter than "normal," being controlled by mass loss rather than by hydrogen burning. Stars with such a short lifetime (compared with the "turnover" time of a cloud) could not escape from the molecular cloud in which they were formed and we would hear about such stars (if at all) not in reviews on the initial

*This research was supported, in part, by a grant from the National Science Foundation, no. AST 78-20708.

0077–8923/82/0395–0284 $1.75/1 © 1982, NYAS

mass function but only in symposia like this one. Put another way—for the same observed overall luminosity function (with $t'_{MS} \ll t_{MS}$), ξ would be much larger and such stars would have a strong effect on molecular clouds. I think (but am not sure) that not having heard any answers on this point at this symposium may already disprove the hypothesis that these "rapidly evaporating" massive stars exist: We've heard of surveys in essentially all parts of the infrared spectrum and these massive stars would presumably have a very large total luminosity. Nevertheless, I hope that, at the next Orion Symposium, we shall have more direct observational upper limits for such objects.

As mentioned, $M\xi$ (in logarithmic units) peaks at the low-mass end of the stellar mass spectrum ($\sim 0.25\ M_\odot$) and ξ has an even stronger peak there. The question has been raised whether this strong weighting towards low mass in star numbers, N, for the Galactic disk as a whole also applies to star formation in molecular clouds. What evidence has been advanced against low-mass stars forming in the same regions as massive stars is quite indirect[3] and, in my opinion, not very convincing. On the other hand, Herbig gave us some pretty direct evidence for the presence of many low-mass stars in the Orion region, so the best (or at least simplest) working hypothesis is

TABLE 1

SOME APPROXIMATE STATISTICS FOR STARS OF LOW, INTERMEDIATE, AND HIGH MASS

$M/M_\odot$	~ 0.25	~ 1	~ 5–10
Spectral type	M	G	B
$L_{MS}(\text{bol})/L_\odot$	10^{-2}	1	10^3
$\sqrt{GM/R_{MS}}$	$400\ \text{km s}^{-1}$	$500\ \text{km s}^{-1}$	$700\ \text{km s}^{-1}$
N fraction	0.5	0.05	0.002
(NL) fraction	0.001	0.01	0.5
(NGM^2/R) fraction	0.3	0.2	0.1
$t_{K.H.} \propto GM^2/R_{MS}L_{MS}$	10^9 y	10^7 y	10^5 y

probably to assume a "standard" initial mass function for the Orion star-forming region. I'll return to the possible importance of low-mass stars below but they are not important in regard to overall radiative luminosity (as shown schematically in TABLE 1), because luminosity is such a strong function of stellar mass.

THE INEFFICIENCY OF RADIATION PRESSURE

I want to advocate models for the high-velocity CO emission where mass outflow feeds the buildup of bulk momentum in a very direct manner, but the question of how much radiation pressure can do indirectly remains. The direct momentum, p, imparted to matter by luminosity, L (per unit time), is uncontroversial: $p = \tau L/c$, where τ is the optical depth and c is the speed of light. If $\tau \lesssim 1$, this is an exceedingly inefficient way to pump energy from a primary source into bulk momentum because c, occurring in the denominator, is so enormous compared with the speed matter can acquire. This inefficiency has already been pointed out in several of the papers in this volume, but one has to check whether or not the optical depth τ could be enormously large. The nominal optical depth of a molecular cloud for the uv emitted directly by O stars is indeed very large, but that is not the relevant quantity: The ratio of scattering to absorption cross-section of dustgrains is only moderately large, so that a photon is transformed to longer and longer wavelengths, where the matter is less and less opaque, after only a few scatterings at each stage. I therefore feel that the effective optical depth, τ, although larger than unity, is not enormously large and that radiation pressure is not important for accelerating matter in molecular clouds, except in very special circumstances.

However, an indirect "rocket effect" due to uv from massive stars can impart momentum much more efficiently. Long before molecular clouds were known, Oort and Spitzer had already pointed out that Lyman continuum photons ionizing atomic hydrogen on one side of a cloud can provide such a rocket effect: With V_{el} a typical speed of the ejected electrons, the absorbing matter could acquire momentum on the order of $p \simeq L/V_{el}$, which is much more efficient than direct radiation pressure because V_{el} occurs in the denominator instead of the much larger c. More recently,[4] Harwit and his colleagues have pointed out that a still more efficient rocket effect is possible in a molecular cloud where dissociation of molecular hydrogen can be used instead of ionization of atomic hydrogen. The efficiency is at least potentially higher for two reasons—uv photons somewhat below the Lyman edge can already dissociate H_2 and, even more important, momenta (per unit time) on the order of $p \simeq L/V_{atom}$ are possible where V_{atom}, the velocity of an ejected hydrogen atom, can be much smaller still than V_{el}. It is too early to tell[4] how well this mechanism can explain any particular observed phenomenon, but uv from O and B stars must certainly dissociate H_2 somewhere in the Orion Molecular Cloud and one must worry about the associated momentum transfer.

HIGH-VELOCITY CO EMISSION AND LOW-MASS STARS

Most (though not all) authors agree that the material showing the intermediate-velocity (20–50 km s^{-1}) and high-velocity (50–100 km s^{-1}) emission in CO and other molecules was accelerated by some kind of mass outflow from some central source. There are, nevertheless, a number of different controversies, some of them connected with the "clumpiness" of the molecular cloud. For simple "first attempt" theoretical models, one dilemma is deciding which phenomenon to explain first—the intermediate- or the high-velocity emission. As emphasized by Draine and Hollenbach, this choice can make a lot of difference, since magnetic fields are likely to lead to C-type shocks at the intermediate velocities, but have much less effect at the highest velocities. The emission spectrum is such that, if the emission intensity is directly proportional to the emitting mass, the intermediate-velocity material contains most of the mass and the high-velocity material contains most of the kinetic energy. I will discuss only models in which the high-velocity material has been accelerated by some kind of spherical blast wave in an ambient medium of relatively low density. In such models, the intermediate velocity material is "an afterthought," *i.e.*, the effect of the blast wave on high-density condensations inside the medium, which can contain most of the mass but do not absorb most of the energy because they fill only a small fraction of the volume.

Even for models that invoke spherical shocks in an originally uniform density medium there are various possibilities, *e.g.,* a single instantaneous explosion (as produced by a supernova) or continuous mass outflow (stellar wind). Furthermore, for a stellar wind, some of the results depend on whether the wind is very fast or only moderately fast (compared with the present shock speed). The three basic parameters of a given model are the original density, n, the present age, t, and the total energy, E_o (up to the present); the calculated quantities are then the present radius, R, and speed, V_{shock}, of the shock front. It is of interest to compare the values of E_o and t required for different models[5] to give the same values for R and V_{shock} (for an assumed value of n). Rough values for three models are given in TABLE 2.[5] The most important results are (1) that the values obtained do not depend extremely strongly on the model, (2) that the required energy input E_o is quite small compared with the energy output of a supernova and—most important—(3) that the present age t is remarkably small compared to the likely lifetime of a molecular cloud core. To put it crudely—the average energy input rate E_o/t is not very remarkable (even though E_o will be raised by the energy given to the intermediate-velocity material with $n \gg 5 \times 10^4$), but the "packaging" into many small chunks is remarkable. A secondary result of TABLE 2 is the fact that a stellar wind model (especially for a small wind speed) has slightly less

TABLE 2

THREE MODELS, ALL ASSUMING $n \simeq 5 \times 10^4$ CM^{-3}, $R \simeq 0.1$ PC, AND $V_{shock} = 100$ KM S^{-1}

	Total Energy E_o (erg)	Present Age t (y)
Explosion	5×10^{48}	700
Fast wind	1.4×10^{48}	800
200 km s^{-1} wind	0.8×10^{48}	1200

stringent requirements on "packaging" (larger t) and energy (smaller E_o) than an explosion model.

We have good evidence for the presence of massive, very luminous stars in the pre-main-sequence stage in the molecular cloud core (and on the main sequence elsewhere in Orion). One might then be inclined to invoke a supernova explosion or a stellar wind from an O star to produce the high-velocity material. This could certainly provide the mean rate E_o/t, but the "packaging" seems badly wrong, since O stars process very large chunks of energy but very rarely. We have therefore proposed[5] outflow from pre-main-sequence stars in the most common mass range (0.2 to 0.5 $M_\odot$, say) to power the high-velocity material in Orion and in other molecular clouds (where similar phenomena with $t < 10^4$ y are observed). The data Herbig presented here and the standard initial mass function make it likely that such stars form often enough to make the packaging plausible, but our proposal involves some slightly radical requirements on energetics: It is clear that pre-main-sequence stars can have episodes of mass outflow ("generalized T Tauri phases"), but the question is whether the kinetic energy of the outflow can come only from a part of the radiative luminosity, L, or can also come from the gravitational energy released by that material which permanently settles onto the star. If energy can be fed only *via* L, then stars of very low mass would not work for two related reasons: L is so extremely small that the time required to reach the main sequence (a more or less orthodox "Hayashi track") is longer than the age of a molecular cloud core and L is so small that it cannot supply the required E_o/t. Our proposal will work only if there is a direct hydrodynamic way for the matter permanently flowing onto the surface of the forming star to transfer some of its gravitational energy release to more material that "flows through near the surface" and is then ejected. As shown in TABLE 1, the total gravitational energy of all stars (unlike the total luminosity) is dominated by stars of low mass. If this hydrodynamical E_o/t is indeed much larger than L, then the time required to reach the main sequence is also shortened drastically. Statistical data on T Tauri–like stars[6] suggest deviations from orthodox Hayashi track timescales but it is not clear how much shorter the times can be.

The proposal that a low-mass star causes the high-velocity flow in the Orion Molecular Cloud Core is somewhat radical from an observational point of view: We have argued about which of the six or seven strong infrared sources can be responsible for what, but our hypothetical low-mass star would not be any one of the seven (its radiative luminosity would be too low), even though its blast wave has engulfed several of these much more luminous and massive objects. The geometry here is qualitatively similar to that described earlier in this volume by Spitzer and Cowie for "supershells," but scaled down in radius and up in density by very many orders of magnitude. We have heard that the high- and intermediate-velocity regions have roughly similar diameters, but more detailed measurements of size and geometry would be very helpful: This picture of high-density clumps being engulfed and pushed and bypassed predicts that the intermediate-velocity material should have a slightly smaller average extent than the high-velocity material.

Incidentally, better angular resolution in infrared observations will also be helpful in resolving a different theoretical controversy regarding star formation—the relative importance of fragmentation and coalescence: If we knew the density distribution of the six or seven (or more) lumps indicated by the infrared emission, we could tell directly whether they are likely to collide and coalesce in 10^5 y, say, or whether they are collapsing.

ACKNOWLEDGMENTS

I am indebted to Drs. S. Beckwith, M. Harwit, G. Knapp, and A. Natta for interesting discussions.

REFERENCES

1. MILLER, G. E. & J. M. SCALO. 1979. Astrophys. J. Suppl. Ser. **41:** 513.
2. LEQUEUX, J. 1979. Astron. Astrophys. **80:** 35.
3. TALBOT, R. J. 1980. Astrophys. J. **235:** 821.
4. HARWIT, M. & J. SCHMID-BURGK. 1982. CRSR Report No. 778. Cornell University. Ithaca, N.Y.
5. BECKWITH, S., A. NATTA & E. SALPETER. 1982. CRSR Report No. 779. Cornell University, Ithaca, N.Y.
6. COHEN, M. & L. V. KUHI. 1979. Astrophys. J. Suppl. Ser. **41:** 743.

———————◆———————

DISCUSSION OF THE PAPER

H. ZINNECKER (*Max-Planck-Institut für Physik und Astrophysik, Garching, F.R.G.*): A comment concerning the IMF. I have recently worked out a simple analytical accretion model for the stellar mass spectrum in the KL nebula. It is summarized elsewhere in this volume.

SALPETER: More observations are particularly important in order to obtain more data on the mass spectra of embedded sources in the cores and outer regions of active clouds and in the more quiescent dark clouds. I keep emphasizing that theorists can't do all that much in this game; it's just too complicated, so more observations are particularly important.

SUMMARY OF THE SYMPOSIUM

George Field

Harvard-Smithsonian Center for Astrophysics
Cambridge, Massachusetts 02138

IMAGES OF ORION

Dr. Henry Draper's photograph revealed an object of enduring fascination. The Orion Nebula, with its O and B stars, H II regions, dark clouds, and molecular emission, has become a touchstone for the study of the formation of stars, much as the Crab Nebula is a key to understanding the energetic processes that accompany the death of stars.

Over the years, optical photographs taken with large telescopes and emulsions of greater sensitivity have shown that the Orion Nebula contains a complex and beautiful detailed structure. But this symposium shows that only by using radio, infrared, and, perhaps, x-ray techniques can we penetrate the dark clouds behind the Orion Nebula where stars are forming at the present time. The images of Orion obtained by such methods are quite different from those obtained by Dr. Draper.

In seeking to understand the processes of star formation, accurate images are of paramount importance, as they are the basis for the quantitative theoretical models that tell us that the present appearance of Orion is but a single stage in a process occurring over millions of years. This symposium demonstrates that our images are still inadequate to the task. We need more data of high angular resolution, particularly at infrared, submillimeter, and millimeter wavelengths, that have sufficiently high spectral resolution to resolve individual molecular lines. This will permit us to deduce densities, temperatures, and velocities.

THE MOLECULAR CLOUDS

The Orion Nebula is at the near edge of the Orion Molecular Cloud, which is but one of the several thousand revealed by CO millimeter-line surveys in our Galaxy. As there is evidence that tracers of star formation, including hot CO regions, compact H II regions, OH and H_2O masers, and luminous infrared sources, occur rather often in other molecular clouds as well as in Orion, we can hope that star formation in Orion is representative of star formation throughout the Galaxy, and possibly, of star formation in spiral galaxies throughout the universe. Thus, our studies of Orion touch upon the general problem of the evolution of galaxies.

The Orion Molecular Cloud has been assayed by CO observations. It is many parsecs across, has a mass on the order of $10^5 M_\odot$, and appears to be rotating. If, like other molecular clouds, the Orion Molecular Cloud is shaped like a bar, it is tumbling end over end—an intriguing fact. Equally interesting is the fact that the Orion Molecular Cloud is some 150 pc from the Galactic plane—most molecular clouds are located within 80 pc of the plane.

We have realized for some time that estimates of the efficiency by which stars are

290

0077–8923/82/0395–0290 $1.75/1 © 1982, NYAS

formed from molecular clouds, together with the short (million-year) free-fall time-scales one calculates for them, would lead to a rate of star formation in the Galaxy far greater than that observed if, in fact, most molecular clouds are in a state of free-fall collapse. What, then, keeps them from collapsing? Evidence presented at this symposium supports the suggestion that they are kept from collapsing by supersonic turbulent motions stirred up by young stars.

The gas pressure in molecular clouds, which follows directly from observed densities and temperatures, is much too small to resist gravitation. Rotation may play a role in resisting collapse, but its effect is to cause clouds to flatten and then break up into fragments. Magnetic fields could also retard collapse, but, if they are sufficiently regular, the collapse will proceed along the field until the gravitational force associated with the higher density is sufficient to overcome the magnetic forces exerted perpendicular to the field. On the other hand, the lines of CO and other molecules in Orion are much broader than can be explained by thermal motions alone, showing that supersonic motions do exist in the cloud. The problem is to account for the persistence of such motions, in view of the high rate of dissipation one calculates for motions along the ambient magnetic field. The evidence in the literature, supplemented by that presented at this symposium, indicates that activity associated with star formation in the Orion Molecular Cloud may cause large-scale motions in the surrounding gas, which establish a turbulent pressure that may substantially delay its collapse. Such a phenomenon could explain the persistence of other molecular clouds in the Galaxy, as well.

Geometry and Kinematics of the Orion Nebula

According to a well-established model of the Orion Nebula, the O and B stars of the Trapezium cluster are ionizing and evaporating the neighboring parts of OMC1 (a dense clump of molecular gas within the Orion Molecular Cloud). Such a model was not seriously questioned at this symposium. As the Trapezium lies between OMC1 and the Earth, space observations of ultraviolet absorption lines formed in the H II region between the Trapezium stars and the Earth should show the same expansive motions observed earlier in the emission lines originating throughout the Orion Nebula. In a general way, this seems to be the case, but one also sees evidence that winds originating in the Trapezium stars themselves affect the absorption profiles. It is possible that θ^2 Ori is in front of the Orion Nebula; the spectrum of its H II region indicates that it is ionization bounded, and thus, separate from the Orion Nebula itself.

It was shown some time ago by 21-cm studies that there is a giant shell of atomic hydrogen approximately centered on the Orion Nebula; Barnard's Loop may be emission from the inner part of this shell, which has been ionized by ultraviolet radiation escaping from the Orion Nebula.

The evidence from 21-cm profiles is that the H I shell is expanding at some 10 km s^{-1}. At this symposium, we learned of another shell, this one composed of highly ionized gas detectable in the ultraviolet by Copernicus and IUE, which is expanding at some 100 km s^{-1}; this shell is outside the H I shell. If it is assumed that the high-velocity gas was accelerated by shock waves originating in supernova explosions in Orion, one must try to understand how the shock waves got through and beyond the H I shell. At this symposium we learned that this would be possible if the H I shell is made up of individual clouds, for then the supernova shock waves can propagate in the medium between the H I clouds, accelerating them outward as it does so. Apparently, the observed radial velocity of the H I shell can be explained if the clouds have been accelerated by about ten supernova shock waves. One wonders whether the gas pressure behind the supernova shock would compress the H I clouds; in other such cases, it has been suggested that such a process may initiate cloud collapse and subsequent star formation. It would be interesting if stars were being formed in the H I shell surrounding Orion; as far as I know, they are not.

GAS IN THE ORION NEBULA

There is a vast amount of information on the optical emission-line spectrum of the Orion Nebula. It has recently become possible to detect emission lines from Orion in the far-infrared region. Those originating in the H II region were predicted on the basis of optical data; their observed intensities appear to fit the theoretical predictions quite well. One far-infrared line, a fine-structure transition of C I, has about 100 times the optical depth that had been predicted on the basis of models of the interface between the H II region and the molecular cloud. These models predict that there should be a sharp transition from C II in the H II region to CO in the molecular cloud, with only a thin layer of C I in between. It remains to be seen whether the discrepancy between theory and observation is real, in view of the fact that the observed line is on the saturated part of the curve of growth and, thus, the measured intensity is very insensitive to the value of optical depth. If it is real, it will yield new information on cloud chemistry.

At this symposium we learned that there are bright optical emission-line objects embedded in the H II region, having sizes on the order of the seeing disk. If they are interpreted as globules of pre-existing neutral gas that have been overrun by the advancing ionization front due to the Trapezium stars, one can deduce a timescale of 3×10^4 y for this phenomenon. This timescale is quite consistent with that of the nebula itself.

Theorists have long predicted that H II regions like the Orion Nebula should drive shock waves into the adjacent molecular gas. It is, therefore, of interest that CO observations reveal motions of the gas at the interface with the molecular cloud corresponding to a radial expansion of about 2 km s^{-1}, in rough agreement with the theory. There is great uncertainty as to where the star-formation activity identified with the infrared source known as the Kleinmann-Low (KL) nebula lies along the line of sight. It was suggested that it may lie in the shocked layer at the edge of the molecular cloud; this would be qualitatively consistent with the theoretical suggestion that the high densities in shocked layers of gas are conducive to gravitational collapse and subsequent star formation, as it is believed that KL is ultimately energized by massive young stars. In fact, such a mechanism plays an important role in a model of sequential formation of stellar associations; the hot stars in the last association formed ionize the gas around them, driving a shock into the neighboring molecular gas, thus creating conditions favorable to the formation of another association. However, it was stated at this symposium that the theory requires a column density of about 1 g cm^{-2} behind the shock before the gas becomes gravitationally unstable. An upper limit to the actual postshock column density follows from the observed extinction to the Becklin-

Neugebauer (BN) infrared source, A_v about 30 magnitudes, corresponding (if the grains are similar to those in other parts of the Galaxy) to a column density of gas equal to only about 0.1 g cm^{-2}. Thus, this interesting suggestion appears to fail quantitatively.

STARS IN ORION

The O and B stars of the Trapezium are well known. I had not realized until this symposium how difficult it is, because of the inevitable interference by the extremely bright H II emission from the Orion Nebula, to observe the fainter stars that accompany them. Nevertheless, reddish fainter stars have been observed in the cluster; these stars must be relatively young, because lithium features are still strong in their spectra. The central density of stars down to $M_v = 6.6$ is 560 pc^{-3}, some 100 times greater than the central density found in most other open clusters. The Orion Nebula is far too young an object for these stars to have formed from the ionized gas of the nebula; they must have existed approximately in their present locations before the gas of the nebula was ionized. It was therefore suggested that the fainter stars in the Trapezium cluster are a sample of the stellar population that would be visible throughout the Orion Molecular Cloud if the obscuring gas and dust of that cloud were removed. From the estimated volume of the Orion Molecular Cloud, one would then infer that there are about 10^5 stars inside it that are optically invisible because of the high extinction. If, on the other hand, the region of high star density extends only through OMC1, there would be only about 10^3 stars involved. Presumably, one way to test these ideas would be sensitive observations at a wavelength long enough to penetrate the dust but short enough so that there is a sufficient amount of radiation in the Rayleigh-Jeans portion of the stellar photospheric emission to be observed. In the discussion following this paper, it was reported that observers using facilities on Mauna Kea have detected a number of candidates for dust-embedded stars at a wavelength of 2.2 μm.

Whether the fainter stars in the Trapezium cluster were formed before the Nebula itself relates to another question. That is, do the massive stars in associations form at the same time, and presumably by the same process, as the low-mass stars? Since, according to theory, the formation of the Orion Nebula should occur immediately following the formation of the massive stars of the Trapezium cluster, we infer that the massive stars in Orion formed later, and thus, probably, by a different process than the low-mass stars.

We have already mentioned several ways in which massive stars affect the surrounding gas: They are responsible for ionizing it, and, thus, for driving shock waves into the molecular gas, and they can accelerate it by means of stellar winds (about which more below). They may also heat and accelerate gas *via* supernova explosions. There is evidence for all of these processes in Orion.

It was pointed out at this symposium that supernova explosions are expected to inject substantial amounts of medium-weight chemical elements, including carbon and oxygen, into the surrounding gas. The recent detection of daughter products of ^{26}Al in meteorites shows that, when the solar system was in the process of forming, fresh ^{26}Al was injected into it by a neighboring supernova (because the 7×10^5 y half-life of ^{26}Al is too short to explain its presence otherwise). Calculations suggest that, in a situation like Orion, the neighboring gas can be modified substantially in its chemical abundances; in particular, the carbon:oxygen ratio can be changed by as much as a factor of two. This may be relevant to the fact that abundances in the Orion Nebula appear to differ significantly from solar values. On a broader front, it may be relevant to the existence of stars and interstellar clouds where the evidence is consistent with carbon being more abundant than oxygen, rather than the reverse, as in the Sun. Carbon stars are a well-known example (although they may be explicable solely in terms of nucleosynthesis in the particular star); the interstellar clouds in which long-chain hydrocarbons as heavy as $HC_{11}N$ are found may be another, as it is hard to imagine such molecules forming in the presence of free oxygen (although the oxygen in these clouds could be locked up in OH or H_2O). In both carbon stars and interstellar clouds, if carbon is more abundant than oxygen, all the oxygen can be locked up in CO, and thus unavailable for forming other molecules.

STAR-FORMATION ACTIVITY IN ORION

There is no object in Orion that has been positively identified as a protostar, so the evidence for recent star formation is circumstantial. Much of the attention centers on the KL nebula and the BN infrared object embedded in it; both are located in OMC1. The region is complicated, and additional observations, particularly at millimeter, submillimeter, and infrared wavelengths, will be required to unravel its mysteries.

The CID 10-μm image of KL presented at this symposium shows BN (also known as IRc1) as well as a number of fainter compact IR sources, together with apparently diffuse emission, which, in principle, could be composed of many discrete unresolved sources. The total luminosity involved is on the order of $10^5 L_\odot$. BN itself produces energy of this order of magnitude (see below), but some of the other infrared "sources" in the region may be scattering energy rather than producing it. At this symposium, we heard evidence from studies of linear polarization in the infrared that this is, indeed, the case. Contours of the electric vector in the polarized infrared emission close around both BN and IRc2, indicating that they are primary sources of radiation; other sources, including IRc3 and IRc4, show no such effect, so they may be secondary sources, such as local maxima in the dust density that result in large scattered power. However, for this interpretation to be tenable, there must be a nonuniform distribution (*e.g.,* a disk) of dust in the immediate vicinity of IRc2, which allows radiation to escape and to be scattered in preferred directions, but which hides IRc2 itself by virtue of its geometrical orientation (*e.g.,* by being edge-on if it is a disk).

The BN object is believed to be a hot star, of approximately early B spectral type, which is losing mass and which is embedded in dust. Its spectrum in the microwave region is that of a stationary, optically thick, ionized cloud with a radius of 15 AU. High-resolution Fourier transform spectrometer results in the 2–5 μm region show, however, that CO lines originate as little as 5 AU from the central source. H II Brackett-series lines have cores about 35 km s^{-1} wide and wings up to 300 km s^{-1} wide. The infrared spectrum suggests optically thick dust about 30 AU from the source. While there is no unique model that reconciles all this data, one model envisages a circumstellar disk seen at an oblique angle. A high-velocity ($\sim$100 km s^{-1}) wind of ionized gas streams out along the axis of the disk to form a bipolar H II region, while low-velocity ($\sim$10 km s^{-1}) ionized gas streams out in the disk, recombining a few AU from the star to form a molecular cloud. Dust forms at about 30 AU. The fact that the size inferred from the microwave spectrum is about the same at various frequencies, rather increasing with lower frequency, as is normally observed with stellar winds, is explained by the fact that recombination occurs suddenly, providing a sharp edge to the H II region.

Of course, the underlying B-type star believed to be responsible for the ionization and the wind is not visible optically because of the large extinction ($A_v \simeq 30$ mag) inferred from the infrared observations. However, the extinction at 2 KeV x-ray energies would be only two magnitudes, so that, if BN were an x-ray source like some other B-type stars, one might hope to detect it. Observations with the high-resolution imager on the Einstein x-ray observatory reveal, in fact, a weak source not far from BN; however, two nearby stars are equally good candidates, so it is not clear whether or not BN is an x-ray source.

One of the most intriguing phenomena in the KL infrared source is the high-velocity molecular gas, observed both in the millimeter region *via* molecules such as CO, SO, SO_2, SiO, and NH_3 and in the infrared region *via* the H_2 vibrational spectrum. The observations indicate that, over a region about 0.05 pc in radius centered somewhere between IRc2, the center of KL, and BN, there is a molecular flow with velocities ranging up to 100 km s^{-1}. If one interprets the intensities in various lines as due to a temperature gradient from about 2000 K inside to about 50 K outside the region, the flow can be one of expansion, with a velocity approximately proportional to the radius. The corresponding "age" of the Hubble-like flow is about 500 y. It contains about 5 $M_\odot$ of gas, and its kinetic energy divided by 500 y corresponds to an average power of about 300 $L_\odot$. There are about 25 mag of visual extinction from front to back. High angular resolution (6 arcsec or 0.013 pc) millimeter-wave observations in the SO line find evidence that the turbulent velocity over the whole region tends to mask any systematic expansion effect. If the latter effect is real, however, it appears to be centered on IRc2.

The whole region contains SiO, OH and H_2O masers; when the H_2O maser positions are tracked with VLBI, they are found to be moving away from a position near that of the high-velocity gas emission. There appear to be two populations of H_2O masers, one with a spread of radial velocities of ~35 km s^{-1}, the other of ~100 km s^{-1}. The proper motions observed within both populations are explained statistically if their distance is 480 ± 80 pc; this constitutes the first reliable geometric parallax of objects as distant as Orion.

The spectra of the OH masers appear to have Zeeman splitting corresponding to magnetic fields on the order of 3×10^{-3} Gs. Each maser source is ~10 AU across.

Various authors in this volume wondered how H_2 molecules could survive in an environment involving such high flow speeds. One theory is that, while shocks would surely form under these conditions, the large temperatures normally expected behind such shocks would be reduced if there were magnetic fields present and, therefore, the H_2 molecules would not be dissociated. Thus, a high-speed shock propagating in a slightly ionized, largely molecular gas would be mediated by momentum transfer from the field to the ions and thence to the molecules. The shock would be much wider and the temperature rise much less than in a field-free shock of the same velocity. Fields of the magnitude of 3×10^{-4} to 10^{-3} Gs will suffice; it is interesting that fields of 3×10^{-3} Gs are inferred from observations of the maser sources. Detailed models envision a strong wind from a central object, perhaps IRc2, which encounters the surrounding molecular gas and drives a shock wave into it, accelerating it to the observed velocities. An alternative model would use the radiation pressure from the central star to accelerate the gas, but it appears to be difficult to satisfy the efficiency required for conversion from radiative energy to kinetic energy.

At this symposium, we learned that the intense high-velocity molecular gas in Orion is not the only such object in the Galaxy. There are five other regions in Orion (NGC 2071, HH25–26, NGC 2024, OMC2, and HH1–2) that show similar characteristics, as well as another half-dozen in other regions. Typically, these objects involve 0.5–30 $M_\odot$ moving at velocities of 10–300 km s^{-1}. Altogether, a typical source in Orion has supplied a momentum of about 300 $M_\odot$ km s^{-1} to the surrounding gas. If the lifetime of such sources is 10^4 y, then the rate at which momentum is supplied is $\dot{P} = 3 \times 10^{-2} M_\odot$ km s^{-1}y^{-1}. If the mass of the Orion Molecular Cloud is $\sim 10^5 M_\odot$, then the observed turbulent velocity of about 2 km s^{-1} implies that the total turbulent momentum is $P \simeq 2 \times 10^5 M_\odot$ km s^{-1}. Hence, the characteristic time for the supply of new momentum is $P/6\dot{P} = 1.1 \times 10^6$ y. This is comparable to the free-fall time, leading to the suggestion that star formation within molecular clouds results in stellar winds that generate enough turbulent pressure to resist further gravitational collapse. Such a mechanism is appealing, because it embodies negative feedback: If a cloud starts to collapse, the rate of star formation goes up, turbulence increases, and the cloud expands again.

Thus, although a precise interpretation of the complex phenomena associated with the birth of stars in Orion is not yet possible, Orion has given birth to a new concept—that of self-generated turbulence in molecular clouds. If this turns out to be the explanation for the unexpectedly long lives of such clouds—the fact that they form so few stars—it will be a big step in our understanding of star formation.

ACKNOWLEDGMENTS

I am grateful for comments and criticism from Phil Solomon, Jim Moran, Jack Welch, Dave Schramm, Bruce Draine, Mike Werner, Dave Hollenbach, Ben Zuckerman, and Al Glassgold.

DISCUSSION OF THE PAPER

M. W. WERNER (*Ames Research Center, Moffet Field, Calif.*): Many of the problems we're having with the study of star formation have to do with our selection of objects. When the IRAS all-sky survey is launched in 1982 to survey the whole sky down to the flux levels with which we've studied Orion, we'll be relieved of the burden of these selection effects. In addition, we may be able to detect the population of stars postulated by Herbig to be within the Orion Molecular Cloud by virtue of their heating effect on the ambient dust within the cloud.

FIELD: Would someone be willing to describe the discovery of young objects in the core of OMC1 made by the UKIRT in Hawaii?

I. GATLEY (*United Kingdom Infrared Telescope, Hilo, Hawaii*): Searches at 2.2 μm of a 40 acrsec field centered on BN have found 26 sources brighter than $K = 12$ mag. These objects are young stars embedded in the core of OMC1 of the type predicted by Herbig.

T. MOUSCHOVIAS (*University of Illinois, Urbana, Ill.*): On the suggestion that stellar winds support molecular clouds, there is at least one observational difficulty. Some dark molecular clouds and Bok globules, which show no evidence of star formation, are too dense not to collapse, yet show significant linewidths. Therefore, a mechanism other than winds is needed to support them, *e.g.*, rotation and/or magnetic fields.

FIELD: A plausible case can be made for the support of Bok globules by magnetic fields or rotation.

MOUSCHOVIAS: Why can't these mechanisms be made to work for Orion without resorting to exotic mechanisms like winds?

FIELD: I agree. Has anyone produced a model of a molecular cloud of this size that is supported by rotation and magnetic fields?

A. STARK (*Bell Telephone Laboratories, Holmdel, N.J.*): Observations made at Bell Labs of some dark clouds in optically thin lines show that they are made up of single or, at most, a few clumps.

W. LANGER (*Plasma Physics Laboratory, Princeton, N.J.*): The dark cloud B335 is spherical, not rotating, and has a narrow line width (0.3 km s^{-1}); it also has an imbedded bipolar source. On the other hand, B5 is elliptical and appears to be supported by both rotation and magnetic fields. I suspect that all the mechanisms that have been mentioned contribute in different ways in different clouds.

P. THADDEUS (*Goddard Institute for Space Studies, New York, N.Y.*): The reason for believing that rotation does not affect the dynamics of large clouds is that there is no other apparent difference between clouds that rotate and clouds that do not.

C. LEUNG (*Rensselaer Polytechnic Institute, Troy, N.Y.*): Kutner and I have recently found a connection between dark globules and molecular clouds, in that they follow a similar velocity dispersion-size correlation. If the velocity dispersion is due to turbulence, this suggests that the dark globules may have formed from molecular clouds.

R. DICKMAN (*University of Massachusetts, Amherst, Mass.*): I have some comments to make regarding Dr. Field's suggestion that rotation may play a significant role in supporting globules against gravitational collapse. Last year, a rather extensive observational program was begun with the 45-ft FCRAO antenna to study a number of dark clouds in the CO $J = 1 \rightarrow 0$ line at high spatial and velocity resolution. While I would hesitate to regard this work as sufficiently complete to be conclusive, my feeling at this point is that rotation probably does not play a widespread role in stabilizing most dark clouds. For a number of reasons, turbulent stabilization mechanisms would appear to deserve serious scrutiny. Recent work by Arny and Fleck, among others, has revived the old notion that one way around the highly dissipative nature of such flows is the possibility that turbulent energy may be fed more or less continuously into interstellar clouds. Arny has shown that the correlation between internal velocity dispersion and cloud size determined by Dr. Larson follows on essentially dimensional grounds, if energy enters through the surfaces of interstellar clouds and is convertible to bulk random motion. In addition, McNally and Settle's recent demonstration that nonradial perturbations of collapsing clouds lead to the growth of vorticity in the flow suggests that a sufficiently random set of such perturbations could produce substantial incoherent rotational motions superposed on a more regular underlying infall. I am not sure that such motions would be observationally or hydrodynamically distinct from what is commonly termed turbulence by radio astronomers. Both processes mentioned above could produce motions whose length scale is an appreciable fraction of the cloud size. In this context, Richard Arquilla of the University of Massachusetts and I have recently begun an analysis of the velocity field of the Bok globule L1523. We have used structure function techniques to search for correlated line-of-sight motions and to establish their scale. While this work is still in its early stages, we appear to be seeing more-or-less random motions whose characteristic length scale is on the order of the cloud size. I think that the issue of dark cloud support mechanisms is clearly still very much an open one and that stochastic motions generated by a variety of mechanisms may well play a significant role.

FIELD: Is the origin of the acoustic energy input from outside the cloud due to supernova shocks?

DICKMAN: Yes.

HENRY DRAPER
THE UNITY OF THE UNIVERSE

E.L. Schucking

*Department of Physics
New York University
New York, New York 10003*

My qualifications to talk about Henry Draper and his work are slight. I am not experienced in astrophotography, mirror grinding, telescope making, spectroscopy, and the arts of the laboratory in which this great man so excelled. Neither am I a historian of science who has pored over Draper's papers, celestial photographs, memoirs, observation-journals, spectrograms, and letters to add to our knowledge of him and his work. All I shall tell you will not be new to the experts—unless it is wrong.

I shall try to retell briefly the story of this remarkable scientist who graced the University of the City of New York and whom we honor at this symposium. The only personal touch I can supply is a century of hindsight from a theoretical physicist.

To see HD in perspective one has to see him first as a member of a glorious scientific family, the Drapers, a clan like the Bernoullis, Cassinis, Struves, Darwins, and Huxleys. His father, John William, was one of the great intellects of nineteenth-century America. He was a photochemist of renown whose work contributed to the discovery of spectral analysis, a pioneer in photography, a historian, and a philosopher. All his sons were scientists. The oldest, John Christopher, was a chemist; Henry, the astro-physiologist, was the second; Daniel, the third, was a meteorologist who built the first weather station in Central Park. Henry's wife Anna, née Palmer, was an observational astronomer, Henry's "National Science Foundation," and the mother of the HD catalogue. Henry's sister's child, Antonia Maury, found in the spectra of stars the first luminosity criteria. Her work on c-stars led E. Hertzsprung to the single most important plot in astrophysics, the H-R diagram, which is still the astronomer's principal clue for finding the measure of the universe in space and in time. Antonia Maury remembers about the Drapers

> Happiest were the days when Henry could be home. Then he sat facing his father; and when the two carried on the conversation, the rest sat silent. For it was well known that everything they said was too important for any word of it to get lost. To his father, in whose face a serene repose mingled with sustained intensity of intellectual penetration, Henry was the perfect foil, his dark eyes flashing electrically over the latest discoveries in physics or astronomy, or when relating some humorous incident, they brimmed with laughter to the point of tears.[1]

HD's work, inspired and initially guided by his great father, must also be seen against the background of nineteenth-century science. This century proclaimed two central ideas about our cosmos: evolution and the unity of the universe. These two themes—which were extended and united in our century into the evolution of space, time, and matter—were promulgated in November and October, 1859, and are commonly associated with the names Charles Darwin and Gustav Kirchhoff.

Darwin's ideas and their impact are well appreciated. Kirchhoff's revolution is not.

299

0077–8923/82/0395–0299 $1.75/1 © 1982, NYAS

FIGURE 1. Henry Draper

One reason is that Kirchhoff's grand discovery is not described by what it means but by the method with which it reaches its conclusion for the solar system: the application of spectrum analysis to the sun. The theme of the unity of the universe was shrouded in technical gobbledygook. Kirchhoff wrote

> I conclude that the dark lines of the solar spectrum which are not evoked by the atmosphere of the earth exist in consequence of the presence, in the incandescent atmosphere of the sun, of those substances which in the spectrum of a flame produce bright lines at the same place.
> ... From the occurrence of these lines (D), the presence of sodium in the atmosphere of the sun may therefore be concluded.[2]

But some contemporaries saw the consequences. Kirchhoff's onetime colleague at Heidelberg, Hermann von Helmholtz, said about Kirchhoff's work,[3] "It has excited the admiration and stimulated the fancy of men as hardly any other discovery has done, because it has permitted an insight into worlds that seemed forever veiled for us." Kirchhoff had demonstrated that the earth and we ourselves are made of sun-stuff, to use the kindergarten language of the world's most popular astronomer from Brooklyn. Chemical analysis had been extended from the chemist's smelly little cabinet to our lofty sun 93 million miles high in space. But could one extend it many million times higher into the sky into the realm of the stars; could one answer the question, What is the universe is made of; Is the universe everywhere the same; Is the sun physically a star among stars? A grandiose theme that Huggins called[4] "the scientific epic of the present century."

Two couples of scientists led the quest to answer the age-old question, What are the stars made of? Their names: William and Margaret Huggins and Henry and Anna

FIGURE 2. Anna Draper, née Palmer

Draper. Huggins had a headstart of some ten years. The Drapers' work led to the still most comprehensive answer: the HD catalogue.

Looking back in 1899, William Huggins described the origin of the spectral analysis of starlight as follows

> I soon became a little dissatisfied with the routine character of ordinary astronomical work, and in a vague way sought about in my mind for the possibility of research upon the heavens in a new direction, or by new methods. It was just at this time, when a vague longing after newer methods of observation for attacking many of the problems of the heavenly bodies filled my mind, that the news reached me of Kirchhoff's great discovery of the true nature and chemical constitution of the sun from his interpretation of the Fraunhofer lines.
>
> Here at last presented itself the very order of work for which in an idenfinite way I was looking—namely, to extend his novel methods of research upon the sun to the other heavenly bodies. This was especially work for which I was to a great extent prepared, from being already familiar with the chief methods of chemical and physical research.
>
> Then it was that an astronomical observatory began, for the first time, to take on the appearance of a laboratory. Primary batteries, giving forth noxious gases, were arranged outside one of the windows: a large induction coil stood mounted on a stand on wheels, so as to follow the positions of the eye-end of the telescope, together with a battery of several Leyden jars; shelves with Bunsen burners, vacuum tubes, and bottles of chemicals, especially of specimens of pure metals, lined its walls.
>
> The observatory became a meeting-place where terrestrial chemistry was brought into direct touch with celestial chemistry. The characteristic light-rays from earthly hydrogen shone side by side with the corresponding radiations from starry hydrogen, or else fell upon the dark lines due to the absorption of hydrogen in Sirius or in Vega. Iron from our mines was line-matched, light for dark, with stellar iron from opposite parts of the celestial sphere. Sodium, which upon the earth is always present with us, was found to be widely diffused through the celestial spaces.
>
> The time was, indeed, one of strained expectation and of scientific exaltation for the astronomer, almost without parallel; for nearly every observation revealed a new fact, and almost every night's work was red-lettered by some discovery.[5]

Fraunhofer had observed absorption lines in stellar spectra as early as 1823.[6] Huggins' observations were visual too. When he attempted to determine radial velocities of stars he wrote, in 1868,

> Unless the air is very steady the lines are seen too fitfully to permit of any certainty in the determination of coincidences of the degree of delicacy which is attempted in the present investigation. I have passed hours in the attempt to determine the position of a single line.[7]

Huggins' measurements of radial velocities of stars turned out to be largely spurious. Visual stellar spectroscopy as practiced by Huggins, Secchi, Rutherfurd, Vogel, and others was an uncertain art. The uncertainty, subjectivity, and often fallacy of visual observations—later vividly demonstrated by Lowell's martial artwork—had to be replaced by objective, reliable records, open to check and countercheck, independent scrutiny, error analysis—in brief, the methods of experimental science. Huggins had seen that and tried to photograph spectra. He wrote:

> In February 1863 the strictly astronomical character of the Observatory was further encroached upon by the erection, in one corner, of a small photographic tent, furnished with baths and other appliances for the wet collodion process. We obtained photographs, indeed, of the spectra of Sirius and Capella; but from want of steadiness and more perfect adjustment of the instruments, the spectra, though defined at the edges, did not show the dark lines, as we expected. The dry collodion plates then available were not rapid enough; and the wet process was so inconvenient for long exposures, from irregular drying and

draining back from the position in which the plates had often to be put, that we did not persevere in our attempts to photograph the stellar spectra.[5]

The man who succeeded in turning astrophysics from an art into a science was Henry Draper. In the summer night of August 1, 1872, with his 28-in. mirror telescope, he and his wife photographed the spectrum of α Lyrae, showing four Balmer lines in absorption. It took Huggins four years to reproduce this spectacular feat. Spectacular indeed, because it was HD's method of spectral line photography applied 57 years later that led to the greatest discovery in science, the discovery that our universe is not eternal and unchanging like a diamond but an explosively evolving organism born billions of years ago.

Three lines of technical developments converged in HD's success: construction and use of large telescopes, invention of precise guiding systems making long exposure times feasible, and application of photography to the extremely low intensities of starlight. Draper excelled in all these fields and his work also ushered in a century of cosmography dominated by large mirror telescopes and astrophotography, our century.

In the late 1830's photography had become the rage of the western world. To illustrate how fantastic this new process appeared I quote one of its pioneers, Fox Talbot,

The phenomenon appears to me to partake of the marvellous ... the most transitory of things, a shadow, may be fettered by the spells of our Natural Magic.[8]

Astrophotography was begun by HD's father, John William, a founder of the NYU Medical School and professor of chemistry. According to the minutes of the Lyceum of Natural History:

March 23, 1840, Dr. Draper announced that he had succeeded in getting a representation of the moon's surface by daguerrotype. The time occupied was 20 minutes, and the size of the figure about one inch in diameter. Daguerre had atttempted the same thing, but did not succeed. This is the first time that anything like a distinct representation of the moon's surface had been obtained.[1]

The place where this feat was performed is probably not far from the site of this symposium, the Washington Square campus of NYU. Draper's house was near Washington Square, and three-year-old Henry must have seen his father experimenting with his heliostat, and his aunt, Dorothy Draper, photographing celestial and human faces on the rooftop.

John William Draper took photographs of the solar spectrum and discovered absorption lines in its infrared and ultraviolet regions. He was the first to photograph the solar spectrum with a diffraction grating. He also pioneered the application of photography to microscopy and in this field he was helped in 1850 by his thirteen-year-old son Henry, who took photomicrographs to illustrate his father's book on human physiology. Henry studied at NYU and wrote his thesis, at the age of twenty, on changes of blood cells in the spleen, illustrated with photomicrographs.

Throughout his life Henry Draper was one of the world's leading scientific photographers and was so honored. Alex S. Lyman writes about him, in 1882,

In 1874, when the transit of Venus across the disc of the sun was imminent, Professor Draper was appointed Superintendent of its Photographic Department by the Commission created

FIGURE 3. The Draper 28-in. mirror telescope

by Congress to observe the transit. In this capacity he performed his duties with such signal success and gratification to scientific men, that Congress ordered a special gold medal to be struck in his honor at the Philadelphia Mint. It bears the inscription, *Decori Decus Addit Avito*—[He adds lustre to ancestral glory]. This was the first time in the history of the nation that Congress has awarded so public a recognition to any man of science.[9]

His interest in large telescopes and their guiding mechanisms began in 1857, when he attended a meeting of the British Association for the Advancement of Science (fondly known as the British ASS) in Dublin. The twenty-year-old Henry Draper was invited by the Earl of Rosse to go with a party to Birr Castle, Parsonstown, to see the famous six-foot reflector, for many years the largest telescope on Earth.

Upon his return to New York he began to build mirror telescopes. By 1867, he has polished more than 100 mirrors, working with seven polishing machines. The power source was a treadwheel he operated himself and on which he reckoned he sometimes walked over ten miles in five hours.[11] After ten years of hard work done in his spare time and at night he had become a master in making silver-on-glass mirrors, the forerunners of the 200-in. on Mount Palomar. His memoir[10] on this art became the bible of modern telescope makers.

In 1867, he married Anna Palmer, daughter of a wealthy New York businessman, who became his scientific collaborator. On their wedding trip, they selected the glass for the 28″ mirror that was to become the centerpiece of the most powerful photographic telescope. For more than a year, Draper ground and polished the primary mirror forty-one times. It took another three years of relentless work to complete the telescope and its driving mechanisms and to take the first successful spectrogram of α Lyrae.

Annie Jump Cannon describes the work of Henry and Anna Draper as follows:

In 1867 a reflector of 28-inch diameter was placed in his private observatory at Hastings-on-Hudson. In the summer, Dr. and Mrs. Draper resided at Dobbs Ferry, two miles distant, and it was their custom to drive together to the observatory for the evening work. So great was her interest that she never went to the observatory without him, and in the days of the wet plate, she herself always coated the glass with collodion. Mrs. Draper told how sometimes after they had been to the observatory and returned to Dobbs Ferry on account of clouds, they would find the sky clearing, and would drive back again two miles to the observatory and recommence work. During the early years of their married life, Dr. Draper was experimenting with the photographs of stellar spectra with his reflector, and in May and August, 1872, he succeeded in photographing the spectrum of Vega, showing four dark lines. This was four years before Huggins obtained a photograph of the dark lines in the spectrum of this star. In 1878, Dr. Draper organized an expedition to go to Rawlins, Wyoming for the purpose of observing the total solar eclipse of July 29. Mrs. Draper not only went with him, but also assisted in various ways. Her special duty was to count the seconds during the eclipse and lest the vision might unnerve her, she was put within a tent and therefore saw nothing at all of the wonderful phenomena. Here she sat patiently and accurately calling out the seconds while the glorious and awe-inspiring spectacle was unfolded.

In the winter, Dr. and Mrs. Draper resided on Madison Avenue, New York City. Here he established a laboratory, connected with the residence by a covered passageway, where his work not dependent on the telescope could be carried on, and where his photographs could be studied. The house, which is between Thirty-ninth and Fortieth Streets, is spacious and well adapted to elaborate entertaining. When originally built by Mr. Palmer, it was the last house in New York City, and he was cautioned by his business friends against investing in property so far away from the center. Mrs. Draper remembered when the old omnibus running on Fifth Avenue went only as far as Thirty-ninth Street, so that when any one alighted and started to walk in their direction they were sure of a visitor.

In November, 1882 when the National Academy of Sciences was meeting in New York

City, Dr. and Mrs. Draper entertained the members at a dinner said to have been one of the most brilliant ever given there. As a novelty, Dr. Draper lighted the table with Edison incandescent lights, some of which were immersed in bowls of water. About fifty were present, and at the close of the dinner, Dr. Draper, although suffering from a severe cold, moved about and talked with several of the guests, among others, Professor E. C. Pickering, director of the Harvard Observatory. They discussed in particular the photographs of stellar spectra Dr. Draper had obtained. Professor Pickering expressed to Dr. Draper his great interest in that work and offered to measure these photographs if they could be sent to Cambridge. Almost immediately after the dinner Dr. Draper was seized with a congestive chill, followed by pneumonia which proved fatal a few days later.[11]

Our universe, or, to be more precise, our presently observable universe, is immensely vast, incredibly rich, and beautifully diverse. If astronomers lived to the ripe old age of a billion years, trillions of such Methusalahs would not run out of subject matter. Our answer to the question What is the universe made of? is always a tentative guess. But perhaps we do know a fair sample. And what this sample suggests is a breathtaking unity, a oneness of us and the stars. When Henry and Anna Draper rode home in their carriage from Hastings-on-Hudson to their house at Dobbs Ferry on the night of the first of August, 1872, they had just obtained the final proof, the demonstrably objective evidence, that one star, 250 thousand billion kilometers high in the heavens, more than a million times further away than the sun, was made of the same atoms that are most abundant in man. A grand theme in the symphony of science, the unity of the universe, had begun to sound.

It is a pleasure to thank the Draper family, Brenda Biram, Owen Gingerich, Alfred Glassgold, Anne Kinney, Howard Plotkin, and Bayrd Still for help in preparation of this paper.

NOTES

1. An extensive bibliography of Henry Draper's writings can be found at the end of the entry "Henry Draper" in the article by Charles A. Whitney in the Dictionary of Scientific Biography, Vol. 4, pp. 180–81. The article erroneously gives the name of HD's older brother John Christopher as Daniel.
2. Some correspondence of Henry Draper is published in Nathan Reingold, 1966. Science in Nineteenth-Century America, Macmillan, London, pp. 251–62.
3. For further information on Henry Draper's work see Owen Gingerich, 1980, The first photograph of a nebula, Sky and Telescope, November, pp. 364–66.
4. The most extensive study of Henry Draper was done by Howard Plotkin in his dissertation at Johns Hopkins University. See also Reference 1.
5. For the background of Draper's work see Henry C. King, 1979, The History of the Telescope, Dover, New York.
6. For the work of Henry's father, John William, see Donald Fleming, 1950, John William Draper and the Religion of Science, University of Pennsylvania Press, Philadelphia, which lists many sources.

REFERENCES

1. PLOTKIN, H. 1981. Henry Draper, Edward C. Pickering and the Birth of American Astrophysics. Colloquium at NYU Physics Department. Personal communication. See this volume.
2. KIRCHHOFF, G. Read to Berlin Academy October 27, 1859. [G. Stokes, Trans.] Philos. Mag. **19** (4th series): 194.

3. HELMHOLTZ, H. 1888. A Memoir of Gustav Robert Kirchhoff. Dtsch. Rundschau. **14:** 232 [Trans. 1889. Smithson. Year Annu. Rep. Smithson. Inst.: 527].
4. HUGGINS, W. 1891. Address of the President to the British Association for the Advancement of Science, Cardiff, 1891. Nature (London) **44:** 2.
5. HUGGINS, W. 1899. An Atlas of Representative Stellar Spectra. Publications of Sir William Huggins's Observatory, London.
6. KAHLBAUM, G. W. A. 1888. Aus der Vorgeschichte der Spectranalyse: 12. Basel.
7. HUGGINS, W. 1868. Philos. Trans. R. Soc. **158:** 546.
8. TALBOT, F. 1964. Quoted in B. C. SAUNDERS & R.E.D. CLARK. Atoms and Molecules: 251. Dover. New York.
9. LYMAN, A. S. 1882. New York University Quarterly **5:** 122.
10. DRAPER, H. 1864. On the Construction of a Silvered Glass Telescope. Smithson. Contrib. **14:** 21.
11. CANNON, A. J. 1915. Science **41:** 381.

HENRY DRAPER'S SCIENTIFIC LEGACY

Owen Gingerich

Harvard-Smithsonian Center for Astrophysics
Cambridge, Massachusetts 02138

THE ORION NEBULA

"The phenomenon of milky nebulosity is certainly of a most interesting nature," William Herschel wrote in 1802 in a statement that holds true today, as this symposium attests. The eminent English observer went on to say, "It is probably of two different kinds, one of them being deceptive, namely, such as arises from widely-extended regions of closely connected clustering stars, contiguous to each other, like the collections that construct our milky-way; the other, on the contrary, being real and possibly at no very great distance from us. The changes I have observed in the great milky nebulosity of Orion, 23 years ago, ... cannot permit us to look upon this phenomenon as arising from immensely distant regions of fixed stars. . . . Its light is not like that of the milky-way, composed of stars."

Herschel's surmise, that the Orion Nebula was something other than a great congeries of stars, was indeed correct, though he arrived at his conclusion partly for spurious reasons, since the changes he supposedly observed were certainly not real. Precisely what generated the nebula's milky light remained a source of bafflement. "To attempt a guess at what this light may be would be presumptuous," he wrote, and Herschel added that it would be all too easy to propose the existence of an effect without its cause, something he dismissed as "unphilosophical."

Astronomers throughout the entire nineteenth century puzzled over the nature of nebulae in general and the Orion Nebula in particular; even the spectroscopic key provided by Kirchhoff in 1859 did not immediately unlock the mystery—the nebulae were too elusively faint for their spectra to be obtained with any ease. Furthermore, the ongoing analysis of purported changes in the nebulae kept astronomers on a false scent that was not to be abandoned until the photographic plate replaced the intricately detailed but subjective drawings made by telescopic observers.

The first of these photographic plates of the Orion Nebula was made by Henry Draper on September 30, 1880. It was a great feat to track the nebula with some precision for a 51-min exposure on a very slow plate, but the results were hardly spectacular. Yet it was a turning point of considerable consequence, a first step that was soon to put the visual observers of nebulae out of business. Had Draper lived longer, he probably would have obtained the spectrum of the Orion Nebula as well, and would thereby have been well on the way to a "philosophical" solution for the physical nature of the nebula.

In that same year, Edward Holden, then still an astronomer at the U.S. Naval Observatory, was assembling a monograph entirely devoted to the visual description of the Orion Nebula and to the apparent changes in its structure. Published in 1882 as a long appendix to the *Washington Observations* for 1878, it marked the end of the visual era and the beginning of the photographic era. Draper's much more successful photograph of 1882 was included as an addendum, and Holden compared it with

0077-8923/82/0395-0308 $1.75/1 © 1982, NYAS

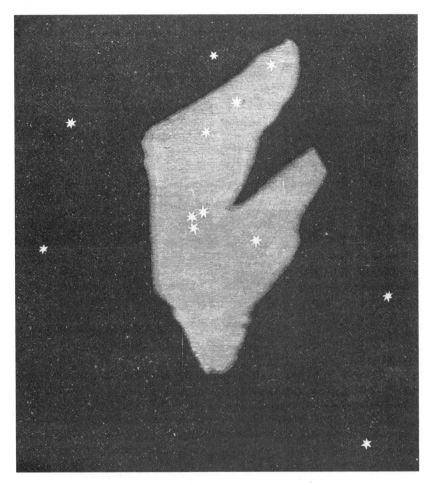

FIGURE 1. The first depiction of the Orion Nebula, from Christiaan Huygens' *Systema Saturnium,* The Hague, 1659.

George Bond's drawing of the Orion Nebula, which was without question the finest visual representation ever made of any nebula.

Holden wrote

Although it is still too soon to give a final discussion to the photographic results attained by Dr. Draper, I cannot refrain from pointing out some of the conclusions which may be drawn from this marvelously perfect representation of the nebula. . . .

Bond's engraving is the most accurate drawing that has been made, even as a map, and as a picture it is decidedly the best representation of a single celestial object which we have by the old methods. The work of observing alone extended over years and consumed many precious hours.

Dr. Draper's negative was made in 137 minutes, and for nearly every purpose is incomparably better than the other. The color and tint of the nebula, which is wonderfully preserved in Bond's engraving, is lost in the photograph; and yet, if the latter is held up between the eye and a window, the pictorial effect is most striking. . . .

This photograph comes as the beginning of a new epoch in such observations.

Draper's accomplishment can be placed in a broader perspective by briefly noting some of the earlier observations of the Orion Nebula and, for this purpose, Holden's

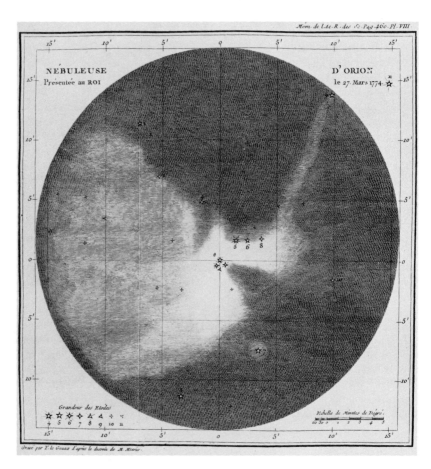

FIGURE 2. Charles Messier's careful drawing of what has become M42 and M43 (below, around the star marked "7"), from the *Mémoires de l'Academie Royale des Sciences* for 1771, Paris, 1774.

monograph provides our guide. The nebula was apparently first noted by Nicholas Peiresc in 1610, in the year after Galileo's first use of the telescope, and an independent discovery was made in 1618 by the Jesuit astronomer Johann Baptist Cysat, who mentioned it in his book on the comet of that year. Cysat described the nebula by saying, "Another [example] of this phenomenon in the heavens is the congeries of stars at the last star in the sword of Orion, for there one can distinguish (with a telescope) a similar congestion of some stars in a very narrow space, and all around and between the stars themselves is a diffused light like a radiant white cloud."

The earliest published drawing of the nebula was included by Christiaan Huygens in his *Systema Saturnium* of 1656. Huygens made a somewhat similar drawing again in 1694, his very last astronomical note. Both depictions clearly show the deep, dark indentation toward the Trapezium that was later designated *Sinus magnus,* the "great bay."

The most spectacular of the early engravings was published by Charles Messier in 1771. Made with a 3½-in. Dollond refractor, it showed the large nebulous arm to the east-southeast that was subsequently named the *Brachium Messieri.* By the time Holden's monograph was published, at least a dozen more features carried picturesque Latin names such as *Pons Schroeteri* ("Schroeter's bridge"), *Palus Bondii* ("Bond's swamp"), and *Lacus Lassellii* ("Lassell's lake"). These names commemorated the observations of Schroeter in the 1790s, the delicate and detailed drawing (already alluded to) made by George Bond from 1859–63, and the large-scale drawing made by Miss Caroline Lassell in 1862 during her father's "celebrated astronomical expedition to Valetta" (that is, with the 48-in. reflector taken to Malta).

By chance, the brilliant edge to the southeast, now interpreted as a shock front, received the name *Frons* ("brow" or "forehead"); the coincidence of *Frons* and "front" is so striking that perhaps the Latin nomenclature should be revived!

Some of the curious and beautiful structures of the nebula never seem to have received their own names. On Bond's drawing, the three bright stars in an east-west line southeast of the Trapezium (including θ^2 Orionis) show distinct cometary tails, and these show up strikingly in the rather cubistic depiction made by D'Arrest in 1872; these unnamed streamers are easily visible in modern photographs, but they seem to elicit very little comment today.

"Probably no object outside the solar system has received more attention from the best observers than the nebula of Orion," Holden wrote in his monograph. He added that "the main object of this memoir is to leave such measures and descriptions of the brightest parts of the nebula of Orion as shall enable another person observing in after years with the same telescope, under like conditions, to say with certainty whether or no changes have occurred in these parts of this nebula." But at the very moment he was preparing his account, events were transpiring along the Hudson River that would reduce his monograph to mere antiquarian interest.

HENRY DRAPER

Dr. Henry Draper's private observatory at Hastings-on-Hudson boasted an 11-in. Clark refractor, with a special correcting lens for photographic work, as well as a 28-in. silver-on-glass reflector of his own construction. This latter achievement would by itself have established him as an outstanding astronomer. Draper had a thorough background in the rapidly emerging field of photography, and the thesis for his medical degree had been illustrated with daguerreotype microphotographs.

Without photography, astronomy could not have expanded to its modern dimensions. Today, even a short exposure records more than the eye can hope to see, but in Draper's lifetime this was never the case. Photography's principal advantage then was permanency—indeed, it was precisely this advantage that was to render Holden's

FIGURE 3. George Bond's drawing of the Orion Nebula made with Harvard's 15-in. refractor, 1859–63, from *Annals of the Observatory of Harvard College*, vol. 8, 1877.

monograph obsolete. But this advantage was counterbalanced by the comparative nuisance of the technique and its novelty.

Henry Draper's success in recording the Orion Nebula must be seen against this background to be appreciated. The wet photographic plates of his day were very slow and long exposures had been impossible because the emulsion dried too quickly. But, as the result of a visit to William Huggins in England in the spring of 1879, Draper

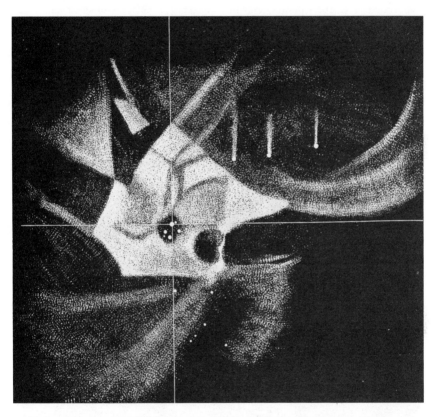

FIGURE 4. H. L. d'Arrest's drawing of the Orion Nebula, made with the 11-in. refractor at Copenhagen in 1872, from Edward S. Holden's *Monograph of the Central Parts of the Nebula of Orion,* Washington, 1882. Note the cometary plumes from the three stars in the upper right quadrant.

discovered the new dry gelatino-bromide plates with their somewhat increased sensitivity. Furthermore, these dry plates made the long exposures required for faint astronomical objects possible for the first time. Draper was able to take a 51-min exposure, long enough to capture some of the faint nebulosity.

Draper promptly sent off a brief description of his success to England, France, and the *American Journal of Science*. He remarked only that his accomplishment had been achieved with the 11-in. Clark triplet, using his own mounting and clock drive, that his exposure was fifty minutes, and that he intended "at an early date to publish a detailed description of the negatives."

But how could he adequately describe his really very fuzzy negative? Following the old adage that a picture is worth a thousand words, Draper decided to distribute actual enlargements of his plate rather than publish a technical description. The prints were mounted on a card approximately six by eight inches, with a printed caption reading "First photograph of the nebula in Orion. Taken by Professor Henry Draper, M.D." Unfortunately, Draper appears never to have recorded the brand of emulsion, though

FIGURE 5. Henry Draper's photograph of the Orion Nebula, a 137-min. exposure on March 14, 1882, reproduced from Holden's monograph. Draper's first photograph, taken 18 months earlier, is reproduced as the frontispiece to this volume.

earlier he had specifically praised those made by Wratten and Wainwright of London as being the most sensitive.

Perhaps another reason why Draper never published further details of his pioneering achievement was that he soon surpassed it with an exposure twice as long, of 104 min. After his initial success, he had written to E. S. Holden saying, "The exposure of the Orion Nebula required was 50 minutes; what do you think of that for a test of my driving clock?" By March, 1882, he had obtained an exposure of 137 min, which, of course, showed far richer detail. This remarkable picture was reproduced by photolithography as an addendum to Holden's 1882 *Monograph of the Central Parts of the Nebula of Orion*. Again, the 11-in. Clark and gelatino-bromide plates were used. "The night was clear but cold and windy. The mean temperature was 27° Fahr.; the wind NNW. and in gusts, the strongest pressure being 5 pounds per square foot about nine o'clock; the whole travel of the wind during the exposure was 35 miles. The variation in the force of the wind is one reason why the stars show some ellipticity under this magnifying power; the gusts of course displaced the telescope somewhat, though the mounting is firm and the clock-work strong."

Two months after taking his 137-min exposure, Draper wrote to Holden, "I think we are by no means at the end of what can be done. If I can stand 6 hours exposure in midwinter another step forward will result." With this plan in mind Draper undertook to build a new form of mounting to permit tracking across the meridian for a continuous exposure of six hours.

But Draper was never to try that experiment. On November 15, 1882, the Drapers held a gala dinner for the National Academy of Sciences. Among those in attendance were Alexander Agassiz, S. P. Langley, Asaph Hall, Willard Gibbs, Charles A. Young, and Edward C. Pickering. In the huge dining room of the Madison Avenue residence forty academicians and other friends sat at a table illuminated with the new Edison incandescent bulbs, some of which were immersed in bowls of water. During the dinner Draper suffered a severe chill, but he continued to converse with his friends until the end of the evening. Just five days later Henry Draper was dead, of pneumonia, at age 45.

DRAPER'S SCIENTIFIC LEGACY

The scientific and photographic legacy of Dr. Draper was great indeed, although it did not directly involve the Orion Nebula. Besides his pioneering nebular photography, Draper had been much concerned with photographing stellar spectra, and it was particularly this part of his work that his devoted wife, Anna Palmer Draper, resolved to continue.

From Harvard, E. C. Pickering offered his sympathy and encouragement. In February, Mrs. Draper carried 21 photographs of stellar spectra to Cambridge for Pickering's microscopic examination and, twelve months later, these results, together with a larger discussion of Draper's spectrographic researches, were published in the *Proceedings of the American Academy of Arts and Sciences*. But in the year that followed, Mrs. Draper's plans to establish her own observatory got nowhere.

At this point Pickering came forward with a rather tentative proposal for a cooperative endeavor that would link Henry Draper's name with the photographic

FIGURE 6. Anna Palmer Draper on a visit to Harvard College Observatory in 1890. At the far left is her niece, Antonia Maury. Consulting with Mrs. Draper is Williamina Fleming, who first developed the Draper spectral classification system. (Photograph courtesy of Harvard College Observatory.)

researches at Harvard, but not until January of 1886 did Pickering make his ideas explicit. He wrote:

> A remark in your note of Jan. 1 'that it seems impracticable to think of doing anything at the Hastings Observatory' has given me much concern. The various difficulties you have encountered in establishing the Henry Draper Physical Observatory have met with my hearty sympathy. In the hope that I may aid you I venture to suggest the following scheme, which at least has the advantage that it leaves you free to alter your plans at any time. Suppose that you should appropriate the interest only, instead of the principal, of a fund to astronomical physics, with the understanding that it will be continued only so long as

FIGURE 7. Cecilia Payne-Gaposhkin in conversation with Annie Jump Cannon, probably around 1938. (Courtesy of *Sky and Telescope*.)

satisfactory progress is maintained. It is very certain that no money would be wasted under these circumstances, and that every effort would be made to secure the greatest results. I can imagine no greater spur to all concerned, than the expectation that the continuance of a work would depend on its successful conduct. How would you like to begin at once to carry on some research in this way? In other words to do now, what would be done if your Observatory were already in operation. If this plan could be tried in Cambridge I would gladly give the necessary supervision.

This plan moved ahead swiftly, and within a year the observatory at Hastings-on-Hudson was abandoned as a research station. The 28-in. reflector was transferred to Cambridge, where its mirror still remains, in the Collection of Historic Scientific Instruments, as a silent testimonial to Draper's achievements as a telescope builder and experimentalist. An even more eloquent memorial is the *Henry Draper Catalogue of Stellar Spectra,* a vast project whose publication was not completed until 1924, after Pickering's death.

This catalogue reached its monumental conclusion in two stages. The initial classification scheme was placed in the hands of Pickering's former housekeeper, Williamina Fleming. Without any special scientific training, Mrs. Fleming soon proved to be one of his ablest and hardest-working assistants, and a person with a real flair for the detail of categorizing the photographic spectra. The scheme, presumably worked out in close consultation with Pickering, employed the letters from A to O, omitting J but adding Q for those spectra that didn't fit in any other category. *The Draper Catalogue of Stellar Spectra* (as distinguished from the later *Henry* Draper Catalogue) was published in 1890 as volume 27 of the *Harvard Annals.*

Meanwhile, Pickering had hired Henry Draper's niece, Antonia C. Maury, a recent graduate of the physics and astronomy program at Vassar, to analyze the spectra of the brighter stars in more detail. Miss Maury was determined to let no detail escape her eyes, and she evolved an alternative classification with 22 categories, which were each further subdivided into three divisions. It was these subdivisions that enabled Ejnar Hertzsprung to verify his discovery of two distinct varieties of stars—dwarfs (divisions *a* and *b*) and giants (*c*). Pickering was not pleased with Miss Maury's categories, however, for he was keen to get the Draper system adopted internationally, and he feared that her clumsier arrangement might not fare well in competition with rival European schemes.

Hence, in 1896, Pickering brought a young Wellesley graduate, Annie J. Cannon, to his staff, and encouraged her to extend and revise Mrs. Fleming's spectral categories. This Miss Cannon accomplished with skill and efficiency, dropping many of the groups as unnecessary, rearranging the order, and splitting the categories into finer decimalized subgroups. When Pickering adopted this system in the Harvard Revised Photometry, Hertzsprung promptly protested:

> In my opinion the separation by Antonia C. Maury of the c- and ac-stars is the most important advancement in stellar classification since the trials by Vogel and Secchi.... To neglect the c-properties in classifying stellar spectra, I think, is nearly the same thing as if the zoologist, who has detected the deciding differences between a whale and a fish, would continue in classifying them together.

Nevertheless, Pickering held out for the simpler arrangement, and his strategy was eminently successful, for the resulting Henry Draper classification scheme became the standard. On the other hand, the acceptance of the Draper classifications may owe as much to the sheer magnitude of the project: the nine volumes, published between 1918 and 1924, contained over 225 000 spectral types determined by Miss Cannon and, in subsequent work, including the *Henry Draper Extension,* she brought the total to over 350 000.

Pickering seemed to eschew speculation about the meaning of this great trove of data. The observatory publications certainly leave us with the impression that he would scarcely allow his staff to suggest that the arrangement of stars in the now-familiar OBAFGKM series was a temperature classification, and it remained for a young researcher from the next generation to show this definitively.

Cecilia Payne came from the English Cambridge to work under Harlow Shapley, who had succeeded Pickering as director in 1921. Shapley assigned her to work on spectra under the knowledgeable Miss Cannon, and he proposed that her work should be turned in as a thesis despite the fact that there was not yet any formal graduate program in astronomy at Harvard. The result was what Otto Struve called "the most brilliant Ph.D. thesis ever written in astronomy." Miss Payne showed that the apparent differences in the spectral classes resulted almost entirely from their differences in surface temperature, and she established such a convincing scale of stellar temperatures that she essentially finished off what had been one of the outstanding astrophysical problems of the first quarter of our century. Almost as an afterthought, under the heading "additional deductions," she laid the groundwork for the astrophysical problem of the next half century, how to find the actual chemical compositions of the stars. And here her results, based on the Draper spectra at Harvard, were literally astonishing beyond belief. She found hydrogen to be a million times more abundant than iron, which was commonly believed by the terrestrial analogue to be the most abundant element. Hence, she rejected her own results, saying that this anomalous abundance must be due to some astrophysical peculiarity of this lightest of all elements.

Several years ago, I asked Professor Payne-Gaposchkin why she had, in fact, declared that the high hydrogen abundance was surely unreal. "I suppose Henry Norris Russell talked me out of it," she replied. Last month I visited the Princeton University Archives to investigate her claim, and there among the papers was the smoking gun:*

January 14, 1925

My dear Miss Payne:

 Here, at last, are your notes on relative abundance which you were so good as to send me some time ago. . . .

 You have some very striking results which appear to me, in general, to be remarkably consistent. Several of the apparent discrepancies can be easily cleared up. [Here Russell discusses Mg, Mg+, and K in some detail.]

 There remains one very much more serious discrepancy, namely, that for hydrogen, helium and oxygen. Here I am convinced that there is something seriously wrong with the present theory. It is clearly impossible that hydrogen should be a million times more abundant than the metals, and I have no doubt that the number of hydrogen atoms in the two quantum state is enormously greater than is indicated by the theory of Fowler and Milne. Compton and I sent a little note to 'Nature' about metastable states, which may help to explain the difficulty. . . .

Very sincerely yours,
Henry Norris Russell

It was not until four years later that Russell reversed himself and declared hydrogen to be the most abundant material in stellar atmospheres, although it took several years for astronomers to be convinced that this result applied to the entire universe.

When we look back at Henry Draper's scientific legacy, it would not be an exaggeration to say that the discovery of the high hydrogen abundance in the universe was an ultimate consequence of his pioneering work and his astrophysical goals.

---

*Quoted by permission of the Princeton University Library.

BIBLIOGRAPHY

GINGERICH, O. 1980. The first photograph of a nebula. Sky and Telescope **60:** 364–66.
HOLDEN, E. S. 1882. Monograph of the Central Parts of the Nebula of Orion. Appendix I of the Washington Astronomical Observations for 1878. Washington, D.C.
JONES, B. Z. & L. G. BOYD. 1971. The Harvard College Observatory 1839–1919. Harvard University Press. Cambridge.
JONES, K. G. 1968. Messier's Nebulae and Star Clusters. Faber & Faber. London.
WHITNEY, C. A. 1971. Henry Draper. *In* Dictionary of Scientific Biography **4:** 179–81. Charles Scribner's Sons. New York.

HENRY DRAPER, EDWARD C. PICKERING, AND THE BIRTH OF AMERICAN ASTROPHYSICS

Howard Plotkin

Department of History of Medicine and Science
University of Western Ontario
London, Ontario N6A 5C1

The great period of rapid advancement in astrophysics began in 1859, when the Berlin Academy of Sciences published the results of the close collaboration between the chemist Robert Bunsen and the physicist Gustav Kirchhoff. During the next three years, Kirchhoff's laboratory experiments allowed him to make the important generalizations that generally came to be known as Kirchhoff's laws of radiation. On the basis of these laws, he postulated the existence of an extensive gaseous atmosphere above the photosphere of the sun that gave rise to the dark Fraunhofer absorption lines. By comparing the Fraunhofer lines in the solar spectrum with the spectra of incandescent gases in the laboratory (the method of spectrum analysis), it was possible to identify the chemical elements present in the solar atmosphere. By the mid-1870s, thirty-three solar elements had been identified in this manner.

These solar discoveries, however, were not the only triumphs of the fledgling science of astrophysics. In 1872, Henry Draper, one of America's pioneer astrophysicists, succeeded in obtaining the first photograph of a stellar spectrum that showed Fraunhofer lines. This photograph led to various attempts to classify the stars on the basis of photographs of their spectra. The culmination of these endeavors was the *Henry Draper Catalogue* published by Edward C. Pickering, the director of the Harvard College Observatory, between 1918 and 1924, in which nearly a quarter-million stellar spectra were measured and classified. The classification scheme embodied in the catalogue was unanimously adopted by the International Union for Cooperation in Solar Research, and is still in use today; indeed, it is heavily relied on by practically all astrophysicists. The *Henry Draper Catalogue* played a major role in the development of astrophysics, and was responsible for Harvard's—and America's— early pre-eminence in this new and exciting field.

As background to the early history of American astrophysics, it will be helpful to first examine some of the experiments in photography and spectrum analysis of John William Draper, Henry Draper, and Edward C. Pickering.

When news of Louis Jacques Mandé Daguerre's successful process for fixing camera images reached New York City in September, 1839, several persons immediately began experimenting with the new technique. Within a week, the first American daguerreotype—of St. Paul's Church and the Astor House Hotel—was made by D. W. Seager. But the real challenge was in taking a portrait; Daguerre himself doubted it could be done. At least three men raced each other to show he was wrong: Alexander S. Wolcott; Samuel F. B. Morse, professor of fine arts at New York University (then called the University of the City of New York); and his new colleague John William Draper, professor of chemistry and botany.

Even before Daguerre had announced his success, Draper had been experimenting

321

0077–8923/82/0395–0321 $1.75/1 © 1982, NYAS

in making temporary copies of objects. For some time he had been interested in the effects of the solar spectrum on the silver salts—bromide, chloride, and nitrate—researches that clearly bore on the problem of fixing camera images.[1] Spurred on by Daguerre's challenge, he now attempted portraitures. Although Wolcott apparently won the race in October, Draper succeeded in taking a striking portrait of his sister Dorothy early the next year.

In April, 1840, Draper and Morse joined forces and opened a portrait parlor, but Draper soon began to apply photography in ways that would take him out of the parlor and back into the mainstream of scientific research: fixing images from the flame of a gas light and from a lime-pea in an oxy-hydrogen blowpipe and taking the first known daguerreotype of the moon that showed any indication of the lunar maria.[2] This daguerreotype, in a modest way, launched the age of astronomical photography. Draper sent it, or a similar one, to the New York Academy of Sciences (then called the Lyceum of Natural History):

> *March 23d*, 1840. Dr. Draper announced that he had succeeded in getting a representation of the moon's surface by the Daguerreotype. . . . The time occupied was 20 minutes, and the size of the figure about 1 inch in diameter. Daguerre had attempted the same thing, but did not succeed. This is the first time that anything like a distinct representation of the moon's surface has been obtained.[3]

Unfortunately, this historic daguerreotype was lost in a fire that destroyed many of the Academy's treasures. Draper clearly found these investigations more interesting than portrait taking and left Morse by the spring of 1841.

In that year, Draper joined Martyn Paine, Valentine Mott, John Revere, Granville Pattison, and Gunning Bedford in founding the New York Medical School. Judging from his publications, however, he did not allow his new position to seriously interfere with his physical and chemical experiments. Early in 1843, he obtained daguerreotypes of the solar spectrum that revealed new Fraunhofer lines. He had found that the plate was more sensitive when one allowed weak, diffuse light to fall on the photographic plate during exposure. By this process he discovered three prominent lines in the infrared which he named α, β, and γ—a discovery later confirmed by Léon Foucault and Hippolyte Fizeau.[4] He also discovered new Fraunhofer lines in the ultraviolet, but here he was preceded slightly by Edmond Becquerel (a fact that Draper apparently did not know).

Draper was not entirely satisfied with these prismatic spectrum photographs, however, as prisms do not yield a normal spectrum. So, in May, 1843, he had Joseph Saxton, an eminent mechanician at the U.S. Mint in Philadelphia, rule a diffraction grating for him. With this grating, measuring ⅝ inch by ⅓ inch, Draper took the first diffraction spectrum daguerreotype of the sun in 1844.[5]

Draper was clearly an exceptional experimentalist. Although some of his pet theories have since been disproved, and some of his best work has to be separated from much of the theory in which it is framed, his pioneering photographic experiments—especially those in spectroscopy—were of fundamental importance.

His eminence as a teacher, author, philosopher, and investigator helped create about him an atmosphere of scientific culture of the highest kind, and he had a profound intellectual influence on all of his children. Although all three of his sons would go on to become distinguished scientists, Henry, the middle, was clearly his favorite:

Happiest were the days when Henry could be home. Then he sat facing his father; and when they two carried on the conversation, the rest sat silent. For it was well known that everything they said was too important for any word of it to be lost. To his father, in whose face a serene repose mingled with sustained intensity of intellectual penetration, Henry was the perfect foil, his dark eyes flashing electrically over the latest discoveries in physics or astronomy, or when relating some humorous incident, they brimmed with laughter to the point of tears.[6]

From an early age Henry was taken into his father's confidence in scientific matters, and was often permitted to assist him not only in his university lectures, but also in his investigations.

In 1850, John taught Henry the art of taking photographs through a microscope and had his thirteen-year-old son take photomicrographs to illustrate his book *Human Physiology, Statical and Dynamical,* which was published six years later. From this period dates Henry's interest in photography, a field in which he later attained great prominence.

When Henry was fifteen he enrolled as an undergraduate at New York University. He did not graduate with his class, however; partly on his father's advice (his father was then Dean), and partly on account of his frail constitution, he left the classical course he was pursuing at the end of his sophomore year and entered the Medical School. By 1857 he had completed all his medical studies, had written a thesis on the changes of blood cells in the spleen, had illustrated it with daguerreotype photomicrographs, and had passed all of his examinations.[7] But he was then only twenty, one year shy of the minimum age required for graduation, so he decided to go abroad with his older brother, John Christopher, for a year of study and recreation.

During this sojourn he attended the annual meeting of the British Association for the Advancement of Science, held in August in Dublin. After the meeting he was invited by the Earl of Rosse to visit his famous six-foot reflecting telescope at Birr Castle, Parsonstown. There Draper saw the "Leviathan of Parsonstown," as the telescope was commonly called, the machinery and methods by which it had been produced, and had the opportunity of observing several celestial objects through it. It was there that he conceived of the possibility of combining photography and astronomy, and resolved to construct a telescope himself to pursue that line of investigation:

On returning home in 1858, I determined to construct a similar, though smaller instrument; which, however, should be larger than any in America, and be especially adapted for photography.[8]

Accordingly, in September, 1858, immediately after his return from Europe, he began the construction of a telescope and observatory on land belonging to his father's estate at Hastings-on-Hudson, a country village about twenty miles north of New York City. Draper could only work on this task part-time, however, for, after receiving his medical diploma that year, he joined the staff of Bellevue Hospital and worked there for the next eighteen months. At the end of that period, early in 1860, he was appointed professor of natural science at his *alma mater,* New York University. By that time, he had completed the grinding and polishing of a 15½-in. speculum mirror out of an alloy of copper and tin in the proportion given by Lord Rosse. Draper's tedious efforts on the mirror came to an abrupt halt in February, when he awoke one morning to find that the speculum had split in half overnight due to the expansion in freezing of a few drops of water that had gotten into the mirror's supporting case.

That summer, when John was abroad, he told this tale to Sir John Herschel.

Herschel informed him that glass was preferable to speculum for astronomical purposes since it was easier to figure, weighed only one-eighth as much, and reflected more than ninety percent of the light falling on it. John forwarded this information to Henry, who immediately began experimenting with silvered-glass mirrors, grinding and polishing more than one hundred ranging from $\frac{1}{4}$ inch to nineteen inches in diameter.

Satisfied that he had mastered the technique, he began to simultaneously grind three 15½-in. glass discs (he had found that it was usually necessary to grind three mirrors of the same aperture and focal length, since two would be so likely to be similar that a third would be necessary in order to make any advance upon them). The mirror was completed in the fall of 1862, and pronounced to be "as good as we know how to make it."

When it was installed at the Hastings-on-Hudson observatory, Draper launched his astronomical career in earnest. The greater part of 1863 was spent taking photographs of the moon, sun, and planets; some fifteen hundred negatives of the moon were taken that year. About 1¼ inches in diameter, they bore enlargement to three feet, and, in one case, to fifty inches, with excellent results. They were among the best photographs of the moon ever taken up to that time.

The year 1867 marked two significant events in Draper's life. The first was his marriage to Anna Palmer, the daughter of Courtlandt Palmer, a very prominent and wealthy New York businessman. The Draper home, at 271 Madison Avenue between 39th and 40th Streets (then the last house in the city), was spacious and particularly well adapted to the elaborate entertaining for which it became well known—especially for entertaining the members of the National Academy of Sciences during its New York meetings after Henry's election in 1877.

During the summer months, the Drapers lived in Dobb's Ferry, two miles from the Hastings observatory, and would drive together to the observatory each clear night. During the days of wet plate photography she always coated the glass with the collodion herself. Indeed, so great was her interest in Henry's work that he never went to the observatory without her.

The second significant event of 1867 was the beginning of construction of a glass mirror with a 28-in. aperture and a 12½-ft focal length. Anna accompanied Henry on an "expedition" to choose the glass, and they afterwards referred to this venture as "our wedding trip." Most of Henry's spare time during the next eighteen months was spent grinding and polishing the mirror. This tedious task was completed in June, 1869, and the mirror was mounted in August.

The following winter, while Henry was working on the driving clock of the telescope, John again travelled abroad. In England he bragged about Henry's telescope to various members of the Royal Astronomical Society, ordered books and scientific apparatus for his son, and spied the land to see what kind of research would make him famous in Europe in the least amount of time:

> From what I see here your proper course is to use your telescope first in getting some good lunar photographs. After what I have said depicting it they will look for something of the kind here . . . that done, try your hand at the stellar spectra.[9]

To make sure that Henry would not lose any valuable time in beginning this line of investigation, he exhorted him to push work on completing the observatory:

By a little pushing you might have it all in readiness by the time I get back early in April. You will have nothing to do at the University as I will take charge there and so might get to work without interruption or delay. Push the thing a little and you can do it.[10]

Henry quickly acted on his father's advice. After constructing no fewer than seven complete driving clocks, repolishing the mirror, and finishing all the accessories, the large telescope was finally completed in May, 1872.

Draper immediately tested his telescope to see if it could be used to photograph stellar spectra. His initial attempt was successful, and he obtained a photograph of the spectrum of α Lyrae (Vega) ½-in. long and ⅟₃₂-in. wide. The photograph, however, did not show any spectrum lines.[11] This was due, undoubtedly, to the low sensitivity of the wet collodion plates and the difficulty of keeping the image of the stellar spectrum motionless on the plate. But repeated experiments enabled him to improve his technique and, on August 1, 1872, he succeeded in obtaining the first photograph of a star (again Vega) that showed distinct Fraunhofer lines.[12] For this historic photograph, he employed a quartz prism placed just inside the focus of the small diagonal mirror of the telescope, no slit or collimating lenses being used. No other astronomers could then duplicate this feat. It was another four years before his closest competitor, William Huggins, could obtain a photograph of stellar spectrum lines—even though he had begun this line of investigation ten years before Draper.

During the following year, Draper employed a diffraction grating made for him by Lewis M. Rutherfurd, another prominent New York pioneer astrophysicist, to photograph the solar spectrum. His resulting photograph was far more detailed than the current standard: "Between wavelengths 3925 and 4205, Ångström shows 118 lines, while my original photograph has at least 293."[13] Draper used this photograph to establish a reference scale for the determination of the wavelengths of the solar Fraunhofer lines.

In the summer of 1879, he went to England in an attempt to garner support among the members of the Royal Astronomical Society for his alleged discovery of oxygen in the sun.[14] While there, he visited Huggins, who told him that the new dry photographic plates had become more sensitive than the wet collodion plates that he was using. Draper brought some of these new plates back with him, along with Huggins' star spectroscope, and immediately used both with his 12-in. Clark refractor. The convenience and success of the dry plates was excellent and, in the next three years, he obtained over eighty high-quality spectrum photographs of the moon, Mars, Jupiter, Comet 1881 III, bright stars and the Orion Nebula. He confirmed Huggins' discovery of hydrogen in Vega, and noted that all the stars he had so far observed could be placed in Angelo Secchi's first two spectral classes.[15]

Draper devoted special attention to the Great Nebula of Orion for eighteen months, with two main objectives:

First, to ascertain whether any changes are taking place in that body by making a series of photographs to be compared in the future with a similar series; and second, to photograph the spectrum of the Nebula in various parts so as to see whether the composition is uniform throughout.[16]

For these photographs, he used a specially corrected 11-in. Clark photographic refractor. His first successful photograph of the nebula was taken on September 30, 1880, with an exposure of fifty minutes.[17] A much better photograph was obtained on March 14, 1882. On that night, a 137-min exposure recorded stars down to magnitude

14.7 and clearly and beautifully revealed the nebula's faint outlying regions.[18] This photograph was the most brilliant success achieved by celestial photography up to that time.

In short, Draper had succeeded in making photography—especially spectrum photography—the best means of studying the sky. He noted:

> It is only a short time since it was considered a feat to get the image of a ninth magnitude star, and now the light of a star of one magnitude less may be photographed even when dispersed into a spectrum.[19]

He was confident that prolonged exposures during the clear nights of the following winter would result in another large step forward.

Draper never got the chance to take that step, however. That fall, while on a two-month hunting trip in the Rockies, he was caught in a blizzard and forced to spend the night high on a mountain without shelter. As a result, he developed double pleurisy and died in November, 1882, at the age of forty-five.

Only five days before his death, at a dinner party at his home for some forty members of the National Academy of Sciences, Draper discussed his recent photographs of stellar spectra with Edward C. Pickering, the director of the Harvard College Observatory. Pickering expressed great interest in the work, and offered to reduce the photographs if he would send them to Cambridge. Two months after Draper's death, Pickering restated his offer to his widow. She visited Pickering in Cambridge, bringing with her twenty-one photographs of stellar spectra, none greater than ¼-in. long. Pickering examined them with a microscope, and suggested that they publish a list of all of Draper's spectra, together with measurements of the relative positions of the lines and calculations of their wavelengths. This paper, with a detailed introduction by Charles A. Young, pictures of the Hastings observatory, and excerpts from Draper's research notebooks, was published in February, 1884.[20] It contained data on seventy-eight spectra, including reduced wavelengths for twenty-one.

Mrs. Draper informed Pickering that she wished to purchase the Hastings observatory from John's estate (he also had died in 1882), move Henry's New York laboratory there, and "endow the whole as an institution for original research in astronomical physics to be called the Henry Draper Astronomical & Physical Observatory." Pickering approved of this plan, and even suggested various men as possible directors for the proposed observatory. None, however, seemed exactly suitable for Mrs. Draper. She lamented: "It is not likely that I will find any one person with the varied scientific knowledge that was peculiar to Henry."[21]

For three years Pickering maintained a lively correspondence with Mrs. Draper, but no concrete plans were ever proposed. Pickering clearly sympathized with her desire to perpetuate her late husband's name and work, and her genuine interest in and love for astronomy. At the same time, however, he recognized that he lacked sufficient means to pursue his own research as far as he wished. He therefore boldly suggested a plan for her consideration: that she appropriate a fund to Harvard for research in astrophysics to be carried out under his supervision. Further, she could choose the area to be investigated and he would publish the results in a single volume of the observatory *Annals* as a memorial volume to Henry Draper.

To his great delight, Mrs. Draper agreed to this plan. As to the area of investigation, she wrote:

I would wish to have it in one of the directions that Dr. Draper intended to pursue. . . . His plans were: At the Observatory to get his instruments into the best order, and then by means of photography to catalogue and classify the stars by their spectra.[22]

This proposal fit in well with Pickering's interests and plans, for he was one of the first observatory directors in America to devote his institution principally to research in the new science of astrophysics. Soon after assuming the directorship in 1877, he had initiated a long-term photometric investigation of the brightness of every star visible from Cambridge. Moreover, he had just recently received an appropriation from the Bache Fund of the National Academy of Sciences to chart all of the visible stars by means of photography. He saw Mrs. Draper's proposal as a means to combine spectroscopy with his stellar photography plans, and seized the opportunity.

Pickering's original intention in 1886 was to photograph the spectra of all stars down to −30° latitude.[23] The instrument used for this employed an 8-in. lens that had been reground and mounted by the Clarks and paid for by the Bache Fund.[24] By placing a large prism in front of the objective, Pickering obtained spectra that could bear enlargement to unprecedented size—four inches wide by twenty-four inches long. Mrs. Draper was greatly pleased with the success of these photographs, and was receptive to Pickering's proposal the following year to "preempt the entire field" by extending the investigation to include special studies of the spectra of faint stars (especially those that were banded or variable) and the comparison of stellar and laboratory spectra.[25]

The expanded Henry Draper Memorial consisted of three main investigations. The first was a general survey of stellar spectra for all stars north of −25° and brighter than the sixth magnitude. The 8-in. Bache telescope was used for this, photographing regions 10° square with exposures of about five minutes. The second was a study of the spectra of fainter stars. The Bache telescope was used for this, also, but with exposures of about an hour. The third was a study of the spectra of brighter stars, using Draper's 11-in. telescope.[26]

By 1888, Pickering announced that these three investigations would be completed in about a year and that he would be sending observers to the southern hemisphere that autumn to complete the photographic work down to the South Celestial Pole.[27] The following year he announced plans to still further broaden the scope of the Henry Draper Memorial by securing a series of standards of stellar photographic magnitudes, which would have a special value in connection with the photometric measures of the spectra that were then being undertaken. With the necessary equipment, personnel, and funds to carry out these broad and fundamental investigations, Pickering was justifiably enthusiastic:

The field of work of the Henry Draper Memorial, as now extended, is almost boundless. The problems to be investigated relate to the fundamental laws regulating the formation of the stellar system.[28]

The first volume of the Henry Draper Memorial, the *Draper Catalogue of Stellar Spectra,* was published as Volume 27 of the observatory *Annals* in 1890. It contained a catalogue of over 28 000 spectra of more than 10 000 stars on 633 plates photographed with the Bache telescope. The classification scheme it employed was developed by Williamina Paton Fleming, a Scotswoman who headed a large staff of woman computers and who had overall responsibility for the preparation of the catalogue.[29]

Later additions to the Henry Draper Memorial included a volume describing the preparation and some of the findings of the catalogue,[30] two volumes of miscellaneous investigations,[31,32] and two volumes devoted to the special investigation of the spectra of the brighter stars—one for northern stars, prepared by Antonia C. Maury, a niece of Henry Draper,[33] and one for southern stars, prepared by Annie Jump Cannon.[34] Although Miss Cannon basically followed Mrs. Fleming's classification scheme (with a modification so that intermediate classes could be indicated), Miss Maury evolved one of her own to better account, in her opinion, for the great number and variety of spectral lines. The result of this was that, by 1912, there were six separate publications forming parts of the Henry Draper Memorial embodying two very different classification schemes.

Pickering therefore felt a need to bring together all the separate findings into a single catalogue, embodying a single classification scheme. The ensuing work, the *Henry Draper Catalogue,* was published as Volumes 91–99 of the *Annals* between 1918 and 1924 (with two extensions up to 1949). It contained more than 242 000 spectra of about 222 000 stars on 2118 plates. The herculean task of classifying nearly a quarter million spectra was undertaken by Miss Cannon, who began the job in October, 1911, and completed it in October, 1915; for this accomplishment, Oxford University awarded her an honorary D.Sc. degree. In her classification scheme, she retained the system employed in both the original *Draper Catalogue of Stellar Spectra* and her own catalogue of bright southern stars (with a further modification for the representation of intermediate spectra). This system is in use today by astrophysicists throughout the world.

The *Henry Draper Catalogue* has thus had a profound impact on the development of astrophysics. In the estimation of a later prominent astrophysicist, Otto Struve:

> The Harvard classification is the greatest single work in the field of stellar spectroscopy.... After nearly 50 years, every astrophysicist still relies heavily upon the *Henry Draper Catalogue,* its more recent Harvard extensions, and several other catalogues built on the Harvard system of classification.[35]

In conclusion, our examination has shown the extent to which various members of the Draper family and Edward C. Pickering played roles in the birth of American astrophysics. In a general way, John William and Henry Draper's pioneering investigations in photography and spectrum analysis provided the foundation for Pickering's researches in these areas. More specifically, Mrs. Draper's desire to memorialize her late husband by establishing a fund to carry out his plan "by means of photography to catalogue and classify the stars by their spectra" matched perfectly with Pickering's grandiose desire to photograph, measure, and classify the spectra of all the stars visible in both the northern and the southern hemispheres. This resulted in a memorial that not only exceeded her greatest expectations, but probably his as well. The Henry Draper Memorial enabled Harvard to become the foremost institution in the United States for astrophysical research, and has itself become a famous milestone in the history of that science.

REFERENCES

1. DRAPER, J. W. 1837. Experiments, solar light. J. Franklin Inst. **20:** 116–25.
2. DRAPER, J. W. 1840. On the process of daguerreotype, and its applications to taking portraits from the life. Philos. Mag. **17:** 222.

3. DRAPER, H. 1864. On the construction of a silvered-glass telescope 15½ inches in aperture and its use in celestial photography. Smithson. Contrib. Knowl. **14:** 33.
4. DRAPER, J. W. 1843. On a new system of inactive tithonographic spaces in the solar spectrum analogous to the fixed lines of Fraunhofer. Philos. Mag. **22:** 360–64.
5. DRAPER, J. W. 1845. On the interference spectrum, and the absorption of the tithonic rays. Philos. Mag. **26:** 465–78.
6. MAURY, A. C. Manuscript. Recollections of my grandfather, John William Draper. Quoted in D. FLEMING. 1950. John William Draper and the Religion of Science: 138. University of Pennsylvania Press. Philadelphia.
7. DRAPER, H. 1858. On the changes of blood cells in the spleen. N.Y. Med. J. **3:** 182–89.
8. DRAPER, H. 1864. Smithson. Contrib. Knowl. **14:** 1.
9. DRAPER, J. W. 1870. Letter. Henry and Anna Palmer Draper Papers. Dec. 19. New York Public Library. New York.
10. DRAPER, J. W. 1871. Letter. Henry and Anna Palmer Draper Papers. Feb. 9. New York Public Library. New York.
11. YOUNG, C. A. & E. C. PICKERING. 1883. Researches upon the photography of planetary and stellar spectra, by the late Henry Draper, M.D., LL.D. With an introduction by Professor C. A. Young, a list of the photographic plates in Mrs. Draper's possession, and the results of the measurements of these plates by Professor E. C. Pickering. Proc. Am. Acad. Arts Sci. **19:** 237.
12. DRAPER, H. 1877. Photographs of the spectrum of Venus and α Lyrae. Philos. Mag. **3:** 238.
13. DRAPER, H. 1873. On diffraction-spectrum photography. Philos. Mag. **46:** 418.
14. DRAPER, H. 1877. Discovery of oxygen in the sun by photography, and a new theory of the solar spectrum. Proc. Am. Philos. Soc. **17:** 74–80.
15. DRAPER, H. 1879. On photographing the spectra of the stars and planets. Am. J. Sci. **18:** 424.
16. DRAPER, H. 1882. On photographs of the spectrum of the nebula in Orion. Am. J. Sci. **23:** 339.
17. DRAPER, H. 1880. Photographs of the nebula in Orion. Am. J. Sci. **10:** 388.
18. DRAPER, H. 1882. On photographs of the nebula in Orion, and of its spectrum. Mon. Not. R. Astron. Soc. **42:** 368.
19. DRAPER, H. 1882. Am. J. Sci. **23:** 340.
20. YOUNG, C. A. & E. C. PICKERING. 1883. Proc. Am. Acad. Arts Sci. **19:** 231–61.
21. DRAPER, A. P. 1883. Letter. Edward C. Pickering Papers. Jan. 17. Harvard University Archives. Cambridge, Mass.
22. DRAPER, A. P. 1886. Letter. Edward C. Pickering Papers. Jan. 31. Harvard University Archives. Cambridge, Mass.
23. PICKERING, E. C. 1886. Photographic study of stellar spectra. Henry Draper Memorial. Nature (London) **33:** 535.
24. PICKERING, E. C. 1888. Stellar photography. Mem. Am. Acad. Arts Sci. **11:** 184–86.
25. PICKERING, E. C. 1887. Letter. Edward C. Pickering Papers. Jan. 18. Harvard University Archives. Cambridge, Mass.
26. PICKERING, E. C. 1887. The Henry Draper Memorial. First annual report of the photographic study of stellar spectra. Nature (London) **36:** 33–34.
27. PICKERING, E. C. 1888. The progress of the Henry Draper Memorial. Second annual report of the photographic study of stellar spectra conducted at the Harvard College Observatory. Nature (London) **38:** 306–7.
28. PICKERING, E. C. 1889. The Henry Draper Memorial. Third annual report of the photographic study of stellar spectra conducted at the Harvard College Observatory. Nature (London) **40:** 18.
29. PICKERING, E. C. 1893. The constitution of the stars. Astron. Astrophys. **12:** 718–22.
30. PICKERING, E. C. 1891. Preparation and discussion of the Draper Catalogue. Ann. Obs. Harvard Coll. **26**(1).
31. PICKERING, E. C. 1897. Miscellaneous investigations of the Henry Draper Memorial. Ann. Obs. Harvard Coll. **26**(2).
32. CANNON, A. J. 1912. Classification of 1,477 stars by means of their photographic spectra. Ann. Obs. Harvard Coll. **56**(4).

33. MAURY, A. C. 1897. Spectra of bright stars photographed with the 11-inch Draper telescope as a part of the Henry Draper Memorial. Ann. Obs. Harvard Coll. **28**(1).
34. CANNON, A. J. 1901. Spectra of bright southern stars photographed with the 13-inch Boyden telescope as a part of the Henry Draper Memorial. Ann. Obs. Harvard Coll. **28**(2).
35. STRUVE, O. & V. ZEBERGS. 1962. Astronomy of the twentieth century: 190–94. Macmillan. New York and London.

THE COLLECTION OF HENRY DRAPER MEMORABILIA IN THE NEW YORK UNIVERSITY ARCHIVES

T. J. Herczeg

Department of Physics and Astronomy
University of Oklahoma
Norman, Oklahoma 73069

Anne Kinney

Department of Physics
New York University
New York, New York 10003

INTRODUCTION

When Henry Draper died in 1882, his major instruments, together with a number of smaller pieces from his observatory at Hastings-on-Hudson, were transferred to the Harvard College Observatory, where the great volumes of the Memorial and the Henry Draper Catalogue were produced. Numerous items from his laboratory in New York, including many photographic plates, were presented to New York University by Mrs. Antonia Dixon, Draper's sister. Originally at the Heights campus, they were eventually transferred to Washington Square and into the care of Prof. Bayrd Still, the University Archivist. It is a substantial collection and, along with that at the Smithsonian, provides an important footnote to Henry Draper's research.

During the summer of 1981, the material from the Archives was studied and organized. In the fall, the most important pieces, with the addition of some memorabilia on loan, were on exhibit in the Bobst Library of the University, remaining on display for the rest of the academic year. On the occasion of the Orion Nebula Symposium it seems appropriate to give a short description, though not a catalogue, of the collection.

The material divides naturally into the following classes: early lunar photographs, photographs of stellar spectra, materials relating to research for the paper on solar oxygen, and material connected with the Harvard College Observatory.

LUNAR PHOTOGRAPHY

A series of 34 plates, original negatives of lunar photographs taken in 1863–64, illustrates Draper's earliest program of celestial photography. The images are about 4 cm in diameter and, although some of the plates show signs of deteriorating emulsion, most negatives are still in good condition, as shown by the example reproduced in FIGURE 1.

According to Plotkin,[2] Draper had taken some 1500 negatives of the moon in the early 1860s. Most of this material is lost, but this group of 34 plates is a significant witness to the early activity at Hastings-on-Hudson.

0077–8923/82/0395–0331 $1.75/1 © 1982, NYAS

FIGURE 1. Fifteen hundred lunar photographs were taken by Henry Draper at around the time of the Civil War. This one, dated October 15, 1863, is one of a set of 34 plates contained in the Archives. According to George F. Barker, who wrote the biography of Henry Draper for the National Academy of Sciences in 1886, these were "the best photographs of the moon ever taken by anyone up to that time."

STELLAR SPECTRA

Henry Draper was a famous pioneer in photographic studies of stellar spectra. The material in the NYU Archives, which represents only a small section of his work in this field, is essentially divided into two groups—the objective prism spectra and the slit spectra.

Objective Prism Spectra

There are 8 plates containing a number of spectra in stellar fields. Since Draper did not have aperature-size prisms attached to his large telescopes, the prism was applied to the converging beam. Most of these plates are positive copies, made perhaps for the purpose of demonstration: two show the Pleiades and one the Hyades region.

Slit Spectra

In August of 1872, Henry Draper took the first spectrum of a star (Vega) showing absorption lines. Although that spectrum no longer exists, the Archive's collection has a very fine set of eleven slit spectra of the Moon, Jupiter, Vega, Arcturus, and Antares taken from 1879 to 1881. The spectra are carefully preserved between glass plates and have clear labels. The dispersion is nearly constant between the K line and Hγ, at about 90 Å mm^{-1}. Two examples are shown in FIGURE 2.

SOLAR SPECTRA

This part of the Draper collection is the largest and, perhaps, most interesting. It contains a considerable portion of the plates used for his investigation of oxygen lines in the solar spectrum. This was carried out by direct comparison of the solar spectrum with the "spectrum of the air," of CO_2, or various other laboratory spectra, such as hydrogen, nitrogen, and cyanogen. The solar and the laboratory spectra were photographed through the same spectrograph and on the same plate. Although later studies did not substantiate Draper's claim that he had found emission lines of oxygen in the photospheric spectrum, it posed a problem that was much discussed for twenty years, and contemporary astrophysicists considered this investigation a crucially important piece of research.*

The Archives contain 33 plates taken for this investigation between February and May, 1877. Several of them have a paper strip pasted on the plate, with the stronger lines of the solar spectrum and the lines of the laboratory spectrum carefully delineated in pencil with the coincidences marked. Remarks such as "noncoincident 5, doubtful 3, coincident 16, total 24" can be found. Other remarks, made in ink on the emulsion, indicate the date and also give the conditions under which the plate was made; for example, "No 8 April 7, 1877 spectrum of air w Leyden 10^m ex. Sun 10^s Engine 90." One of them bears the remark, "Used for enlarging, April 24 1877"; another glass plate, protected by a second, is identified by a typewritten remark (perhaps added later) as the copy used for demonstration in the Philosophical Society lecture. One spectrum reaches unusually far into the ultraviolet (the H and K lines are shifted to the middle of the region in focus) and contains wavelength indications in 10^{-8} cm. With few exceptions, the plates are in good condition.

This method of "visual" coincidences would hardly be applied today, especially when looking for emission features in a spectrum crowded with absorption lines. Draper's procedure, however, seems to be justified by the then still somewhat imperfect state of wavelength tables and perhaps even more by the fact that, in his

---

*When Draper's paper, read before the Americal Philosophical Society on June 20, 1877, was published in Volume 74 of The *Journal of the Franklin Institute,* the editors felt justified in adding the remark, "This paper has been pronounced, by able judges, to be the most important contribution to solar physics since Kirchhoff's great discovery."

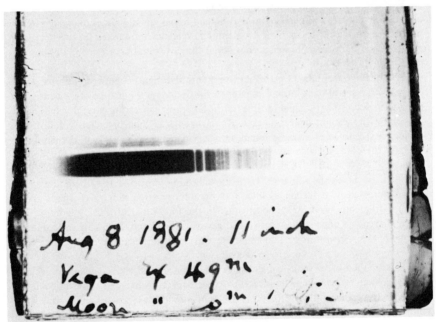

FIGURE 2. These two spectra of Vega and the Moon were taken in 1879 and 1881, with a dispersion of about 90 Ångströms per millimeter. The Balmer series is seen in the two stellar spectra and the calcium doublet is seen in the lunar spectra.

superbly equipped laboratory, he was able to make this type of comparison far more accurately than most of his colleagues.

Spectra of this type were also taken in the years following the 1877 publication. In two boxes are twenty spectra (taken from 1878–79) of the Sun next to a comparison spectrum of air and, in one case, iron. Several are at a much higher dispersion than those from 1877; one bears the note: "Camera crooked and so this must not be used for measurement. Definition bad at ends."

THE SOLAR ECLIPSE OF 1878

The eclipse of July 29, 1878, was photographed by Henry Draper on an expedition to Battle Lake, Wyoming, where he was accompanied by Thomas Edison and Mrs.

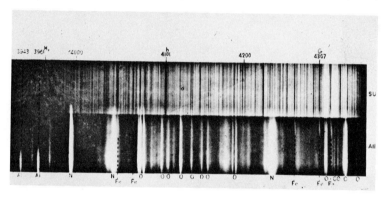

FIGURE 3. This spectrum was published by Henry Draper in his paper of 1877 in an attempt to prove the existence of oxygen emission lines in the sun. It is an example of his method of exposing both a solar spectrum and a laboratory-produced comparison spectrum on the same plate. The upper part of the photograph is the solar spectrum, the lower part is the spectrum of air (produced with a Leyden spark with terminals of iron and aluminium). Lines of oxygen, nitrogen, iron, and aluminium are marked.

Draper, among others. He succeeded in taking a photograph of the corona, two copies of which are part of the collection.

THE HENRY DRAPER MEMORIAL

After Henry Draper's death in 1882, a memorial was set up at the Harvard University Observatory by Mrs. Draper, with the help of E. C. Pickering. A set of nine plates in the collection is accompanied by an inventory that begins, "Lantern slides sent to Mrs. Draper, December 23, 1887, as examples of those which were sent in her name to the Royal Astronomical Society." These plates include a view of the Harvard Observatory with the Draper domes, the Pleiades, the Orion nebula through an 8-in. telescope, spectra of six first-magnitude stars, spectra of β Persei at full brightness and at minimum, with notes concerning the difference, of α Lyrae and o Ceti, and of three other stars, with remarks about the details lost in the printing process.

Additionally, a few plates taken for the Harvard Observatory in the course of regular work found their way into this collection. They were presumably sent to Mrs. Draper as illustrations of further research. One of them is the globular cluster Omega Centauri, with bright stars remarkably well resolved right down to the center.

Some miscellaneous personal memorabilia include stereoscopic New York city-scapes, group pictures, and an unidentified building with colonnades.

REFERENCES

1. DRAPER, H. 1877. Discovery of oxygen in the sun by photography, and a new theory of the solar spectrum. J. Franklin Institute **74:** 81.
2. PLOTKIN, H. N. 1972. Henry Draper, A Scientific Biography.: 63. Diss. Johns Hopkins Univ. Baltimore.

Index of Contributors

(Italic page numbers refer to comments made in discussion.)